Introduction to
Spatial Econometrics

T0304221

STATISTICS: Textbooks and Monographs

Recent Titles

Introduction to
Spatial Econometrics

James LeSage
Texas State University-San Marcos

San Marcos, Texas, U.S.A.

R. Kelley Pace
Louisiana State University

Baton Rouge, Louisiana, U.S.A.

CRC Press
Taylor & Francis Group
Boca Raton   London   New York

CRC Press is an imprint of the
Taylor & Francis Group, an **informa** business

A CHAPMAN & HALL BOOK

Chapman & Hall/CRC
Taylor & Francis Group
6000 Broken Sound Parkway NW, Suite 300
Boca Raton, FL 33487-2742

First issued in paperback 2022

© 2009 by Taylor & Francis Group, LLC
CRC Press is an imprint of Taylor & Francis Group, an Informa business

No claim to original U.S. Government works

ISBN 13: 978-1-03-247774-9 (pbk)
ISBN 13: 978-1-4200-6424-7 (hbk)

DOI: 10.1201/9781420064254

Library of Congress Cataloging-in-Publication Data

LeSage, James P.
  Introduction to spatial econometrics / James LeSage, Robert Kelley Pace.
    p. cm. -- (Statistics : a series of textbooks and monographs ; 196)
  Includes bibliographical references and index.
  ISBN-13: 978-1-4200-6424-7 (alk. paper)
  ISBN-10: 1-4200-6424-X (alk. paper)
  1. Space in economics--Econometric models. 2. Space in
economics--Mathematical models. I. Pace, Robert Kelley. II. Title. III. Series.

HT388.L47 2009
330.01'5195--dc22   2008038890

Visit the Taylor & Francis Web site at
http://www.taylorandfrancis.com

and the CRC Press Web site at
http://www.crcpress.com

Contents

List of Figures

List of Tables

Preface

This text provides an introduction to spatial econometric modeling along with numerous applied illustrations of the methods. It is intended as a text for students and researchers with a basic background in regression methods interested in learning about spatial regression models. There has been a surge of interest in these modeling methods in recent years, yet there exists no comprehensive up-to-date text that discusses the variety of approaches available in a consistent manner. This text would be appropriate for an advanced undergraduate or graduate level course in the subject.

When producing a text, there are always trade-offs between breadth and depth of coverage and we have attempted to cover a wide range of alternative topics including: maximum likelihood and Bayesian estimation, different types of spatial regression specifications such as the spatial autoregressive and matrix exponential, applied modeling situations involving different circumstances including origin-destination flows, limited dependent variables, and space-time data samples. This breadth of coverage comes at the expense of detailed derivations in some parts of the text. In these cases, we provide a host of references to the growing body of spatial econometric literature.

Readers interested in implementing the methods discussed here should find useful MATLAB code that is publicly available at: spatial-econometrics.com and spatial-statistics.com. Toolboxes are the name given by the MathWorks Inc. to related sets of MATLAB functions aimed at solving a particular class of problems. The two web sites are the home of the *Spatial Econometrics Toolbox* and *Spatial Statistics Toolbox*, which contain a number of functions useful for spatial econometric estimation. All of the applied examples presented in the text were constructed using these toolbox functions. We have chosen not to discuss details regarding MATLAB computer codes for the methods presented in the text, but have modified the documentation for the toolbox code to reference various sections in this text.

One of our goals in writing the text was to provide a number of different motivations for the phenomena known as *simultaneous spatial dependence*. This is a central concept that justifies use of spatial autoregressive processes that have become a mainstay of spatial econometrics. Luc Anselin in his influential 1988 text on spatial econometrics provides a strong argument for use of models capable of addressing simultaneous spatial dependence that arises in spatial data samples. However, this concept has made the field somewhat mysterious, and we believe the alternative motivations provided here for use of spatial regression models involving spatial lags of the dependent variable will help demystify the concept.

Another goal of the text was to aid practitioners with interpretation of spatial regression models, especially those that include spatial lags of the dependent variable. The applied literature contains a number of studies that misinterpret regression results from these models. We provide new methods that produce useful summary measures of the direct and indirect or spatial spillover impacts that arise in these models in response to changes in the explanatory variables. A number of applied illustrations are provided that should help practitioners with this task.

Another important issue is the relationship between spatiotemporal processes and long-run equilibrium states that are characterized by simultaneous spatial dependence. We devote a chapter of the text to motivating how spatiotemporal processes are related to a host of spatial models characterized by simultaneous and conditional spatial dependence. Using spatiotemporal processes of the type explored here would ensure that space-time panel model specifications could be justified as arising from underlying space-time interactions. This may help improve current space-time panel data specifications.

The views expressed regarding spatial econometric modeling represent a consensus that has arisen from almost daily phone conversations between the authors over the ten year period of our collaborative research. Due to the rapidly evolving nature of the field, much of the material reflects recent ideas that have not appeared elsewhere. For example, the chapter on limited dependent variable modeling provides a comprehensive development of new ideas that differ from some past work, and extensions to the case of multinomial spatial autoregressive probit models. The chapter on matrix exponential spatial specifications elaborates in a number of ways on our *Journal of Econometrics* article on this topic. Our scalar summary measures of spatial impact estimates have been the subject of conference presentations but have not appeared in print. The same is true of the numerous motivations for spatial regression models that include spatial lags of the dependent variable.

Acknowledgements

Many years ago, Luc Anselin encouraged Jim LeSage to produce a text describing Bayesian spatial econometric methods, and has been a source of encouragement along the way.

Interaction at conferences and work on joint projects with a number of colleagues over the years has provided a welcome opportunity to discuss and debate spatial econometric issues. Some of the ideas in this text have been stimulated by joint research with colleagues: Corrine Autant-Bernard, Ron Barry, Eric Blankmeyer, Cem Ertur, Manfred Fischer, Wilfried Koch, Carlos Llano, Julie LeGallo, Olivier Parent, Wolfgang Polasek, Tony Smith, and Christine Thomas-Agnan. Other ideas arose from conference sessions and discussions involving: Sudipto Banerjee, Badi Baltagi, Roger Bivand, David Brasington, J. Paul Elhorst, Bernard Fingleton, Alan Gelfand, Art Getis, Daniel Griffith, Carter Hill, Garth Holloway, James Kau, Harry Kelejian, Kara Kockelman, Donald Lacombe, Ingmar Prucha, Dek Terrell, C. F. Sirmans, Carlos Slawson, and Michael Tiefelsdorf.

During preparation of the manuscript, we received a great deal of proof-reading assistance from: Shuang Zhu, Mihaela Craioveanu, Olivier Parent, Garth Holloway, and Donald Lacombe.

We would like to thank David Grubbs and Taylor & Francis for proposing the idea of a text on spatial econometrics, and Jessica Vakili for editorial assistance.

The authors would like to thank the McCoy family and Jerry D. and Linda Gregg Fields for their generous support of the McCoy College of Business Administration at Texas State University-San Marcos, and the Louisiana Real Estate Commission for support of the E.J. Ourso College of Business at Louisiana State University and the Real Estate Research Institute.

Finally, the authors would like to thank the Louisiana and Texas Sea Grant programs and especially the National Science Foundation for their support of our research on spatial econometric methods through the following grants BSC-0136193, BSC-0136229, BCS-0554937, SES-0729259, and SES-0729264.

Symbol Description

$\odot$	represents Hadamard or element-by-element multiplication
$\otimes$	represents a Kronecker product
ι_n	denotes an $n \times 1$ vector of ones
iid	stands for independent and identically distributed
∂	denotes a partial derivative
$|A|$	is the determinant of the matrix A
(a,b)	open interval that excludes the endpoints a and b
$[a,b]$	closed interval that includes the endpoints a and b
diag	extracts the main diagonal from a matrix
tr	trace operator for matrices
abs	absolute value operator
plim	probability limit operator
$\delta()$	is an indicator function, $\delta(A) = 1$ for outcomes where A occurs, $\delta(A) = 0$ otherwise.
$\propto$	proportionality symbol
vec	an operator that stacks columns of a matrix to form a vector
$\pi()$	denotes prior distributions
$N(a,b)$	represents a normal distribution with mean a and variance b
$IG(a,b)$	represents an inverse gamma distribution with parameters a,b
NIG	represents a combination of normal and inverse gamma distributions
$TMNV$	represents a truncated multivariate normal distribution
$\chi^2(r)$	represents a chi-squared distribution with parameter r
$\mathcal{B}(a,b)$	represents a beta distribution with parameters a, b
$\mathcal{D}$	represents the set of model data $\{y, X, W\}$
$\mathcal{P}()$	represents the Poisson distribution
κ	denotes a real constant
$\Gamma(a)$	represents the Gamma function, $\int_0^\infty t^{a-1}e^{-t}dt$
$Beta()$	represents the Beta function, $\int_0^1 t^{a-1}(1-t)^{b-1}dt$

Chapter 1

Introduction

Section 1.1 of this chapter introduces the concept of *spatial dependence* that often arises in cross-sectional spatial data samples. Spatial data samples represent observations that are associated with points or regions, for example homes, counties, states, or census tracts. Two motivational examples are provided for spatial dependence, one based on spatial spillovers stemming from congestion effects and a second that relies on omitted explanatory variables. Section 1.2 sets forth *spatial autoregressive* data generating processes for spatially dependent sample data along with *spatial weight matrices* that play an important role in describing the structure of these processes. We provide more detailed discussion of spatial data generating processes and associated spatial econometric models in Chapter 2, and spatial weight matrices in Chapter 4. Our goal here is to provide an introduction to spatial autoregressive processes and spatial regression models that rely on this type of process. Section 1.3 provides a simple example of how congestion effects lead to spatial spillovers that impact neighboring regions using travel times to the central business district (CBD) region of a metropolitan area. Section 1.4 describes various scenarios in which spatial econometric models can be used to analyze spatial spillover effects. The final section of the chapter lays out the plan of this text. A brief enumeration of the topics covered in each chapter is provided.

1.1   Spatial dependence

Consider a cross-sectional variable vector representing observations collected with reference to points or regions in space. Point observations could include selling prices of homes, employment at various establishments, or enrollment at individual schools. Geographic information systems typically support *geocoding* or *address matching* which allow addresses to be automatically converted into locational coordinates. The ability to geocode has led to vast amounts of spatially-referenced data. Observations could include a variable like population or average commuting time for residents in regions such as census tracts, counties, or metropolitan statistical areas (MSAs). In contrast to point observations, for a region we rely on the coordinates of an interior point representing the center (the *centroid*). An important point is that in spatial regression models each observation corresponds to a location or region.

The *data generating process* (DGP) for a conventional cross-sectional non-spatial sample of n *independent* observations $y_i, i = 1, \ldots, n$ that are linearly related to explanatory variables in a matrix X takes the form in (1.1), where we have suppressed the intercept term, which could be included in the matrix X.

$$y_i = X_i\beta + \varepsilon_i \tag{1.1}$$
$$\varepsilon_i \sim N(0, \sigma^2) \quad i = 1, \ldots, n \tag{1.2}$$

In (1.2), we use $N(a, b)$ to denote a univariate normal distribution with mean a and variance b. In (1.1), X_i represents a $1 \times k$ vector of covariates or explanatory variables, with associated parameters β contained in a $k \times 1$ vector. This type of data generating process is typically assumed for linear regression models. Each observation has an underlying mean of $X_i\beta$ and a random component ε_i. An implication of this for situations where the observations i represent regions or points in space is that observed values at one location (or region) are independent of observations made at other locations (or regions). Independent or *statistically independent* observations imply that $E(\varepsilon_i\varepsilon_j) = E(\varepsilon_i)E(\varepsilon_j) = 0$. The assumption of independence greatly simplifies models, but in spatial contexts this simplification seems strained.

In contrast, *spatial dependence* reflects a situation where values observed at one location or region, say observation i, depend on the values of *neighboring* observations at nearby locations. Suppose we let observations $i = 1$ and $j = 2$ represent neighbors (perhaps regions with borders that touch), then a data generating process might take the form shown in (1.3).

$$y_i = \alpha_i y_j + X_i\beta + \varepsilon_i \tag{1.3}$$
$$y_j = \alpha_j y_i + X_j\beta + \varepsilon_j$$
$$\varepsilon_i \sim N(0, \sigma^2) \quad i = 1$$
$$\varepsilon_j \sim N(0, \sigma^2) \quad j = 2$$

This situation suggests a simultaneous data generating process, where the value taken by y_i depends on that of y_j and vice versa. As a concrete example, consider the set of seven regions shown in Figure 1.1, which represent three regions to the west and three to the east of a central business district (CBD).

For the purpose of this example, we will consider these seven regions to constitute a single metropolitan area, with region $R4$ being the central business district. Since the entire region contains only a single roadway, all commuters share this route to and from the CBD.

We might observe the following set of sample data for these regions that relates travel times to the CBD (in minutes) contained in the dependent vari-

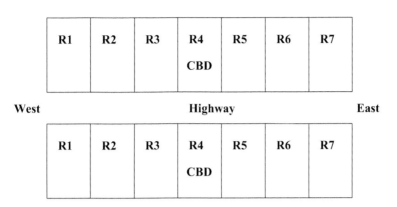

FIGURE 1.1: Regions east and west of the Central Business District

able vector y to distance (in miles) and population density (population per square block) of the regions in the two columns of the matrix X.

$$
y = \begin{pmatrix} \text{Travel times} \\ 42 \\ 37 \\ 30 \\ 26 \\ 30 \\ 37 \\ 42 \end{pmatrix} \qquad X = \begin{pmatrix} \text{Density} & \text{Distance} \\ 10 & 30 \\ 20 & 20 \\ 30 & 10 \\ 50 & 0 \\ 30 & 10 \\ 20 & 20 \\ 10 & 30 \end{pmatrix} \begin{array}{ll} \text{ex-urban areas} & R1 \\ \text{far suburbs} & R2 \\ \text{near suburbs} & R3 \\ \text{CBD} & R4 \\ \text{near suburbs} & R5 \\ \text{far suburbs} & R6 \\ \text{ex-urban areas} & R7 \end{array}
$$

The pattern of longer travel times for more distant regions $R1$ and $R7$ versus nearer regions $R3$ and $R5$ found in the vector y seems to clearly violate independence, since travel times appear similar for neighboring regions. However, we might suppose that this pattern is explained by the model variables *Distance* and *Density* associated with each region, since these also appear similar for neighboring regions. Even for individuals in the CBD, it takes time to go somewhere else in the CBD. Therefore, the travel time for intra-CBD travel is 26 minutes despite having a distance of 0 miles.

Now, consider that our set of observed travel times represent measurements taken on a particular day, so we have travel times to the CBD averaged over a 24 hour period. In this case, some of the observed pattern might be explained by congestion effects that arise from the shared highway. It seems plausible that longer travel times in one region should lead to longer travel times in neighboring regions on any given day. This is because commuters pass from one region to another as they travel along the highway to the CBD. Slower times in $R3$ on a particular day should produce slower times for this day in regions $R2$ and $R1$. Congestion effects represent one type of spatial spillover, which do not occur simultaneously, but require some time for the traffic delay to arise. From a modeling viewpoint, congestion effects such as these will not be explained by the model variables *Distance* and *Density*. These are dynamic feedback effects from travel time on a particular day that impact travel times of neighboring regions in the short time interval required for the traffic delay to occur. Since the explanatory variable distance would not change from day to day, and population density would change very slowly on a daily time scale, these variables would not be capable of explaining daily delay phenomena. Observed daily variation in travel times would be better explained by relying on travel times from neighboring regions on that day. This is the situation depicted in (1.3), where we rely on travel time from a neighboring observation y_j as an explanatory variable for travel time in region i, y_i. Similarly we use y_i to explain region j travel time, y_j.

Since our observations were measured using average times for one day, the measurement time scale is not fine enough to capture the short-interval time dynamic aspect of traffic delay. This would result in observed daily travel times in the vector y that appear to be simultaneously determined. This is an example of why measured spatial dependence may vary with the time-scale of data collection.

Another example where observed spatial dependence may arise from omitted variables would be the case of a hedonic pricing model with sales prices of homes as the vector y and characteristics of the homes as explanatory variables in the matrix X. If we have a cross-sectional sample of sales prices in a neighborhood collected over a period of one year, variation in the characteristics of the homes should explain part of the variation in observed sales prices. Consider a situation where a single home sells for a much higher price than would be expected based solely on its characteristics. Assume this sale took place at the mid-point of our 12 month observation period, shortly after a positive school quality report was released for a nearby school. Since school quality was not a variable included in the set of explanatory variables representing home characteristics, the higher than expected selling price might reflect a new premium for school quality. This might signal other sellers of homes served by the same school to ask for higher prices, or to accept offers that are much closer to their asking prices during the last six months of our observation period. This would lead to a situation where use of selling prices from neighboring homes produce improved explanatory power for homes served by the high quality school during the last six months of our sample. Other omitted variables could be accessibility to transportation, nearby amenities such as shopping or parks, and so on. If these were omitted from the set of explanatory variables consisting solely of home characteristics, we would find that selling prices from neighboring homes are useful for prediction.

An illustration that non-spatial regression models will ignore spatial dependence in the dependent variable is provided by a map of the ordinary least-squares residuals from a production function regression: $\ln(Q) = \alpha \iota_n + \beta \ln(K) + \gamma \ln(L) + \varepsilon$, estimated using the 48 contiguous US states plus the District of Columbia. Gross state product for the year 2001 was used as Q, with labor L being 2001 total non-farm employment in each state. Capital estimates K for the states are from Garofalo and Yamarik (2002). These residuals are often referred to as the *Solow residual* if constant returns to scale are imposed so that $\beta = \phi, \gamma = (1 - \phi)$. In the context of a Solow growth model, they are interpreted as reflecting economic growth above the rate of capital growth, or that not explained by growth in factors of production. In the case of our production function model, these would be interpreted as total factor productivity, so they reflect output attributable to regional variation in the technological efficiency with which these factors are used.

Figure 1.2 shows a *choropleth map* of total factor productivity (the residuals from our production function regression). A choropleth map relies on shaded or patterned areas to reflect the measured values of the variable being displayed on the map. It provides a visual depiction of how values of a variable differ over space. Figure 1.3 displays an associated legend for the map taking the form of a histogram showing the frequency distribution of states according to the magnitude of their residuals. We see negative residuals for 12 states, including the cluster of 7 neighboring states, Texas, Oklahoma, Louisiana, Mississippi, Tennessee Arkansas and Alabama. A negative residual would indicate that observed output Q was lower than output predicted by the regression based on labor and capital available to these states. From the legend in Figure 1.3 we see that blue, green and purple states represent positive residuals. Of the 11 green states we see a cluster of these states in the northeast, indicating that observed output for these states was above that predicted by our regression model, reflecting higher than expected total factor productivity.

If the residuals were randomly distributed with regard to location, we would not see clusters of red and green states that are indicative of negative and positive residuals associated with neighboring states. This type of clustering represents a visual depiction of spatial dependence in the residuals or factor productivity from the non-spatial regression model.

A question arises — what leads to the observed spatial dependence in total factor productivity? There is a role for spatial econometric modeling methods to play in answering this question. As we will see, different model specifications suggest different theoretical justifications, and vice versa. In traditional econometrics there are three uses of empirical models: 1) estimation and inference regarding parameters, 2) prediction or out-of-sample forecasting and 3) model comparison of alternative specifications.

We can use spatial econometric models in the same three ways to answer the

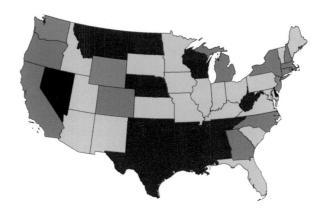

FIGURE 1.2:   Solow residuals, 2001 US states (see color figure on the insert following page 24)

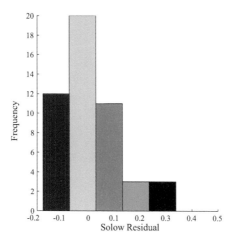

FIGURE 1.3:   Solow residuals map legend (see color insert)

question regarding observed spatial dependence in dependent variables from our models as well as residuals. For example, there has been some theoretical work on extending neoclassical growth models to provide a justification for a *spatially lagged dependent variable* (Lopez-Bazo et al., 2004; Ertur and Koch, 2007) in our production function model. A spatial lag of the dependent variable is an explanatory variable vector constructed using an average of values from neighboring regions. These theoretical models posit physical and human capital externalities as well as technological interdependence between regions, which leads to a reduced form regression that includes a spatial lag of the dependent variable.

Spatial econometric model comparison methods could be used to test these theories by comparing models that include a spatial lag of the dependent variable to other model specifications that do not. Predictions or out-of-sample forecasts from models including a spatially lagged dependent variable could be compared to models that do not include these terms to provide evidence in favor of these theories. Finally, estimates and inferences regarding the significance of the parameter associated with the spatially lagged variable could be used to show consistency of these theories with the sample data.

There are other possible explanations for the observed pattern of spatial dependence. Since we are mapping residuals that reflect total factor productivity, these are conditional on capital and labor inputs. There is a great deal of literature that examines regional production from the standpoint of the new economic geography (Duranton and Puga, 2001; Autant-Bernard, 2001; Autant-Bernard, Mairesse and Massard, 2007; Parent and LeSage, 2008). These studies point to spatial spillovers that arise from technological innovation, measured using regional patents as a proxy for the stock of knowledge available to a region. In Chapter 3 we will provide an applied illustration of this total factor productivity relationship that is used to quantify the magnitude of spatial spillovers arising from regional differences in technical innovation.

In time series, lagged dependent variables can be justified by theoretical models that include costly adjustment or other behavioral frictions which give rise quite naturally to time lags of the dependent variable. As we saw with the travel time to the CBD example, a similar motivation can be used for spatial lags. Another justification often used in the case of time series is that the lagged dependent variable accounts for variation in the dependent variable that arises from unobserved or latent influences. As we have seen in the case of our hedonic home sales price example, a similar justification can be used for a spatial lag of the dependent variable. Latent unobservable influences related to culture, infrastructure, or recreational amenities can affect the dependent variable, but may not appear as explanatory variables in the model. Use of a spatial regression model that includes a spatial lag of the dependent variable vector can capture some of these influences.

1.2   The spatial autoregressive process

We could continue in the fashion of (1.3) to generate a larger set of observations as shown in (1.4).

$$y_i = \alpha_{i,j}y_j + \alpha_{i,k}y_k + X_i\beta + \varepsilon_i \qquad (1.4)$$
$$y_j = \alpha_{j,i}y_i + \alpha_{j,k}y_k + X_j\beta + \varepsilon_j$$
$$y_k = \alpha_{k,i}y_i + \alpha_{k,j}y_j + X_k\beta + \varepsilon_k$$
$$\varepsilon_i \sim N(0,\sigma^2) \qquad i = 1$$
$$\varepsilon_j \sim N(0,\sigma^2) \qquad j = 2$$
$$\varepsilon_k \sim N(0,\sigma^2) \qquad k = 3$$

It is easy to see that this would be of little practical usefulness, since it would result in a system with many more parameters than observations.

Intuitively, once we allow for dependence relations between a set of n observations/locations, there are potentially $n^2 - n$ relations that could arise. We subtract n from the potential n^2 dependence relations because we rule out dependence of an observation on itself.

The solution to the over-parameterization problem that arises when we allow each dependence relation to have relation-specific parameters is to impose structure on the spatial dependence relations. Ord (1975) proposed a parsimonious parameterization for the dependence relations (which built on early work by Whittle (1954)). This structure gives rise to a data generating process known as a *spatial autoregressive process*. Applied to the dependence relations between the observations on variable y, we have expression (1.5).

$$y_i = \rho \sum_{j=1}^{n} W_{ij}y_j + \varepsilon_i \qquad (1.5)$$
$$\varepsilon_i \sim N(0,\sigma^2) \qquad i = 1,\ldots,n$$

Where we eliminate an intercept term by assuming that the vector of observations on the variable y is in deviations from means form. The term: $\sum_{j=1}^{n} W_{ij}y_j$ is called a *spatial lag*, since it represents a linear combination of values of the variable y constructed from observations/regions that neighbor observation i. This is accomplished by placing elements W_{ij} in the $n \times n$ *spatial weight matrix* W, such that $\sum_{j=1}^{n} W_{ij}y_j$ results in a scalar that represents a linear combination of values taken by neighboring observations.

As an example, consider the seven regions shown in Figure 1.1. The single *first-order neighbor* to region $R1$ is region $R2$, since this is the only region that has borders that touch region $R1$. Similarly, region $R2$ has 2 first-order neighbors, regions $R1$ and $R3$. We can define *second-order neighbors* as regions that are neighbors to the first-order neighbors. Second-order neighbors to region $R1$ would consist of all regions having borders that touch the first-order neighbor (region $R2$), which are: regions $R1$ and $R3$. It is important to note that region $R1$ is a second-order neighbor to itself. This is because region $R1$ is a neighbor to its neighbor, which is the definition of a second-order neighboring relation. If the neighboring relations are symmetric, each region will always be a second order neighbor to itself. By nature, contiguity relations are symmetric, but we will discuss other definitions of neighboring relations in Chapter 4 that may not result in symmetry.

We can write a matrix version of the spatial autoregressive process as in (1.6), where we use $N(0, \sigma^2 I_n)$ to denote a zero mean disturbance process that exhibits constant variance σ^2, and zero covariance between observations. This results in the diagonal variance-covariance matrix $\sigma^2 I_n$, where I_n represents an n-dimensional identity matrix. Expression (1.6) makes it clear that we are describing a relation between the vector y and the vector Wy representing a linear combination of neighboring values to each observation.

$$y = \rho W y + \varepsilon \tag{1.6}$$
$$\varepsilon \sim N(0, \sigma^2 I_n)$$

To illustrate this, we form a 7×7 spatial weight matrix W using the first-order contiguity relations for the seven regions shown in Figure 1.1. This involves associating rows of the matrix with the observation index i, and columns with the index j representing neighboring observations/regions to region i. We begin by forming a first-order contiguity matrix C shown in (1.7). For row 1 we place a value of 1 in column 2, reflecting the fact that region $R2$ is first-order contiguous to region $R1$. All other elements of row 1 receive values of zero. Similarly, for each row we place a 1 in columns associated with first-order contiguous neighbors, resulting in the matrix C shown in (1.7).

$$
C =
\begin{pmatrix}
 & R1 & R2 & R3 & R4 & R5 & R6 & R7 \\
R1 & 0 & 1 & 0 & 0 & 0 & 0 & 0 \\
R2 & 1 & 0 & 1 & 0 & 0 & 0 & 0 \\
R3 & 0 & 1 & 0 & 1 & 0 & 0 & 0 \\
R3 & 0 & 0 & 1 & 0 & 1 & 0 & 0 \\
R5 & 0 & 0 & 0 & 1 & 0 & 1 & 0 \\
R6 & 0 & 0 & 0 & 0 & 1 & 0 & 1 \\
R7 & 0 & 0 & 0 & 0 & 0 & 1 & 0
\end{pmatrix}
\tag{1.7}
$$

We note that the diagonal elements of the matrix C are zero, so regions are not considered neighbors to themselves. For the purpose of forming a spatial lag or linear combination of values from neighboring observations, we can normalize the matrix C to have row sums of unity. This *row-stochastic* matrix which we label W is shown in (1.8), where the term row-stochastic refers to a non-negative matrix having row sums normalized so they equal one.

$$W = \begin{pmatrix} 0 & 1 & 0 & 0 & 0 & 0 & 0 \\ 1/2 & 0 & 1/2 & 0 & 0 & 0 & 0 \\ 0 & 1/2 & 0 & 1/2 & 0 & 0 & 0 \\ 0 & 0 & 1/2 & 0 & 1/2 & 0 & 0 \\ 0 & 0 & 0 & 1/2 & 0 & 1/2 & 0 \\ 0 & 0 & 0 & 0 & 1/2 & 0 & 1/2 \\ 0 & 0 & 0 & 0 & 0 & 1 & 0 \end{pmatrix} \tag{1.8}$$

The 7×7 matrix W can be multiplied with a 7×1 vector y of values taken by each region to produce a *spatial lag* vector of the dependent variable vector taking the form Wy. The matrix product Wy works to produce a 7×1 vector representing the value of the spatial lag vector for each observation $i, i = 1, \ldots, 7$. We will provide details on various approaches to formulating spatial weight matrices in Chapter 4, which involve alternative ways to defining and weighting *neighboring observations*. For now, we note that use of the matrix W which weights each neighboring observation equally will result in the spatial lag vector being a simple average of values from neighboring (first-order contiguous) observations to each region. The matrix multiplication process is shown in (1.9), along with the resulting spatial lag vector Wy.

$$Wy = \begin{pmatrix} 0 & 1 & 0 & 0 & 0 & 0 & 0 \\ 1/2 & 0 & 1/2 & 0 & 0 & 0 & 0 \\ 0 & 1/2 & 0 & 1/2 & 0 & 0 & 0 \\ 0 & 0 & 1/2 & 0 & 1/2 & 0 & 0 \\ 0 & 0 & 0 & 1/2 & 0 & 1/2 & 0 \\ 0 & 0 & 0 & 0 & 1/2 & 0 & 1/2 \\ 0 & 0 & 0 & 0 & 0 & 1 & 0 \end{pmatrix} \begin{pmatrix} y_1 \\ y_2 \\ y_3 \\ y_4 \\ y_5 \\ y_6 \\ y_7 \end{pmatrix}$$

$$= \begin{pmatrix} y_2 \\ (y_1 + y_3)/2 \\ (y_2 + y_4)/2 \\ (y_3 + y_5)/2 \\ (y_4 + y_6)/2 \\ (y_5 + y_7)/2 \\ y_6 \end{pmatrix} \tag{1.9}$$

The scalar parameter ρ in (1.6) describes the strength of spatial dependence in the sample of observations. Use of a single parameter to reflect an average level of dependence over all dependence relations arising from observations $i = 1, \ldots, n$, is one way in which parsimony is achieved by the spatial autoregressive structure. This is in stark contrast to our starting point in (1.3) and (1.4), where we allowed each dependency to have its own parameter.

We can graphically examine a scatter plot of the relation between the observations in the vector y (in deviation from means form) and the average values of neighboring observations in the vector Wy using a *Moran scatter plot*. An example is shown in Figure 1.4, where we plot total factor productivity of the states, constructed using the residuals from our 2001 production function regression on the horizontal axis, and the spatial lag values on the vertical axis. By virtue of the transformation to deviation from means, we have four Cartesian quadrants in the scatter plot centered on zero values for the horizontal and vertical axes. These four quadrants reflect:

Quadrant I (red points) states that have factor productivity (residuals) above the mean, where the average of neighboring states' factor productivity is also greater than the mean,

Quadrant II (green points) states that exhibit factor productivity below the mean, but the average of neighboring states' factor productivity is above the mean,

Quadrant III (blue points) states with factor productivity below the mean, and the average of neighboring states' factor productivity is also below the mean,

Quadrant IV (purple points) states that have factor productivity above the mean, and the average of neighboring states' productivity is below the mean.

From the scatter plot, we see a positive association between factor productivity observations y on the horizontal axis and the spatially lagged observations from Wy shown on the vertical axis, suggesting the scalar parameter ρ is greater than zero. Another way to consider the strength of positive association is to note that there are very few green and purple points in the scatter plot. Green points represent states where factor productivity is below average and that of neighboring states Wy is above average. The converse is true of the purple points, where above average factor productivity coincides with below average factor productivity Wy from neighboring states. In contrast, a large number of points in quadrants II and IV with few points in quadrants I and III would suggest negative spatial dependence so that $\rho < 0$.

Points in the scatter plot can be placed on a map using the same color coding scheme, as in Figure 1.5. Red states represent regions with higher than average (positive) factor productivity where the average of neighboring states' factor productivity is also above the mean. The map makes the clustering of northeast and western states with above average factor productivity levels where neighboring states also have above average factor productivity quite clear. Similarly, clustering of states with lower than average factor productivity levels and surrounding states that are also below the mean is evident in the central and southern part of the US.

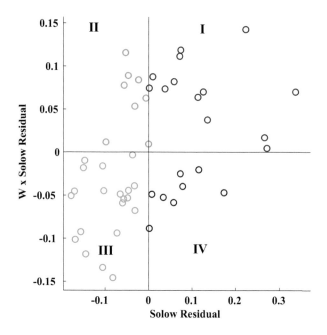

FIGURE 1.4:   Moran scatter plot of 2001 US states factor productivity (see color insert)

It is tempting to interpret the scalar parameter ρ in the spatial autoregressive process as a conventional correlation coefficient between the vector y and the *spatial lag* vector Wy. This temptation should be avoided, as it is not entirely accurate. We will discuss this point in more detail in Chapter 2, but note that the range for correlation coefficients is $[-1, 1]$, whereas ρ cannot equal one.

1.2.1   Spatial autoregressive data generating process

The spatial autoregressive process is shown in (1.10) using matrix notation, and the implied data generating process for this type of process is in (1.11). We introduce a constant term vector of ones ι_n, and associated parameter α to accommodate situations where the vector y does not have a mean value of zero.

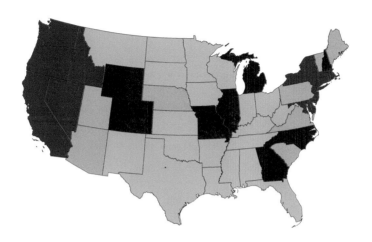

FIGURE 1.5: Moran plot map of US states 2001 factor productivity (see color insert)

$$y = \alpha \iota_n + \rho W y + \varepsilon \qquad (1.10)$$
$$(I_n - \rho W)y = \alpha \iota_n + \varepsilon$$
$$y = (I_n - \rho W)^{-1} \iota_n \alpha + (I_n - \rho W)^{-1} \varepsilon \qquad (1.11)$$
$$\varepsilon \sim N(0, \sigma^2 I_n)$$

The $n \times 1$ vector y contains our dependent variable and ρ is a scalar parameter, with W representing the $n \times n$ spatial weight matrix. We assume that ε follows a multivariate normal distribution, with zero mean and a constant scalar diagonal variance-covariance matrix $\sigma^2 I_n$.

The model statement in (1.10) can be interpreted as indicating that the expected value of each observation y_i will depend on the mean value α plus a linear combination of values taken by neighboring observations scaled by the dependence parameter ρ. The *data generating process* statement in (1.11) expresses the *simultaneous* nature of the spatial autoregressive process. To further explore the nature of this, we can use the following infinite series to express the inverse:

$$(I_n - \rho W)^{-1} = I_n + \rho W + \rho^2 W^2 + \rho^3 W^3 + \ldots \qquad (1.12)$$

where we assume for the moment that $\text{abs}(\rho) < 1$. This leads to a spatial autoregressive data generating process for a variable vector y:

$$
\begin{aligned}
y &= (I_n - \rho W)^{-1} \iota_n \alpha + (I_n - \rho W)^{-1} \varepsilon \\
y &= \alpha \iota_n + \rho W \iota_n \alpha + \rho^2 W^2 \iota_n \alpha + \ldots \\
&\quad + \varepsilon + \rho W \varepsilon + \rho^2 W^2 \varepsilon + \rho^3 W^3 \varepsilon + \ldots
\end{aligned}
\qquad (1.13)
$$

Expression (1.13) can be simplified since the infinite series: $\iota_n \alpha + \rho W \iota_n \alpha + \rho^2 W^2 \iota_n \alpha + \ldots$ converges to $(1 - \rho)^{-1} \iota_n \alpha$ since α is a scalar, the parameter $\text{abs}(\rho) < 1$, and W is row-stochastic. By definition, $W \iota_n = \iota_n$ and therefore $W(W \iota_n)$ also equals $W \iota_n = \iota_n$. Consequently, $W^q \iota_n = \iota_n$ for $q \geq 0$ (recall that $W^0 = I_n$). This allows us to write:

$$y = \frac{1}{(1 - \rho)} \iota_n \alpha + \varepsilon + \rho W \varepsilon + \rho^2 W^2 \varepsilon + \rho^3 W^3 \varepsilon + \ldots \qquad (1.14)$$

To further explore the nature of this data generating process, we consider powers of the row-stochastic spatial weight matrices $W^2, W^3, \ldots$ that appear in (1.14). Let us assume that rows of the weight matrix W are constructed to represent *first-order* contiguous neighbors. The matrix W^2 will reflect *second-order* contiguous neighbors, those that are neighbors to the first-order neighbors. Since the neighbor of the neighbor (second-order neighbor) to an observation i includes observation i itself, W^2 has positive elements on the diagonal when each observation has at least one neighbor. That is, higher-order spatial lags can lead to a connectivity relation for an observation i such that $W^2 \varepsilon$ will extract observations from the vector ε that point back to the observation i itself. This is in stark contrast with our initial independence relation in (1.1), where the Gauss-Markov assumptions rule out dependence of ε_i on other observations j, by assuming zero covariance between observations i and j in the data generating process.

To illustrate this point, we show W^2 based on the 7×7 first-order contiguity matrix W from (1.8) in (1.15), where positive elements appear on the diagonal. We see that for region $R1$ for example, the second-order neighbors are regions $R1$ and $R3$. That is, region $R1$ is a second-order neighbor to itself as well as to region $R3$, which is a neighbor to the neighboring region $R2$.

$$W^2 = \begin{pmatrix} 0.50 & 0 & 0.50 & 0 & 0 & 0 & 0 \\ 0 & 0.75 & 0 & 0.25 & 0 & 0 & 0 \\ 0.25 & 0 & 0.50 & 0 & 0.25 & 0 & 0 \\ 0 & 0.25 & 0 & 0.50 & 0 & 0.25 & 0 \\ 0 & 0 & 0.25 & 0 & 0.50 & 0 & 0.25 \\ 0 & 0 & 0 & 0.25 & 0 & 0.75 & 0 \\ 0 & 0 & 0 & 0 & 0.50 & 0 & 0.50 \end{pmatrix} \quad (1.15)$$

Given that $\text{abs}(\rho) < 1$, the data generating process assigns less influence to disturbance terms associated with higher-order neighbors, with a geometric decay of influence as the order rises. Stronger spatial dependence reflected in larger values of ρ leads to a larger role for the higher order neighbors.

The dependence of each observation y_i on disturbances associated with neighboring observations as well as higher-order neighbors suggests a mean and variance-covariance structure for the observations in the vector y that depend in a complicated way on other observations. It is instructive to consider the mean of the variable y that arises from the spatial autoregressive data generating process in (1.13). Note that we assume the spatial weight matrix is exogenous, or fixed in repeated sampling, so that:

$$\begin{aligned} E(y) &= \frac{1}{(1-\rho)}\alpha\iota_n + E(\varepsilon) + \rho W E(\varepsilon) + \rho^2 W^2 E(\varepsilon) + \dots \\ &= \frac{1}{(1-\rho)}\alpha\iota_n \end{aligned} \quad (1.16)$$

It is interesting to note that in social networking (Katz, 1953; Bonacich, 1987) interpret the vector $b = (I_n - \rho P)^{-1}\iota_n$ as a measure of *centrality* of individuals in a social network, where the matrix P is a binary peer matrix, so the vector b reflects row sums of the matrix inverse.[1] The vector b (referred to as Katz-Bonacich Centrality in social networking) measures the number of direct and indirect connections that an individual in a social network has. For example, if the matrix P identifies friends, then P^2 points to friends of friends, P^3 to friends of friends of friends, and so on. In social networking, individuals are viewed as located at nodes in a network, and the parameter ρ reflects a discount factor that creates decay of influence for friends/peers that are located at more distant nodes. These observations merely point out that the spatial autoregressive process has played an important role in other disciplines beside spatial statistics, and will likely continue to grow in use and importance.

Simultaneous feedback is useful in modeling spatial dependence relations where we wish to accommodate spatial feedback effects from neighboring regions to an origin location i where an initial impact occurred. In fact, these

---

[1] The binary peer matrix is defined like our contiguity matrix C, having values of 1 for peers and 0 for non-peers.

models allow us to treat all observations as potential origins of an impact without loss of generality. One might suppose that feedback effects would take time, but there is no explicit role for passage of time in a cross-sectional relation. Instead, we can view the cross-sectional sample data observations as reflecting an equilibrium outcome or steady state of the spatial process we are modeling. We develop this idea further in Chapter 2 and Chapter 7. This is an interpretation often used in cross-sectional modeling and Sen and Smith (1995) provide examples of this type of situation for conventional spatial interaction models used in regional analysis. The goal in spatial interaction models is to analyze variation in flows between regions that occur over time using a cross-section of observed flows between origin and destination regions that have taken place over a finite period of time, but measured at a single point in time. We discuss spatial econometric models for origin-destination flows in Chapter 8.

This simultaneous dependence situation does not occur in time series analysis, making spatial autoregressive processes distinct from time series autoregressive processes. In time series, the *time lag* operator L is strictly triangular and contains zeros on the diagonal. Powers of L are also strictly triangular with zeros on the diagonal, so that L^2 specifies a two-period time lag whereas L creates a single period time lag. It is never the case that L^2 produces observations that point back to include the present time period.

1.3   An illustration of spatial spillovers

The spatial autoregressive structure can be combined with a conventional regression model to produce a spatial extension of the standard regression model shown in (1.17), with the implied data generating process in (1.18). We will refer to this as simply the *spatial autoregressive model* (SAR) throughout the text. We note that Anselin (1988) labeled this model a "mixed-regressive, spatial-autoregressive" model, where the motivation for this awkward nomenclature should be clear.

$$y = \rho W y + X\beta + \varepsilon \qquad (1.17)$$
$$y = (I_n - \rho W)^{-1} X\beta + (I_n - \rho W)^{-1}\varepsilon \qquad (1.18)$$
$$\varepsilon \sim N(0, \sigma^2 I_n)$$

In this model, the parameters to be estimated are the usual regression parameters β, σ and the additional parameter ρ. It is noteworthy that if the scalar parameter ρ takes a value of zero so there is no spatial dependence in the vector of cross-sectional observations y, this yields the least-squares regression model as a special case of the SAR model.

To provide an illustration of how the spatial regression model can be used to quantify spatial spillovers, we reuse the earlier example of travel times to the CBD from the seven regions shown in Figure 1.1. We consider the impact of a change in population density for a single region on travel times to the CBD for all seven regions. Specifically, we double the population density in region $R2$ and make a prediction of the impact on travel times to the CBD for all seven regions.

We use parameter estimates: $\hat{\beta}' = \begin{bmatrix} 0.135 & 0.561 \end{bmatrix}$ and $\hat{\rho} = 0.642$ for this example. The estimated value of ρ indicates positive spatial dependence in commuting times. Predictions from the model based on the explanatory variables matrix X would take the form:

$$\hat{y}^{(1)} = (I_n - \hat{\rho}W)^{-1}X\hat{\beta}$$

where $\hat{\rho}, \hat{\beta}$ are maximum likelihood estimates.

A comparison of predictions $\hat{y}^{(1)}$ from the model with explanatory variables from X and $\hat{y}^{(2)}$ from the model based on $\tilde{X}$ shown in (1.19) is used to illustrate how the model generates spatial spillovers when the population density of a single region changes. The matrix $\tilde{X}$ reflects a doubling of the population density of region $R2$.

$$\tilde{X} = \begin{pmatrix} 10 & 30 \\ 20 & \mathbf{40} \\ 30 & 10 \\ 50 & 0 \\ 30 & 10 \\ 20 & 20 \\ 10 & 30 \end{pmatrix} \tag{1.19}$$

The two sets of predictions $\hat{y}^{(1)}, \hat{y}^{(2)}$ are shown in Table 1.1, where we see that the change in region $R2$ population density has a direct effect that increases the commuting times for residents of region $R2$ by 4 minutes. It also has an indirect or spillover effect that produces an increase in commuting times for the other six regions. The increase in commuting times for neighboring regions to the east and west (regions $R1$ and $R3$) are the greatest and these spillovers decline as we move to regions in the sample that are located farther away from region $R2$ where the change in population density occurred.

It is also of interest that the cumulative indirect impacts (spillovers) can be found by adding up the increased commuting times across all other regions (excluding the own-region change in commuting time). This equals $2.57 + 1.45 + 0.53 + 0.20 + 0.07 + 0.05 = 4.87$ minutes, which is larger than the direct (own-region) impact of 4 minutes. The total impact on all residents of the seven region metropolitan area from the change in population density of

TABLE 1.1: Spatial spillovers from changes in Region $R2$ population density

Regions / Scenario	$\hat{y}^{(1)}$	$\hat{y}^{(2)}$	$\hat{y}^{(2)} - \hat{y}^{(1)}$
$R1$:	42.01	44.58	2.57
$R2$:	37.06	41.06	4.00
$R3$:	29.94	31.39	1.45
$R4$: CBD	26.00	26.54	0.53
$R5$:	29.94	30.14	0.20
$R6$:	37.06	37.14	0.07
$R7$:	42.01	42.06	0.05

region $R2$ is the sum of the direct and indirect effects, or 8.87 minutes increase in travel times to the CBD.[2]

The model literally suggests that the change in population density of region $R2$ would immediately lead to increases in the observed daily commuting times for all regions. A more palatable interpretation would be that the change in population density would lead over time to a new equilibrium steady state in the relation between daily commuting times and the distance and density variables. The predictions of the direct impacts arising from the change in density reflect $\partial y_i / \partial X_{i2}$, where X_{i2} refers to the ith observation of the second explanatory variable in the model. The cross-partial derivatives $\partial y_j / \partial X_{i2}$ represent indirect effects associated with this change.

To elaborate on this, we note that the DGP for the SAR model can be written as in (1.20), where the subscript r denotes explanatory variable r,

$$y = \sum_{r=1}^{k} S_r(W)X_r + (I_n - \rho W)^{-1}\varepsilon \qquad (1.20)$$

$$E(y) = \sum_{r=1}^{k} S_r(W)X_r \qquad (1.21)$$

where $S_r(W) = (I_n - \rho W)^{-1}\beta_r$ acts as a "multiplier" matrix that applies higher-order neighboring relations to X_r. Models that contain spatial lags of the dependent variable exhibit a complicated derivative of y_i with respect to X_{jr}, where i, j denote two distinct observations. It follows from (1.21) that:

$$\frac{\partial E(y_i)}{\partial X_{jr}} = S_r(W)_{ij} \qquad (1.22)$$

where $S_r(W)_{ij}$ represents the ijth element of the matrix $S_r(W)$.

---

[2]Throughout the text we will use the terms *impacts* and *effects* interchangeably when referring to direct and indirect effects or impacts.

As expression (1.22) indicates, the standard regression interpretation of coefficient estimates as partial derivatives: $\hat{\beta}_r = \partial y / \partial X_r$, no longer holds. Because of the transformation of X_r by the $n \times n$ matrix $S_r(W)$, any change to an explanatory variable in a given region (observation) can affect the dependent variable in all regions (observations) through the matrix inverse.

Since the impact of changes in an explanatory variable differ over all observations, it seems desirable to find a summary measure for the own derivative $\partial y_i / \partial X_{ir}$ in (1.22) that shows the impact arising from a change in the ith observation of variable r. It would also be of interest to summarize the cross derivative $\partial y_i / \partial X_{jr} (i \neq j)$ in (1.22) that measures the impact on y_i from changes in observation j of variable r. We pursue this topic in detail in Chapter 2, where we provide summary measures and interpretations for the impacts that arise from changes represented by the own- and cross-partial derivatives.

Despite the simplicity of this example, it provides an illustration of how spatial regression models allow for spillovers from changes in the explanatory variables of a single region in the sample. This is a valuable aspect of spatial econometric models that sets them apart from most spatial statistical models, an issue we discuss in the next section.

An ordinary regression model would make the prediction that the change in population density in region $R2$ affects only the commuting time of residents in region $R2$, with no allowance for spatial spillover impacts. To see this, we can set the parameter $\rho = 0$ in our model, which produces the non-spatial regression model. In this case $\hat{y}^{(1)} = X\hat{\beta}_o$ and $\hat{y}^{(2)} = \tilde{X}\hat{\beta}_o$, so the difference would be $\tilde{X}\hat{\beta}_o - X\hat{\beta}_o = (\tilde{X} - X)\hat{\beta}_o$, where the estimated parameters $\hat{\beta}_o$ would be those from a least-squares regression.

If the DGP for our observed daily travel times is that of the SAR model, least-squares estimates will be biased and inconsistent, since they ignore the spatial lag of the dependent variable. To see this, note that the estimates for $\hat{\beta}$ from the SAR model take the form: $\hat{\beta} = (X'X)^{-1}X'(I_n - \hat{\rho}W)y$, a subject we pursue in more detail in Chapter 2. For our simple illustration where all values of y and X are positive, and the spatial dependence parameter is also positive, this suggests an upward bias in the least-squares estimates. This can be seen by noting that:

$$\hat{\beta} = (X'X)^{-1}X'y - \hat{\rho}(X'X)^{-1}X'Wy$$
$$\hat{\beta} = \hat{\beta}_o - \hat{\rho}(X'X)^{-1}X'Wy$$
$$\hat{\beta}_o = \hat{\beta} + \hat{\rho}(X'X)^{-1}X'Wy$$

Since all values of y are positive, the spatial lag vector Wy will contain averages of the neighboring values which will also be positive. This in conjunction with only positive elements in the matrix X as well as positive $\hat{\rho}$ lead us to conclude that the least-squares estimates $\hat{\beta}_o$ will be biased upward relative to the unbiased estimates $\hat{\beta}$. For our seven region example, the least-squares estimates were: $\hat{\beta}'_o = \begin{bmatrix} 0.55 & 1.25 \end{bmatrix}$, which show upward bias relative to the spatial autoregressive model estimates: $\hat{\beta}' = \begin{bmatrix} 0.135 & 0.561 \end{bmatrix}$. Intuitively, the ordinary least-squares model attempts to explain variation in travel times that arises from spillover congestion effects using the distance and population density variables. This results in an overstatement of the true influence of these variables on travel times.

Least-squares predictions based on the matrices X and $\tilde{X}$ are presented in Table 1.2. We see that no spatial spillovers arise from this model, since only the travel time to the CBD for region $R2$ is affected by the change in population density of region $R2$. We also see the impact of the upward bias in the least-squares estimates, which produce an inflated prediction of travel time change that would arise from the change in population density.

TABLE 1.2: Non-spatial predictions for changes in Region $R2$ population density

Regions / Scenario	$\hat{y}^{(1)}$	$\hat{y}^{(2)}$	$\hat{y}^{(2)} - \hat{y}^{(1)}$
$R1$:	42.98	42.98	0.00
$R2$:	36.00	47.03	11.02
$R3$:	29.02	29.02	0.00
$R4$: CBD	27.56	27.56	0.00
$R5$:	29.02	29.02	0.00
$R6$:	36.00	36.00	0.00
$R7$:	42.98	42.98	0.00

1.4 The role of spatial econometric models

A long-running theme in economics is how pursuit of self interest results in benefits or costs that fall on others. These benefits or costs are labeled externalities. In situations where spillovers are spatial in nature, spatial econometric models can quantify the magnitude of these, as illustrated by the travel time to the CBD example.

There are a host of other examples. Technological innovation that arises as a result of spatial knowledge spillovers from nearby regions is an example of a positive externality or spillover. It is argued that a large part of knowledge is tacit because ideas leading to technical innovation are embodied in persons and linked to the experience of the inventor. This stock of knowledge increases in a region as local inventors discover new ideas and diffuses mostly via face-to-face interactions. We can think of knowledge as a local public good that benefits researchers within a region as well as nearby neighboring regions.

This motivates a spatial specification for unobserved knowledge that would not be included as a model explanatory variable. It is generally believed that tacit knowledge linked to the experience of inventors and researchers does not "travel well," so knowledge spillovers are thought to be local in nature falling only on nearby regions. We can use spatial regression models to quantify the spatial extent of spillovers by examining indirect effects using the series expansion $I_n + \rho W + \rho^2 W^2 + \ldots$ that arises in the partial derivative expression for these effects. Chapter 3 will explore this issue in an applied illustration that relates regional total factor productivity to knowledge spillovers.

Pollution provides another example since these negative externalities or spillovers are likely to be spatial in nature. The ability to quantify direct and indirect effects from pollution sources should be useful in empirical analysis of the classic Pigovian tax and subsidy solutions for market failure.

Regional governments are often thought to take into account actions of neighboring governments when setting tax rates (Wilson, 1986) and deciding on provision of local government services (Tiebout, 1956). Spatial econometric models can be used to empirically examine the magnitude and statistical significance of local government interaction. Use of the partial derivative measures of direct and indirect effects that arise from changes in the explanatory variables should be particularly useful from a public policy perspective. In a model of county government decisions, direct effects estimates pertain to impacts that would be of primary concern to that county's government officials, whereas spillover and total effects reflect the broader perspective of society at large. Much of the public choice literature focuses on situations where private and public, or local and national government interests diverge. In the case of local and national governments, the divergence can be viewed in terms of spatial spillover effects. Again, the ability of spatial regression models to quantify the relative magnitude of the divergence should be useful to those studying public choice issues.

There is a fundamental difference between models containing spatial lags of the dependent variable and those modeling spatial dependence in the disturbances. We explore this using the general error model in (1.23), where $F(W)$ in (1.24) represents a non-singular matrix function involving a spatial weight matrix W.

$$y = X\beta + \epsilon \qquad (1.23)$$
$$\epsilon = F(W)\varepsilon \qquad (1.24)$$

The expectation of y for these error models appears in (1.25).

$$E(y) = X\beta \qquad (1.25)$$

This means that all of the various types of error models have the same expectation as the non-spatial model. Sufficiently large sample sizes using consistent estimators on the various models should yield identical estimates of the parameters β. For small samples the estimates could vary, and using models that differ from the DGP could lead to inconsistent estimates of dispersion for the model parameters. Interpretation of the parameters β from this type of model is the same as for a non-spatial linear regression model.

Anselin (1988) provides a persuasive argument that the focus of spatial econometrics should be on measuring the effects of spillovers. We pay limited attention to error models in this text because these models eliminate spillovers by construction. These could be added by making X more spatially complex, but there are more appealing alternatives that we will explore here.

1.5   The plan of the text

This introductory chapter focused on a brief introduction to spatial dependence and spatial autoregressive processes, as well as spatial weight matrices used in these processes. These processes can be used to produce a host of spatial econometric models that accommodate spatial dependence taking various forms.

Chapter 2 provides more detailed motivations for spatial dependence and the use of spatial regression models. We elaborate on the idea that omitted or excluded variables in our models that exhibit spatial dependence can lead to spatial regression models that contain spatial lags of the dependent variable. Cross-sectional simultaneous spatial regression models are also motivated as a long-run steady-state outcome of non-simultaneous dependence situations. We consider situations where economic agents can observe past actions of neighboring agents, for example county government officials should be aware of neighboring government tax rates or levels of government services provision in the previous period. This type of non-simultaneous space-time dynamic relationship is consistent with a cross-sectional simultaneous spatial regression relationship that represents the long-run steady state outcome of the space-time dynamic relationship. We also provide details regarding interpretation of estimates from these models. An elaboration is provided regarding direct and indirect effects associated with changes in explanatory variables that was introduced in the travel time example of this chapter.

Chapter 3 will focus on a family of spatial regression models popularized by Anselin (1988) in his influential text on spatial econometrics. The implications for estimates and inferences based on least-squares estimates from non-spatial regression in the presence of spatial dependence are discussed. This chapter also provides details regarding computational aspects of maximum likelihood estimation for the family of spatial regression models. Computational methods have advanced considerably since 1988, the year of Anselin's text.

Chapter 4 addresses various computational and theoretical aspects of spatial econometric models. Topics include computation of spatial weight matrices, log-determinants (including numerous special cases such as the matrix exponential, equation systems, multiple weight matrices, and flow matrices), derivatives of log-determinants, diagonals of the variance-covariance matrix, and closed-form solutions for a number of single-parameter spatial models.

Conventional Bayesian methods for analyzing spatial econometric models (Anselin, 1988; Hepple, 1995a,b) as well as more recent *Bayesian Markov Chain Monte Carlo* (MCMC) methods (LeSage, 1997) for estimating spatial regression models are the subject of Chapter 5. The approach set forth in LeSage (1997) allows formal treatment of *spatial heterogeneity* that is motivated in Chapter 2. We show that many of the computational advances described for maximum likelihood estimation in Chapter 3 also work to simplify Bayesian estimation of these models.

Model specification and comparison is the topic of Chapter 6. Specification issues considered include the form of the weight matrix, the usual concern about appropriate explanatory variables, and questions regarding which of the alternative members of the family of spatial regression models introduced in Chapter 3 should be employed. We show how formal Bayesian model comparison methods proposed by LeSage and Parent (2007) can be used to answer questions regarding appropriate explanatory variables for the family of models from Chapter 3. Bayesian model comparison methods can also be used to discriminate between models based on alternative spatial weight matrices as noted by LeSage and Pace (2004a) and different specifications arising from the family of spatial regression models from Chapter 3 (Hepple, 2004).

Chapter 7 is unlike other chapters in the text since it is more theoretical, focusing on spatiotemporal foundations for observed cross-sectional spatial dependence. Starting with the assumption that regions are influenced only by *own* and *other regions* past period values we develop a spatiotemporal motivation for simultaneous spatial dependence implied by the spatial autoregressive process. We elaborate on the discussion in Chapter 2 showing how time dependence on past decisions of neighboring economic agents will lead to simultaneous spatial regression specifications. We show that a strict spatiotemporal framework consistent with a spatial partial adjustment mechanism can result in a long-run equilibrium characterized by simultaneous spatial dependence.

Spatial econometric extensions of conventional least-squares gravity or spatial interaction models described in Sen and Smith (1995) are the topic of Chapter 8. We present spatial regression models similar to those from Chapter 3 introduced by LeSage and Pace (2008) that can be applied to models that attempt to explain variation in flows between origins and destinations. Allowing for spatial dependence at origins, destinations, and between origins and destinations leads to a situation where changes at either the origin or destination will give rise to forces that set in motion a series of events. We explore the notion advanced by Behrens, Ertur and Koch (2007) that spatial dependence suggests a multilateral world where indirect interactions link all regions. This contrasts with conventional emphasis on bilateral flows from origin to destination regions.

Chapter 9 sets forth an alternative approach for spatial econometric modeling that replaces the spatial autoregressive process with a matrix exponential approach to specifying spatial dependence structures (LeSage and Pace, 2007, 2004b). This has both computational as well as theoretical advantages over the more conventional spatial autoregressive process. We discuss both maximum likelihood and Bayesian approaches to estimating models based on this new spatial process specification.

Chapter 10 takes up the topic of spatial regressions involving binary, count or truncated dependent variables. This draws on work regarding binary dependent variables in the context of the family of models from Chapter 3 described in LeSage (2000) and surveyed by Flemming (2004). Use of spatial autoregressive processes as a Bayesian prior for spatially structured effects introduced in Smith and LeSage (2004) for the case of probit models and in LeSage, Fischer and Scherngell (2007) for Poisson count data models are also discussed. This approach to structuring individual effects parameters can be used to overcome problems that typically arise when estimating individual effects (Christensen, Roberts and Sköld, 2006; Gelfand, Sahu and Carlin, 1995).

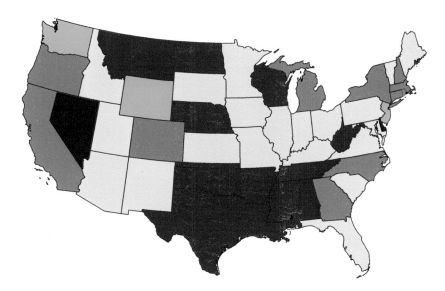

FIGURE 1.2
Solow residuals, 2001 US states.

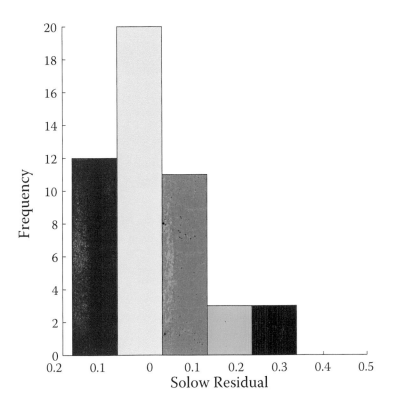

FIGURE 1.3
Solow residuals map legend.

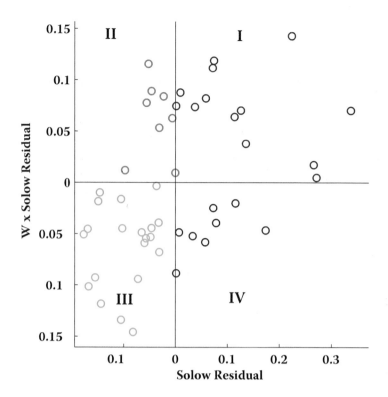

FIGURE 1.4
Moran scatter plot of 2001 US states factor productivity.

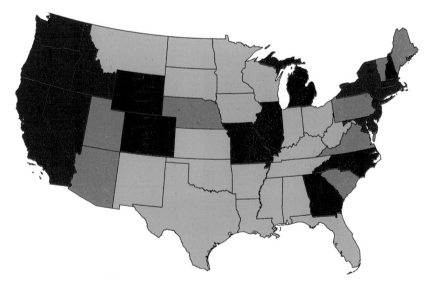

FIGURE 1.5
Moran plot map of US states 2001 factor productivity.

Chapter 2

Motivating and Interpreting Spatial Econometric Models

In the first five sections of this chapter, we provide separate motivations for regression models that include spatial autoregressive processes. These motivations are explored in more detail in later chapters of the text, with the presentation here being less formal. Section 2.1 shows how cross-sectional model relations involving spatial lags of the dependent variable (the SAR model) come from economic agents considering past period behavior of neighboring agents. Section 2.2 provides a second situation where omitted variables that exhibit spatial dependence lead to a model that includes spatial lags of both the dependent as well as independent variables. Sections 2.3 to 2.5 provide additional motivations based on spatial heterogeneity, externalities, and model uncertainty. Taken together, the motivations in Sections 2.1 to 2.5 show how a host of alternative spatial regression structures arise when dependence enters into a combination of the explanatory variables, dependent variables, or disturbances.

Section 2.6 briefly introduces a family of conventional spatial regression models that have appeared in the empirical literature. Section 2.7 is devoted to a discussion of interpreting the parameter estimates from spatial regression models. This issue has been particularly misunderstood in applied studies that have relied on spatial regression models. We introduce some relatively straightforward procedures that simplify analysis of impacts that result from changes in the explanatory variables of these models.

2.1 A time-dependence motivation

Economic agents often make current period decisions that are influenced by the behavior of other agents in previous periods. For example, local governments might set tax rates after observing rates set by neighboring regions in previous time periods. Although the tax rates were set over time by the cross-section of regions representing our sample, the observed cross-sectional tax rates would exhibit a pattern of spatial dependence.

To illustrate this, consider a relation where the dependent variable vector at

time t, denoted y_t, is determined using a spatial autoregressive scheme that depends on *space-time lagged values* of the dependent variable from neighboring
observations. This would lead to a time lag of the average neighboring values
of the dependent variable observed during the previous period, Wy_{t-1}. We
can also include current period own-region characteristics X_t in our model.
In the event that the characteristics of regions remain relatively fixed over
time, we can write $X_t = X$ and ignore the time subscript for this matrix
of regional characteristics. As a concrete example of this type of situation,
consider a model involving home selling prices as the dependent variable y_t,
which depend on past period selling prices of neighboring homes, Wy_{t-1}.
Characteristics of homes such as the number of bedrooms or baths change
very slowly over time. This suggests the following relation as a representation
for the space-time lagged autoregressive process:

$$y_t = \rho W y_{t-1} + X\beta + \varepsilon_t \tag{2.1}$$

Note that we can replace y_{t-1} on the right-hand side above with: $y_{t-1} = \rho W y_{t-2} + X\beta + \varepsilon_{t-1}$ producing:

$$y_t = X\beta + \rho W \left(X\beta + \rho W y_{t-2} + \varepsilon_{t-1} \right) + \varepsilon_t \tag{2.2}$$
$$y_t = X\beta + \rho W X\beta + \rho^2 W^2 y_{t-2} + \varepsilon_t + \rho W \varepsilon_{t-1} \tag{2.3}$$

Recursive substitution for past values of the vector y_{t-r} on the right-hand
side of (2.3) over q periods leads to (2.4) and (2.5).

$$y_t = \left(I_n + \rho W + \rho^2 W^2 +, \dots, +\rho^{q-1} W^{q-1} \right) X\beta + \rho^q W^q y_{t-q} + u \tag{2.4}$$
$$u = \varepsilon_t + \rho W \varepsilon_{t-1} + \rho^2 W^2 \varepsilon_{t-2} +, \dots, +\rho^{q-1} W^{q-1} \varepsilon_{t-(q-1)} \tag{2.5}$$

These expressions can be simplified by noting that $E(\varepsilon_{t-r}) = 0, r = 0, \dots, q-1$, implies that $E(u) = 0$. In addition, the magnitude of $\rho^q W^q y_{t-q}$ becomes
small for large q, under the usual assumption that $|\rho| < 1$ and assuming
that W is row-stochastic, so the matrix W has a principal eigenvalue of 1.
Consequently, we can interpret the observed cross-sectional relation as the
outcome or expectation of a long-run equilibrium or steady state shown in
(2.6).

$$\lim_{q \to \infty} E(y_t) = (I_n - \rho W)^{-1} X\beta \tag{2.6}$$

Note that this provides a dynamic motivation for the data generating process of the cross-sectional SAR model that serves as a workhorse of spatial
regression modeling. That is, a cross-sectional SAR model relation can arise
from time-dependence of decisions by economic agents located at various
points in space when decisions depend on those of neighbors.

2.2   An omitted variables motivation

Omitted variables may easily arise in spatial modeling because unobservable factors such as location amenities, highway accessibility, or neighborhood prestige may exert an influence on the dependent variable. It is unlikely that explanatory variables are readily available to capture these types of latent influences. We explore this situation using a very simple scenario involving a dependent variable y that is completely explained by two explanatory variables x and z with associated scalar parameters β and θ. For simplicity, we assume the $n \times 1$ vectors x and z are distributed $N(0, I_n)$, and we assume x and z are independent.

$$y = x\beta + z\theta \tag{2.7}$$

Given both x and z, solution of the linear system would yield an exact β and θ. The absence of a disturbance term simplifies discovery of the parameters.

Consider the case where the vector z is not observed. Since the unobserved variable z is not correlated with the observed vector x, we can still uncover β. In this case, the vector $z\theta$ acts as the disturbance term, which we label ε in the relation shown in (2.8).

$$y = x\beta + \varepsilon \tag{2.8}$$

Expression (2.8) represents a normal linear model with independent and identically distributed (*iid*) disturbances, where the ordinary least-squares estimator $\hat{\beta} = (x'x)^{-1}x'y$ is known to be the best linear unbiased estimator.

As an alternative scenario, consider a situation where the explanatory variable vector z exhibits zero covariance with the vector x, but follows the spatial autoregressive process shown in (2.9).

$$z = \rho W z + r \tag{2.9}$$
$$z = (I_n - \rho W)^{-1} r \tag{2.10}$$

In (2.9), ρ is a real scalar parameter, r is a $n \times 1$ vector of disturbances distributed $N(0, \sigma_r^2 I_n)$, and W is an $n \times n$ spatial weight matrix with $W_{ij} > 0$ when observation j is a neighbor to observation i, and $W_{ij} = 0$ otherwise. We also set $W_{ii} = 0$, and assume that W has row-sums of unity and that $(I_n - \rho W)^{-1}$ exists. From our discussion of spatial autoregressive processes, each element of Wz would represent a linear combination of elements from the vector z associated with neighboring locations. When working with spatial data samples, it seems intuitively plausible that unobserved latent factors such as location amenities, highway accessibility, or neighborhood prestige would exhibit spatial dependence of the type assigned to the vector z.

Substituting (2.10) into (2.7) yields (2.11), which reflects the generalized normal linear model containing non-spherical disturbances. The effect of θ is to increase the variance of r and in (2.12) we define $u = \theta r$. As is well-known, least-squares estimates for the parameter β in (2.11) are still unbiased, but not efficient.

$$y = x\beta + (I_n - \rho W)^{-1}(\theta r) \tag{2.11}$$
$$y = x\beta + (I_n - \rho W)^{-1}u \tag{2.12}$$
$$E(y) = x\beta \tag{2.13}$$

Given the prevalence of omitted variables in spatial econometric practice, it seems unlikely that x and u are uncorrelated. A simple approach to representing this correlation is to specify that u depends linearly on x, plus a disturbance term v that is independent of x as in (2.14), where the scalar parameter γ and the variance of the disturbance term v (σ_v^2) determine the strength of the relation between x and $z = (I_n - \rho W)^{-1}u$.

$$u = x\gamma + v \tag{2.14}$$
$$v \sim N(0, \sigma_v^2 I_n)$$

In this scenario, the more complicated DGP is shown in (2.16).

$$y = x\beta + (I_n - \rho W)^{-1}(x\gamma + v) \tag{2.15}$$
$$y = x\beta + (I_n - \rho W)^{-1}x\gamma + (I_n - \rho W)^{-1}v \tag{2.16}$$

It is no longer the case that the least-squares estimate $\hat{\beta}$ is unbiased. If we transform expression (2.16) to have *iid* errors, we see that this situation gives rise to a model shown in (2.18) that Anselin (1988) labeled the spatial Durbin model (SDM). This model includes a spatial lag of the dependent variable Wy, as well as the explanatory variable vector x, and a spatial lag of the explanatory variable Wx.

$$(I_n - \rho W)y = (I_n - \rho W)x\beta + x\gamma + v \tag{2.17}$$
$$y = \rho Wy + x(\beta + \gamma) + Wx(-\rho\beta) + v \tag{2.18}$$

In Chapter 3 we will pursue the relation between omitted variables that exhibit spatial dependence and the implied spatial regression models that result. The magnitude of bias that arises in these cases will also be explored more fully.

2.3 A spatial heterogeneity motivation

Specifying models to have an individual effect, usually modeled as a separate intercept for each individual or unit, has become more popular with the prevalence of large *panel data* sets. To give this some form, let the $n \times 1$ vector a in (2.19) represent individual intercepts.

$$y = a + X\beta \qquad (2.19)$$

Typically, panel data sets include multiple observations for each unit, so estimating a vector of parameters such as a is feasible. In a spatial context where we have only a single observation for each region we can treat the vector a as a spatially structured random effect vector. Making an assumption that observational units in close proximity should exhibit effects levels that are similar to those from neighboring units provides one way of modeling spatial heterogeneity. This can be implemented by assigning the spatial autoregressive process shown in (2.20) and (2.21) to govern the vector of intercepts a. For the moment, we assume a is independent of X.

$$a = \rho W a + \varepsilon \qquad (2.20)$$
$$a = (I_n - \rho W)^{-1}\varepsilon \qquad (2.21)$$

Since we have introduced the scalar parameter ρ and a scalar noise variance parameter σ_ε^2 in conjunction with the exogenous sample connectivity information contained in the matrix W, we can feasibly estimate the $n \times 1$ vector of parameters a. Combining (2.19) and (2.21) yields the DGP of the spatial error model (SEM).

$$y = X\beta + (I_n - \rho W)^{-1}\varepsilon \qquad (2.22)$$

Consequently, spatial heterogeneity provides another way of motivating spatial dependence. In this case, the dependence can be viewed as error dependence.

What if a is not independent of X? Suppose that ε in (2.21) is replaced by $X\gamma + \epsilon$ to model the disturbances, where there is a portion that is correlated with the explanatory variables and a portion that is independent noise. In this case, a has the form in (2.23) which leads to the reduced forms for a and y in (2.24) and (2.25) and the empirical model with *iid* disturbances ϵ in (2.26).

$$a = \rho W a + X\gamma + \epsilon \qquad (2.23)$$
$$a = (I_n - \rho W)^{-1} X\gamma + (I_n - \rho W)^{-1}\epsilon \qquad (2.24)$$
$$y = X\beta + (I_n - \rho W)^{-1}(X\gamma + \epsilon) \qquad (2.25)$$
$$y = \rho W y + X(\beta + \gamma) + W X(-\rho\beta) + \epsilon \qquad (2.26)$$

The model in (2.26) takes the form of the SDM. Models involving spatially structured effects parameters are discussed in Chapter 8 in the context of origin-destination flows and Chapter 10 for the case of limited dependent variable models.

2.4 An externalities-based motivation

In a spatial context, externalities (both positive and negative) arising from neighborhood characteristics often have direct sensory impacts. For example, lots with trash provide habitat for rats and snakes that may visit contiguous yards and reduce their property values. On the other hand, homes surrounded by those with beautifully landscaped yards containing fragrant plants would have a positive effect on the house values. In terms of modeling, the spatial average of neighboring home characteristics (WX) could play a direct role in determining house prices contained in the vector y, as shown in (2.27).

$$y = \alpha \iota_n + X\beta_1 + W X\beta_2 + \varepsilon \qquad (2.27)$$

We refer to this as the spatial lag of X model or *SLX*, since the model contains spatial lags (WX) of neighboring home characteristics as explanatory variables.

2.5 A model uncertainty motivation

In applied practice we are often faced with uncertainty regarding the type of model to employ as well as conventional parameter uncertainty and uncertainty regarding specification of the appropriate explanatory variables. As an example, suppose there exists uncertainty regarding use of the autoregressive (SAR) model specification $y = \rho W y + X\beta + \varepsilon$. In particular, we introduce a competing model specification that involves spatial dependence in the disturbances of the model, $y = X\beta + u, u = \rho W u + \varepsilon$, which we refer to as the spatial error model (SEM). The respective DGPs for these two models are shown in (2.28) and (2.29). These are DGPs not estimation models, so we could have identical parameter vectors β and equal values for ρ in each DGP.

$$y_a = (I_n - \rho W)^{-1} X \beta + (I_n - \rho W)^{-1} \varepsilon \tag{2.28}$$
$$y_b = X\beta + (I_n - \rho W)^{-1} \varepsilon \tag{2.29}$$

We will discuss Bayesian model comparison methods in Chapter 6 that can be used to produce posterior model probabilities. Let π_a, π_b represent the weights or probabilities associated with the autoregressive and error models, and further assume that these two models represent the only models considered so that: $\pi_a + \pi_b = 1$. A Bayesian solution to model uncertainty is to rely on *model averaging* which involves drawing inferences from a linear combination of models. Posterior model probabilities are used as weights to produce estimates and inferences based on the combined or averaged model parameters.

It is interesting to consider the DGP associated with a linear combination of the SAR and SEM models, shown in (2.30). As the manipulations show, this leads to the SDM model in (2.31).

$$y_c = \pi_a y_a + \pi_b y_b \tag{2.30}$$
$$y_c = R^{-1} X (\pi_a \beta) + X(\pi_b \beta) + (\pi_a + \pi_b) R^{-1} \varepsilon$$
$$y_c = R^{-1} X (\pi_a \beta) + X(\pi_b \beta) + R^{-1} \varepsilon$$
$$R y_c = X (\pi_a \beta) + R X (\pi_b \beta) + \varepsilon$$
$$R y_c = X \beta + W X(-\rho \pi_b \beta) + \varepsilon$$
$$R y_c = X \beta_1 + W X \beta_2 + \varepsilon$$
$$y_c = \rho W y_c + X \beta_1 + W X \beta_2 + \varepsilon \tag{2.31}$$
$$R = I_n - \rho W$$

Combinations of other models that we introduce in the next section can be used to produce more elaborate versions of the SDM model that contain higher-order spatial lags involving terms such as $W^2 X$.

This suggests that uncertainty regarding the specific character of spatial dependence in the underlying DGP provides another motivation for models involving spatial lags of dependent and explanatory variables. In this example, we have uncertainty regarding the presence of spatial dependence in the dependent variable versus the disturbances. We will address Bayesian model averaging as a solution to model uncertainty in more detail in Chapter 6.

2.6   Spatial autoregressive regression models

As noted in Chapter 1, the spatial autoregressive structure can be combined with a conventional regression model to produce a spatial extension of the linear regression model that we have labeled the SAR model. This model is shown in (2.32), with the implied *data generating process* in (2.33).

$$y = \rho W y + \alpha \iota_n + X\beta + \varepsilon \tag{2.32}$$
$$y = (I_n - \rho W)^{-1}(\alpha \iota_n + X\beta) + (I_n - \rho W)^{-1}\varepsilon \tag{2.33}$$
$$\varepsilon \sim N(0, \sigma^2 I_n)$$

In this model, the parameters to be estimated are the usual regression parameters α, β, σ and the additional parameter ρ. Spatial lags are a hallmark of spatial regression models, and these can be used to provide extended versions of the SAR model. We have already seen one such extension, the spatial Durbin Model (SDM) which arose from our omitted variables motivation. This model includes spatial lags of the explanatory variables as well as the dependent variable. This model is shown in (2.34) along with its associated *data generating process* in (2.35).

$$y = \rho W y + \alpha \iota_n + X\beta + WX\gamma + \varepsilon \tag{2.34}$$
$$y = (I_n - \rho W)^{-1}(\alpha \iota_n + X\beta + WX\gamma + \varepsilon) \tag{2.35}$$
$$\varepsilon \sim N(0, \sigma^2 I_n)$$

We can also use spatial lags to reflect dependence in the disturbance process, which leads to the spatial error model (SEM), shown in (2.36).

$$y = \alpha \iota_n + X\beta + u \tag{2.36}$$
$$u = \rho W u + \varepsilon$$
$$\varepsilon \sim N(0, \sigma^2 I_n)$$

Another member of the family of spatial regression models is one we label SAC, taking the form in (2.37), where the matrix W_1 may be set equal to W_2. This model contains spatial dependence in both the dependent variable and the disturbances.

$$y = \alpha \iota_n + \rho W_1 y + X\beta + u \tag{2.37}$$
$$u = \theta W_2 u + \varepsilon$$
$$\varepsilon \sim N(0, \sigma^2 I_n)$$
$$y = (I_n - \rho W_1)^{-1}(X\beta + \alpha \iota_n) + (I_n - \rho W_1)^{-1}(I_n - \theta W_2)^{-1}\varepsilon \tag{2.38}$$

We note that spatial regression models have been proposed that use a moving average process in place of the spatial autoregressive process. For example, $u = (I_n - \theta W)\varepsilon$ could be used to model the disturbances. This type of process provides a method for capturing *local effects* arising from immediate neighbors, as opposed to the autoregressive process that models *global effects* (Anselin, 2003).

The (local) spatial moving average can be combined with a (global) spatial autoregressive process to produce a model that Anselin and Bera (1998) label a spatial autoregressive moving average model, SARMA. This takes the form in (2.39), with the DGP shown in (2.40), where as in the case of the SAC model, the matrix W_1 might be set equal to W_2.

$$
\begin{aligned}
y &= \alpha \iota_n + \rho W_1 y + X\beta + u \\
u &= (I_n - \theta W_2)\varepsilon \\
\varepsilon &\sim N(0, \sigma^2 I_n) \\
y &= (I_n - \rho W_1)^{-1}(X\beta + \alpha \iota_n) + (I_n - \rho W_1)^{-1}(I_n - \theta W_2)\varepsilon
\end{aligned}
$$
(2.39)

(2.40)

The distinction between the SAC and the SARMA model lies in the differences between the disturbances in their respective DGPs (2.38) and (2.40). The SAC uses $(I_n - \rho W_1)^{-1}(I_n - \theta W_2)^{-1}\varepsilon$ while SARMA uses $(I_n - \rho W_1)^{-1}(I_n - \theta W_2)\varepsilon$. Given the series representation of the inverse in terms of matrix powers, it should be clear that the SAC will place more weight on higher powers of W than SARMA. However, both of these models have: $E(y) = (I_n - \rho W_1)^{-1}(X\beta + \alpha \iota_n)$, which is the same as $E(y)$ for the SAR model. Therefore, these models concentrate on a more elaborate model for the disturbances, whereas the SDM elaborates on the model for spillovers.

In addition, many other models exist such as the matrix exponential, fractional differencing, and other variants of the ARMA specifications. We will deal with some of these in Chapter 9.

2.7   Interpreting parameter estimates

Spatial regression models exploit the complicated dependence structure between observations which represent countries, regions, counties, etc. Because of this, the parameter estimates contain a wealth of information on relationships among the observations or regions. A change in a single observation (region) associated with any given explanatory variable will affect the region itself (a direct impact) and potentially affect all other regions indirectly (an indirect impact). In fact, the ability of spatial regression models to capture these interactions represents an important aspect of spatial econometric models noted in Behrens and Thisse (2007).

A virtue of spatial econometrics is the ability to accommodate extended modeling strategies that describe multi-regional interactions. However, this rich set of information also increases the difficulty of interpreting the resulting estimates. In Section 2.7.1 we describe the theory behind analysis of the impact of changing explanatory variables on the dependent variable in the model. Computational approaches to calculating summary measures of these impacts are the subject of Section 2.7.2, with measures of dispersion for these summary statistics discussed in Section 2.7.3. A partitioning of the summary measures of impact that allows an examination of the rate of decay of impact over space is set forth in Section 2.7.4 and Section 2.7.5 discusses an error model that contains both X and WX which we label (SDEM) as a simplified means of estimating direct and indirect impacts.

2.7.1   Direct and indirect impacts in theory

Linear regression parameters have a straightforward interpretation as the partial derivative of the dependent variable with respect to the explanatory variable. This arises from linearity and the assumed independence of observations in the model: $y = \sum_{r=1}^{k} x_r \beta_r + \varepsilon$. The partial derivatives of y_i with respect to x_{ir} have a simple form: $\partial y_i / \partial x_{ir} = \beta_r$ for all i, r; and $\partial y_i / \partial x_{jr} = 0$, for $j \neq i$ and all variables r.

One way to think about this is that the information set for an observation i in regression consists only of exogenous or predetermined variables associated with observation i. Thus, a linear regression specifies: $E(y_i) = \sum_{r=1}^{k} x_{ir} \beta_r$, and takes a restricted view of the information set by virtue of the independence assumption.

In models containing spatial lags of the explanatory or dependent variables, interpretation of the parameters becomes richer and more complicated. A number of researchers have noted that models containing spatial lags of the dependent variable require special interpretation of the parameters (Anselin and LeGallo, 2006; Kelejian, Tavlas and Hondronyiannis, 2006; Kim, Phipps, and Anselin, 2003; LeGallo, Ertur, and Baumont, 2003).

In essence, spatial regression models expand the information set to include information from neighboring regions/observations. To see the effect of this, consider the SDM model which we have re-written in (2.41).

$$(I_n - \rho W)y = X\beta + WX\theta + \iota_n \alpha + \varepsilon$$

$$y = \sum_{r=1}^{k} S_r(W)x_r + V(W)\iota_n \alpha + V(W)\varepsilon \qquad (2.41)$$

$$S_r(W) = V(W)(I_n \beta_r + W\theta_r)$$

$$V(W) = (I_n - \rho W)^{-1} = I_n + \rho W + \rho^2 W^2 + \rho^3 W^3 + \dots$$

To illustrate the role of $S_r(W)$, consider the expansion of the data generating process in (2.41) as shown in (2.42) (Kim, Phipps, and Anselin, 2003, c.f. equation(4)).

$$
\begin{pmatrix} y_1 \\ y_2 \\ \vdots \\ y_n \end{pmatrix} = \sum_{r=1}^{k} \begin{pmatrix} S_r(W)_{11} & S_r(W)_{12} & \cdots & S_r(W)_{1n} \\ S_r(W)_{21} & S_r(W)_{22} & & \\ \vdots & \vdots & \ddots & \\ S_r(W)_{n1} & S_r(W)_{n2} & \cdots & S_r(W)_{nn} \end{pmatrix} \begin{pmatrix} x_{1r} \\ x_{2r} \\ \vdots \\ x_{nr} \end{pmatrix} \qquad (2.42)
$$

$$
+ V(W)\iota_n \alpha + V(W)\varepsilon
$$

The case of a single dependent variable observation in (2.43) makes the role of the matrix $S_r(W)$ more transparent. We use $S_r(W)_{ij}$ in this equation to denote the i,jth element of the matrix $S_r(W)$, and $V(W)_i$ to indicate the ith row of $V(W)$.

$$
y_i = \sum_{r=1}^{k} [S_r(W)_{i1}x_{1r} + S_r(W)_{i2}x_{2r}+, \ldots, +S_r(W)_{in}x_{nr}]
$$
$$
+ V(W)_i \iota_n \alpha + V(W)_i \varepsilon \qquad (2.43)
$$

It follows from (2.43) that unlike the case of the independent data model, the derivative of y_i with respect to x_{jr} is potentially non-zero, taking a value determined by the i,jth element of the matrix $S_r(W)$. It is also the case that the derivative of y_i with respect to x_{ir} usually does not equal β_r as in least-squares.

$$
\frac{\partial y_i}{\partial x_{jr}} = S_r(W)_{ij} \qquad (2.44)
$$

An implication of this is that a change in the explanatory variable for a single region (observation) can potentially affect the dependent variable in all other observations (regions). This is of course a logical consequence of our SDM model, since the model takes into account other regions dependent and explanatory variables through the introduction of Wy and WX. For a model where the dependent variable vector y reflects say levels of regional per capita income, and the explanatory variables are regional characteristics (e.g., human and physical capital, industrial structure, population density, etc.), regional variation in income levels is modeled to depend on income levels from neighboring regions captured by the spatial lag vector Wy, as well as characteristics of neighboring regions represented by WX.

The own derivative for the ith region shown in (2.45) results in an expression $S_r(W)_{ii}$ that measures the impact on the dependent variable observation i from a change in x_{ir}. This impact includes the effect of *feedback loops* where observation i affects observation j and observation j also affects observation i

as well as longer paths which might go from observation i to j to k and back to i.

$$\frac{\partial y_i}{\partial x_{ir}} = S_r(W)_{ii} \qquad (2.45)$$

Consider the scalar term $S_r(W)_{ii}$ in light of the matrix, $S_r(W) = (I_n - \rho W)^{-1}(I_n \beta_r + W \theta_r)$. Focusing on the inverse term and the series expansion of this inverse from Chapter 1 expression (1.12), neighboring region influences arise as a result of impacts passing through neighboring regions and back to the region itself. To see this, observe that the matrix W^2 from Chapter 1 expression (1.15) reflects second order neighbors and contains non-zero elements on the diagonal. These arise because region i is considered a neighbor to its neighbor, so that impacts passing through neighboring regions will exert a feedback influence on region i itself. The magnitude of this type of feedback will depend upon: (1) the position of the regions in space, (2) the degree of connectivity among regions which is governed by the weight matrix W in the model, (3) the parameter ρ measuring the strength of spatial dependence, and (4) the parameters β and θ. The diagonal elements of the $n \times n$ matrix $S_r(W)$ contain the direct impacts, and off-diagonal elements represent indirect impacts.

There are some situations where practitioners are interested in impacts arising from changes in a single region or the impact of changes on a single region, which would be reflected in one column or row of the matrix $S_r(W)$ as we will motivate shortly. For example, Kelejian, Tavlas and Hondronyiannis (2006) examine the impact of financial contagion arising from a single country on other countries in the model, Anselin and LeGallo (2006) examine diffusion of point source air pollution, and LeGallo, Ertur, and Baumont (2003) and Dall'erba and LeGallo (2007) examine impacts of changing explanatory variables (such as European Union structural funds) in strategic regions on overall economic growth.

In general however, since the impact of changes in an explanatory variable differs over all regions/observations, Pace and LeSage (2006) suggest a desirable summary measure of these varying impacts. A natural scalar summary would be based on summing the total impacts over the rows (or columns) of the matrix $S_r(W)$, and then taking an average over all regions. They label the average of row sums from this matrix as the *Average Total Impact* **to** *an Observation*, and refer to the average of column sums as *Average Total Impact* **from** *an Observation*. An average of the diagonal of matrix $S_r(W)$ provides a summary measure of the *Average Direct Impact*. Finally, a scalar summary of the *Average Indirect Impact* is by definition the difference between the Average Total Impact and Average Direct Impact. Formally, the definitions of these summary measures of impact are:

1. *Average Direct Impact.* The impact of changes in the ith observation of x_r, which we denote x_{ir}, on y_i could be summarized by measuring the

average $S_r(W)_{ii}$, which equals $n^{-1}\operatorname{tr}(S_r(W))$. Note that averaging over the direct impact associated with all observations i is similar in spirit to typical regression coefficient interpretations that represent average response of the dependent to independent variables over the sample of observations.

2. *Average Total Impact to an Observation.* The sum across the ith row of $S_r(W)$ would represent the total impact on individual observation y_i resulting from changing the rth explanatory variable by the same amount across all n observations (e.g., $x_r + \delta\iota_n$ where δ is the scalar change). There are n of these sums given by the column vector $c_r = S_r(W)\iota_n$, so an average of these total impacts is $n^{-1}\iota_n' c_r$.

3. *Average Total Impact from an Observation.* The sum down the jth column of $S_r(W)$ would yield the total impact over all y_i from changing the rth explanatory variable by an amount in the jth observation (e.g., $x_{jr} + \delta$). There are n of these sums given by the row vector $r_r = \iota_n' S_r(W)$, so an average of these total impacts is $n^{-1}r_r\iota_n$.

It is easy to see that the numerical values of the summary measures for the two forms of average total impacts set forth in 2) and 3) above are equal, since $\iota_n' c_r = \iota_n' S_r(W)\iota_n$, as does $r_r\iota_n = \iota_n' S_r(W)\iota_n$. However, these two measures allow for different interpretative viewpoints, despite their numerical equality.

The *from an observation* view expressed in 3) above relates how changes in a single observation j influences all observations. In contrast, the *to an observation* view expressed in 2) above considers how changes in all observations influence a single observation i. Averaging over all n of the total impacts, whether taking the *from an observation* or *to an observation* approaches, leads to the same numerical result. Therefore, the average total impact is the average of all derivatives of y_i with respect to x_{jr} for any i, j. The average direct impact is the average of all own derivatives. Consequently, the average of all derivatives (average total impact) less the average own derivative (average direct impact) equals the average cross derivative (average indirect impact).

The application of Kelejian, Tavlas and Hondronyiannis (2006) examines the impact of financial contagion arising from a single country on other countries in the model, taking the *from an observation* viewpoint expressed in 3) above. On the other hand, Dall'erba and LeGallo (2007) examine impacts of changing explanatory variables (such as European Union structural funds) which apply to all regions on own-region as well as overall economic growth, an example more consistent with the *to an observation* view. We will provide additional examples in our applied illustrations throughout the text.

We need to keep in mind that the scalar summary measures of impact reflect how these changes would work through the simultaneous dependence system over time to culminate in a new steady state equilibrium.

We note that our measures of impact for the SAR model can be derived from (2.46).

$$(I_n - \rho W)y = X\beta + \iota_n\alpha + \varepsilon$$

$$y = \sum_{r=1}^{k} S_r(W)x_r + V(W)\iota_n\alpha + V(W)\varepsilon \qquad (2.46)$$

$$S_r(W) = V(W)I_n\beta_r$$

$$V(W) = (I_n - \rho W)^{-1} = I_n + \rho W + \rho^2 W^2 + \rho^3 W^3 + \dots$$

The summary measure of total impacts, $n^{-1}\iota_n' S_r(W)\iota_n$, for this model take the simple form in (2.47) for row-stochastic W.

$$n^{-1}\iota_n' S_r(W)\iota_n = n^{-1}\iota_n'(I_n - \rho W)^{-1}\beta_r\iota_n$$

$$= (1 - \rho)^{-1}\beta_r \qquad (2.47)$$

In contrast to our approach and nomenclature, Abreu, de Groot, and Florax (2004) consider the (simpler) SAR model and the expression in (2.48). They refer to β_r as a direct effect, $W\rho\beta_r$ as an indirect effect, and the term in brackets is labeled induced effects.

$$\frac{\partial y}{\partial x_r'} = I_n\beta_r + W\rho\beta_r + [W^2\rho^2\beta_r + W^3\rho^3\beta_r + \dots] \qquad (2.48)$$

Under their labeling, as they correctly point out, the direct effects *do not* correspond to the partial derivative of y_i with respect to x_{ir}, and the indirect effects *do not* correspond to the partial derivative of y_i with respect to x_{jr} for $i \neq j$. In contrast, our definitions of direct effect and indirect effects *do* correspond to the own- and cross-partial derivatives respectively. An additional benefit of our approach is that we reduce the number of labels from three (direct, indirect, and induced) to only two (direct and indirect).

Relative to the SAR model, the SDM model total impacts arising from changes in X_r exhibit a great deal of heterogeneity arising from the presence of the additional matrix $W\theta_r$ in the total effects. In particular, this allows the spillovers from a change in each explanatory variable to differ as opposed to the SAR case which has a common, global multiplier for each variable.

For the case of the SAC model, the total impacts take the same form as in the SAR model, since the spatial autoregressive model for the disturbances in this model do not come into play when considering the partial derivative of y with respect to changes in the explanatory variables X. Of course, in applied practice, impact estimates for the SAR model would be based on SAR model estimates for the coefficients ρ, β, whereas those for the SAC would be based on SAC model estimates for these parameters, which would likely be different. It is also the case that the SARMA model introduced in Section 2.6

would have the same total impacts as the SAR and SAC models, again with the caveat that estimates based on the SARMA model for ρ, β would be used in calculating these impacts. Consequently, for larger data sets SAR, SAC, and SARMA should yield very similar estimated impacts (in the absence of misspecification affecting variables other than the disturbance terms).

2.7.2 Calculating summary measures of impacts

We formally define in (2.49) through (2.51), $\bar{M}(r)_{total}$, $\bar{M}(r)_{direct}$, and $\bar{M}(r)_{indirect}$, representing the average total impacts, the average direct impacts, and the average indirect impacts from changes in the model variable X_r.

$$\bar{M}(r)_{direct} = n^{-1} \operatorname{tr}(S_r(W)) \qquad (2.49)$$

$$\bar{M}(r)_{total} = n^{-1} \iota'_n S_r(W) \iota_n \qquad (2.50)$$

$$\bar{M}(r)_{indirect} = \bar{M}(r)_{total} - \bar{M}(r)_{direct} \qquad (2.51)$$

It is computationally inefficient to calculate the summary impact estimates using the definitions above, since this would require inversion of the $n \times n$ matrix $(I_n - \rho W)$ in $S_r(W)$. We propose an approximation to the infinite expansion of $S_r(W)$ based on traces of the powers of W. This of course requires that the highest power considered in the approximation is large enough to ensure approximate convergence. Chapter 4 discusses a linear in n approximation of this type that aids in calculating the scalar summary measures for the direct, indirect and total impacts.

2.7.3 Measures of dispersion for the impact estimates

In order to draw inferences regarding the statistical significance of the impacts associated with changing the explanatory variables, we require the distribution of our scalar summary measures for the various types of impact. Computationally efficient simulation approaches can be used to produce an empirical distribution of the parameters $\alpha, \beta, \theta, \rho, \sigma$ that are needed to calculate the scalar summary measures. This distribution can be constructed using a large number of simulated parameters drawn from the multivariate normal distribution of the parameters implied by the maximum likelihood estimates.

Alternatively, Bayesian Markov Chain Monte Carlo (MCMC) estimation methods set forth in LeSage (1997), discussed in Chapter 5 can be used to produce estimates of dispersion for the scalar impacts. Since MCMC estimation yields samples (draws) from the posterior distribution of the model parameters, these can be used in (2.49) and (2.50) to produce a posterior distribution for the scalar summary measures of impact. As shown by Gelfand et al. (1990), MCMC can yield valid inference on non-linear functions of the parameters such as the direct and indirect impacts in (2.49) and (2.50). All that is required is evaluation and storage of the draws reflecting the non-linear combinations of the parameters. Posterior estimates of dispersion are based on simple variance calculations applied to these stored draws, a topic taken up in Chapter 5.

2.7.4   Partitioning the impacts by order of neighbors

It should be clear that impacts arising from a change in the explanatory variables will influence low-order neighbors more than higher-order neighbors. We would expect a profile of decline in magnitude for the impacts as we move from lower- to higher-order neighbors. In some applications the particular pattern of decay of influence on various order neighbors may be of interest. We provide an example of this in Chapter 3.

Since the impacts are a function of $S_r(W)$, these can be expanded as a linear combination of powers of the weight matrix W using the infinite series expansion of $(I_n - \rho W)^{-1}$. Applying this to (2.49) and (2.50) where we use the definition of $S_r(W)$ for the SAR model allows us to observe the impact associated with each power of W. These powers correspond to the observations themselves (zero-order), immediate neighbors (first-order), neighbors of neighbors (second-order), and so on.

$$S_r(W) \approx (I_n + \rho W + \rho^2 W^2 + \rho^3 W^3 +, \ldots, +\rho^q W^q)\beta_r \qquad (2.52)$$

As an example, Table 2.1 shows both the *cumulative* and *marginal or spatially partitioned* direct, indirect and total impacts associated with orders 0 to 9 for the case of a SAR model where $\beta_r = 0.5$ and $\rho = 0.7$. From the table we see a cumulative direct effect equal to 0.586, which given the coefficient of 0.5 indicates that there is feedback equal to 0.086 arising from each region impacting neighbors that in turn impacts neighbors to neighbors and so on. In this case these feedback effects account for the difference between the coefficient value of $\beta_r = 0.5$ and the cumulative direct effect of 0.586.

The cumulative indirect effects equal to 1.0841 are nearly twice the magnitude of the cumulative direct effects of 0.5860. Based on the t-statistics calculated from a set of 5,000 simulated parameter values, all three effects are significantly different from zero.

The spatial partitioning of the direct effect shows that by the time we reach 9th-order neighbors we have accounted for 0.5834 of the 0.5860 cumulative direct effect. Of note is the fact that for W^0 there is no indirect effect, only direct effect, and for W^1 there is no direct effect, only indirect. To see this, consider that when $q = 0$, $W^0 = I_n$, and we have: $S_r(W) = I_n \beta_r = 0.5 I_n$. When $q = 1$ we have only an indirect effect, since there are zero elements on the diagonal of the matrix W. Also, the row-stochastic nature of W leads to an average of the sum of the rows that takes the form: $\rho \beta_r = 0.7 \times 0.5 = 0.35$, when $q = 1$.

TABLE 2.1: Spatial partitioning of direct, indirect and total impacts

	Cumulative Effects		
	Mean	Std. dev	t-statistic
Direct effect X_r	0.5860	0.0148	39.6106
Indirect effect X_r	1.0841	0.0587	18.4745
Total effect X_r	1.6700	0.0735	22.7302
W-order	Spatially Partitioned Effects		
	Total	Direct	Indirect
W^0	0.5000	0.5000	0
W^1	0.3500	0	0.3500
W^2	0.2452	0.0407	0.2045
W^3	0.1718	0.0144	0.1574
W^4	0.1204	0.0114	0.1090
W^5	0.0844	0.0066	0.0778
W^6	0.0591	0.0044	0.0547
W^7	0.0415	0.0028	0.0386
W^8	0.0291	0.0019	0.0272
W^9	0.0204	0.0012	0.0191
$\sum_{q=0}^{9} W^q$	1.6220	0.5834	1.0386

While cumulative indirect effects having larger magnitudes than the direct effects might seem counterintuitive, the marginal or partitioned impacts make it clear that individual indirect effects falling on first-order, second-order and higher-order neighboring regions are smaller than the average direct effect of 0.5 falling on the "own-region." Cumulating these effects however leads to a larger indirect effect which represents smaller impacts spread over many regions.

We see the direct effects die down quickly as we move to higher-order neighbors, whereas the indirect or spatial spillover effects decay more slowly as we move to higher-order neighbors.

2.7.5 Simplified alternatives to the impact calculations

Spatial regression models such as the SEM that do not involve spatial lags of the dependent variable produce estimates for the parameters β that can be interpreted in the usual regression sense as partial derivatives: $\partial y_i / \partial x_{ir} = \beta_r$ for all i, r; and $\partial y_i / \partial x_{jr} = 0$, for $j \neq i$ and all variables r. Of course, these models do not allow indirect impacts to arise from changes in the explanatory variables, similar to the least-squares situation where the dependent variable observations are treated as independent.

An alternative to the SEM model that we label the *spatial Durbin error model* (SDEM) includes a spatial lag of the explanatory variables WX, as well as spatially dependent disturbances. This model, which augments the SEM model with a spatial lag of the explanatory variables is shown in (2.53), with the model DGP in (2.54).

$$y = X\beta + WX\gamma + \iota_n\alpha + u \qquad (2.53)$$
$$u = R^{-1}\varepsilon$$
$$y = X\beta + WX\gamma + \iota_n\alpha + R^{-1}\varepsilon \qquad (2.54)$$
$$R = I_n - \rho W$$
$$E(y) = \sum_{r=1}^{k} S_r(W)x_r + \iota_n\alpha \qquad (2.55)$$
$$S_r(W) = (I_n\beta_r + W\gamma_r)$$

The SDEM model does not allow for a separate lagged dependent variable effect, but does allow for spatially dependent errors and spatial lags of the explanatory variables. Relative to the more general SDM, it simplifies interpretation of the impacts, since the direct impacts are represented by the model parameters β and indirect impacts correspond to γ. This also allows us to use measures of dispersion such as the standard deviation or t-statistic for these regression parameters as a basis for inference regarding significance of the direct and indirect impacts.

The version of the SDEM in (2.53) uses the same spatial weight matrix for the errors and the spatially lagged explanatory variables, but this could be generalized to allow for different weights and not affect the simplicity of interpreting the direct and indirect impacts as corresponding to the model parameters.

The SDEM replaces the global multiplier found in the SDM with local multipliers that simplify interpretation of the model estimates. However, we note that the SDEM could result in underestimation of higher-order (global) indirect impacts. The SDEM does not nest the SDM and *vice versa*. However, one can devise an extended SDM that nests the SDEM.

2.8   Chapter summary

We provided numerous motivations for why spatial regression relationships might arise that include spatial lags of the dependent variable vector. Models containing spatially lagged dependent variables have been used most often in situations where there is an intuitive or theoretical motivation that y will depend on neighboring values of y. For example, the hedonic housing price literature where it is generally thought that home prices depend on prices of recently sold neighboring homes. This is because appraisers/real estate agents presumably use information on recently sold homes to determine the asking price. Another example is competition between local governments, where it seems intuitive that local governments can react to actions taken by nearby local governments.

There may be a wider role for these models than previously thought since we were able to provide three motivations based on: omitted variables, space-time dependence and model uncertainty that resulted in models involving spatial lags of the dependent variable.

This type of development has wide-ranging implications for the interface of economic theory and spatial econometrics. It suggests that spatial econometric models may be applicable in many situations where they have not previously been employed.

Chapter 3

Maximum Likelihood Estimation

As shown in Chapter 2, estimation of spatial models via least squares can lead to inconsistent estimates of the regression parameters for models with spatially lagged dependent variables, inconsistent estimation of the spatial parameters, and inconsistent estimation of standard errors. In contrast, maximum likelihood is consistent for these models (Lee, 2004). Consequently, this chapter focuses on maximum likelihood estimation of spatial regression models. Historically, much of the spatial econometrics literature has focused on ways to avoid maximum likelihood estimation because of perceived computational difficulties. There have been a great many improvements in computational methods for maximum likelihood estimation of spatial regression models since the time of Anselin's 1988 text. These improvements allow models involving samples containing more than 60,000 US Census tract observations to be estimated in only a few seconds on desktop and laptop computers.

Section 3.1 addresses maximum likelihood estimation for the SAR, SDM, SEM, and other models. Section 3.1 provides a number of techniques that greatly reduce previous computational difficulties that arose in estimation of these models. Section 3.2 turns attention to maximum likelihood estimation of variance-covariance estimates of dispersion for the model parameters required for inference. We provide a new approach that can be used to reduce the computational tasks needed to construct maximum likelihood estimates of dispersion needed for inference.

As already motivated, omitted variables are a likely problem in applied work with regional economic data, and Section 3.3 further explores the empirical impact of spatial dependence on omitted variables bias in spatial regression models. We present theoretical expressions for the bias along with statistical tests and model specifications that mitigate problems posed by omitted variables that are correlated with included explanatory variables.

The chapter concludes with an application in Section 3.4 that illustrates many of the issues discussed in the chapter. We rely on a simple model containing a single explanatory variable used to explain factor productivity differences among European Union regions.

3.1   Model estimation

We address maximum likelihood estimation for the family of spatial regression models including SAR, SDM, SEM and SAC, which were introduced in Section 2.6. The spatial Durbin model (SDM) provides a general starting point for discussion of spatial regression model estimation since this model subsumes the spatial error model (SEM) and the spatial autoregressive model (SAR).

In Section 3.1.1 we discuss maximum likelihood estimation of the SAR and SDM models whose likelihood functions coincide. In Section 3.1.2 we turn attention to the SEM model likelihood function and estimation procedure, and models involving multiple weight matrices are discussed in Section 3.1.3.

3.1.1   SAR and SDM model estimation

The SDM model is shown in (3.1) along with its associated *data generating process* in (3.2),

$$y = \rho W y + \alpha \iota_n + X\beta + W X\theta + \varepsilon \tag{3.1}$$

$$y = (I_n - \rho W)^{-1}\left(\alpha \iota_n + X\beta + W X\theta + \varepsilon\right) \tag{3.2}$$

$$\varepsilon \sim N(0, \sigma^2 I_n)$$

where 0 represents an $n \times 1$ vector of zeros and ι_n represents an $n \times 1$ vector of ones associated with the constant term parameter α. This model can be written as a SAR model by defining: $Z = \begin{bmatrix} \iota_n & X & WX \end{bmatrix}$ and $\delta = \begin{bmatrix} \alpha & \beta & \theta \end{bmatrix}'$, leading to (3.3). This means that the likelihood function for SAR and SDM models can be written in the same form where: $Z = \begin{bmatrix} \iota_n & X \end{bmatrix}$ for the SAR model and $Z = \begin{bmatrix} \iota_n & X & WX \end{bmatrix}$ for the SDM model.

$$y = \rho W y + Z\delta + \varepsilon \tag{3.3}$$

$$y = (I_n - \rho W)^{-1} Z\delta + (I_n - \rho W)^{-1}\varepsilon \tag{3.4}$$

$$\varepsilon \sim N(0, \sigma^2 I_n)$$

From the model statement (3.3), *if* the true value of the parameter ρ was known to be say ρ^*, we could rearrange the model statement in (3.3) as shown in (3.5).

$$y - \rho^* W y = Z\delta + \varepsilon \tag{3.5}$$

This suggests an estimate for δ of $\hat{\delta} = (Z'Z)^{-1}Z'(I_n - \rho^* W)y$. In this case we could also find an estimate for the noise variance parameter $\hat{\sigma}^2 = n^{-1}e(\rho^*)'e(\rho^*)$, where $e(\rho^*) = y - \rho^* W y - Z\hat{\delta}$.

These ideas motivate that we can concentrate the full (log) likelihood with respect to the parameters β, σ^2 and reduce maximum likelihood to a univariate optimization problem in the parameter ρ.

Maximizing the full log-likelihood for the case of the SAR model would involve setting the first derivatives with respect to the parameters β, σ^2 and ρ equal to zero and simultaneously solving these first-order conditions for all parameters.

In contrast, equivalent maximum likelihood estimates could be found using the log-likelihood function concentrated with respect to the parameters β and σ^2. This involves substituting *closed-form solutions* from the first order conditions for the parameters β and σ^2 to yield a log-likelihood that is said to be *concentrated* with respect to these parameters. We label these expressions $\hat{\beta}(\rho), \hat{\sigma}^2(\rho)$, and note that they depend on sample data plus the unknown parameter ρ. In the case of the SAR model, this leaves us with a concentrated log-likelihood that depends only on the single scalar parameter ρ. Optimizing the *concentrated* log-likelihood function with respect to ρ, to find the maximum likelihood estimate $\hat{\rho}$ allows us to use this estimate in the closed-form expressions for $\hat{\beta}(\hat{\rho})$ and $\hat{\sigma}^2(\hat{\rho})$ to produce maximum likelihood estimates for these parameters.

Working with the concentrated log-likelihood yields exactly the same maximum likelihood estimates $\hat{\beta}$, $\hat{\sigma}$, and $\hat{\rho}$ as would arise from maximizing the full log-likelihood (Davidson and MacKinnon, 1993, p. 267-269). The motivation for optimizing the concentrated log-likelihood is that this simplifies the optimization problem by reducing a multivariate optimization problem to a univariate problem. Another advantage of using the concentrated log-likelihood is that simple adjustments to output from the optimization problem (that we describe later) can be used to produce a computationally efficient variance-covariance matrix that we use for inference regarding the parameters. These inferences are identical to those that would be obtained from solving the more cumbersome optimization problem involving the full log-likelihood.

The log-likelihood function for the SDM (and SAR) models takes the form in (3.6) (Anselin, 1988, p. 63), where ω is the $n \times 1$ vector of eigenvalues of the matrix W.

$$\ln L = -(n/2)\ln(\pi\sigma^2) + \ln|I_n - \rho W| - \frac{e'e}{2\sigma^2} \tag{3.6}$$
$$e = y - \rho Wy - Z\delta$$
$$\rho \in (\min(\omega)^{-1}, \max(\omega)^{-1})$$

If ω contains only real eigenvalues, a positive definite variance-covariance matrix is ensured by the condition: $\rho \in (\min(\omega)^{-1}, \max(\omega)^{-1})$, as shown in Ord (1975). The matrix W can always be constructed to have a maximum eigenvalue of 1. For example, scaling the weight matrix by its maximum eigenvalue as noted by Barry and Pace (1999); Kelejian and Prucha (2007). In this case, the interval for ρ becomes $(\min(\omega)^{-1}, 1)$ and a subset of this widely employed in practice is $\rho \in [0, 1)$. We provide more details regarding the admissible values for ρ in Chapter 4. The admissible values can become more complicated for non-symmetric weight matrices W since these may have complex eigenvalues.

As noted, the log-likelihood can be concentrated with respect to the coefficient vector δ and the noise variance parameter σ^2. Pace and Barry (1997) suggested a convenient approach to concentrating out the parameters δ and σ^2, shown in (3.7). The term κ is a constant that does not depend on the parameter ρ, and $|I_n - \rho W|$ is the determinant of this $n \times n$ matrix. We use the notation $e(\rho)$ to indicate that this vector depends on values taken by the parameter ρ, as does the scalar concentrated log-likelihood function value $\ln L(\rho)$.

$$\ln L(\rho) = \kappa + \ln|I_n - \rho W| - (n/2)\ln(S(\rho)) \qquad (3.7)$$
$$S(\rho) = e(\rho)'e(\rho) = e_o'e_o - 2\rho e_o'e_d + \rho^2 e_d'e_d$$
$$e(\rho) = e_o - \rho e_d$$
$$e_o = y - Z\delta_o$$
$$e_d = Wy - Z\delta_d$$
$$\delta_o = (Z'Z)^{-1}Z'y$$
$$\delta_d = (Z'Z)^{-1}Z'Wy$$

To simplify optimization of the log-likelihood with respect to the scalar parameter ρ, Pace and Barry (1997) proposed evaluating the log-likelihood using a $q \times 1$ vector of values for ρ in the interval $[\rho_{\min}, \rho_{\max}]$, labeled as $\rho_1, \ldots, \rho_q$ in (3.8).

$$\begin{pmatrix} \ln L(\rho_1) \\ \ln L(\rho_2) \\ \vdots \\ \ln L(\rho_q) \end{pmatrix} = \kappa + \begin{pmatrix} \ln|I_n - \rho_1 W| \\ \ln|I_n - \rho_2 W| \\ \vdots \\ \ln|I_n - \rho_q W| \end{pmatrix} - (n/2) \begin{pmatrix} \ln(S(\rho_1)) \\ \ln(S(\rho_2)) \\ \vdots \\ \ln(S(\rho_q)) \end{pmatrix} \qquad (3.8)$$

In Chapter 4 we discuss a number of approaches to efficiently calculating the term $\ln|I_n - \rho_i W|$ over a vector of values for the parameter ρ. In our discussion here, we simply assume that these values are available during optimization of the log-likelihood. Given a sufficiently fine grid of q values for the log-likelihood, interpolation can supply intervening points to any desired accuracy (which follows from the smoothness of the log-likelihood function). Note, the scalar moments $e_o'e_o$, $e_d'e_o$, and $e_d'e_d$ and the $k \times 1$ vectors δ_o, δ_d are computed prior to optimization, and so given a value for ρ, calculating $S(\rho)$ simply requires weighting three numbers. Given the optimum value of ρ, this becomes the maximum likelihood estimate of ρ denoted as $\hat{\rho}$. Therefore, it requires very little computation to arrive at the vector of concentrated log-likelihood values.

Given the maximum likelihood estimate $\hat{\rho}$, (3.9), (3.10), and (3.11) show the maximum likelihood estimates for the coefficients $\hat{\delta}$, the noise variance parameter $\hat{\sigma}^2$, and associated variance-covariance matrix for the disturbances.

$$\hat{\delta} = \delta_o - \hat{\rho}\delta_d \tag{3.9}$$

$$\hat{\sigma}^2 = n^{-1}S(\hat{\rho}) \tag{3.10}$$

$$\hat{\Omega} = \hat{\sigma}^2 \left[(I_n - \hat{\rho}W)'(I_n - \hat{\rho}W)\right]^{-1} \tag{3.11}$$

Although the vectorized approach works well, Chapter 4 discusses an alternative closed-form solution technique for ρ. However, we prefer to discuss the vectorized approach here due to its simplicity.

The likelihood function combines a transformed sum-of-squared errors term with the log determinant term acting as a penalty function that prevents the maximum likelihood estimate of ρ from being equal to an estimate based solely on the minimized (transformed) sum of squared errors, $S(\rho)$. The vectorized approach provides the additional advantage of ensuring a global as opposed to a local optimum.

Maximum likelihood estimation could proceed using a variety of univariate optimization techniques. These could include the vectorized approach just discussed based on a fine grid of values of ρ (large q), non-derivative search methods such as the Nelder-Mead simplex or bisection search scheme, or by applying a derivative-based optimization technique (Press et al., 1996). Some form of Newton's method with numerical derivatives has the advantage of providing the optimum as well as the second derivative of the concentrated log likelihood at the optimum $\hat{\rho}$. This numerical estimate of the second derivative in conjunction with other information can be useful in producing a numerical estimate of the variance-covariance matrix for the parameters. We discuss this topic in more detail in Section 3.2.

As shown above, an apparent barrier to implementing these models for large n is the $n \times n$ matrix W. If W contains all non-zero elements, it would require enormous amounts of memory to store this matrix for problems involving large samples such as the US Census tracts where $n > 60,000$. Fortunately, W is usually *sparse*, meaning it contains a large proportion of zeros. For example, if one relies on contiguous regions or some number m of nearest neighboring regions to form W, the spatial weight matrix will only contain mn non-zeros as opposed to n^2 non-zeros for a *dense* matrix. The proportion of non-zeros becomes m/n which falls with n. Contiguity weight matrices have an average of six neighbors per row (for spatially random sets of points on a plane). As an example, using the 3,111 US counties representing the lower 48 states plus the district of Columbia, there are 9,678,321 elements in the $3,111 \times 3,111$ matrix W, but only $3,111 \times 6 = 18,666$ would be non-zero, or 0.1929 percent of the entries. In addition, calculating matrix-vector products such as Wy and WX take much less time for sparse matrices. In both cases, sparse matrices require linear in n operations ($O(n)$) while a dense W would require quadratic in n operations ($O(n^2)$). As shown in Chapter 4, sparse matrix techniques greatly accelerate computation of the log-determinant and other quantities of interest.

To summarize, a number of techniques facilitate calculation of maximum likelihood estimates for the SDM and SAR models. These techniques include concentrating the log-likelihood, pre-computing a table of log-determinants as well as moments such as $e'_o e_d$, and using sparse W. Taken together, these techniques greatly reduce the operation counts as well as computer memory required to solve problems involving large data samples. Chapter 4 provides more detail about these and other techniques that can aid in calculation of maximum likelihood estimates.

3.1.2   SEM model estimation

The model statement for a model containing spatial dependence in the disturbances that we label SEM is shown in (3.12), with the DGP for this model in (3.13), where we define X to be the $n \times k$ explanatory variables matrix that may or may not include a constant term, and β the associated $k \times 1$ vector of parameters.

$$y = X\beta + u \tag{3.12}$$
$$u = \lambda W u + \varepsilon$$
$$y = X\beta + (I_n - \lambda W)^{-1}\varepsilon \tag{3.13}$$
$$\varepsilon \sim N(0, \sigma^2 I_n)$$

The full log-likelihood has the form in (3.14).

$$\ln L = -(n/2)\ln(\pi\sigma^2) + \ln|I_n - \lambda W| - \frac{e'e}{2\sigma^2} \tag{3.14}$$
$$e = (I_n - \lambda W)(y - X\beta)$$

For a given λ, optimization of the log-likelihood function shows (Ord, 1975; Anselin, 1988) that $\beta(\lambda) = (X(\lambda)'X(\lambda))^{-1}X(\lambda)'y(\lambda)$, where $X(\lambda) = (X - \lambda W X)$, $y(\lambda) = (y - \lambda W y)$, and $\sigma^2(\lambda) = e(\lambda)'e(\lambda)/n$ where $e(\lambda) = y(\lambda) - X(\lambda)\beta(\lambda)$. Therefore, we can concentrate the log-likelihood with respect to β and σ^2 to yield the concentrated log-likelihood as a function of λ in (3.15).

$$\ln L(\lambda) = \kappa + \ln|I_n - \lambda W| - (n/2)\ln(S(\lambda)) \tag{3.15}$$
$$S(\lambda) = e(\lambda)'e(\lambda) \tag{3.16}$$

Unlike the SAR or SDM case, $S(\lambda)$ is not a simple quadratic in the spatial parameter. As currently stated in (3.16), evaluating the concentrated log-likelihood for any given value of λ requires manipulation of $n \times 1$ and $n \times$

k matrices for each choice of λ. This becomes tedious for large data sets, optimization techniques that require many trial values of λ, and in simulations. However, variables that require $O(n)$ computations can be pre-computed so that calculating $S(\lambda)$ during optimization only requires working with moment matrices of dimension k by k or smaller. These moment matrices involve the independent and dependent variables as a function of λ.[1]

$$A_{XX}(\lambda) = X'X - \lambda X'WX - \lambda X'W'X + \lambda^2 X'W'WX$$
$$A_{Xy}(\lambda) = X'y - \lambda X'Wy - \lambda X'W'y + \lambda^2 X'W'Wy$$
$$A_{yy}(\lambda) = y'y - \lambda y'Wy - \lambda y'W'y + \lambda^2 y'W'Wy$$
$$\beta(\lambda) = A_{XX}(\lambda)^{-1} A_{Xy}(\lambda)$$
$$S(\lambda) = A_{yy}(\lambda) - \beta(\lambda)' A_{XX}(\lambda)\beta(\lambda)$$

With these moments and a pre-computed grid of log-determinants (coupled with an interpolation routine) updating the concentrated log-likelihood in (3.15) for a new value of λ is almost instantaneous. Applying a univariate optimization technique such as Newton's method to (3.15) to find $\hat{\lambda}$ and substituting this into $\sigma^2(\lambda)$, $\beta(\lambda)$ and $\Omega(\lambda)$ leads to the maximum likelihood estimates (3.17) to (3.19).

$$\hat{\beta} = \beta(\hat{\lambda}) \tag{3.17}$$
$$\hat{\sigma}^2 = n^{-1} S(\hat{\lambda}) \tag{3.18}$$
$$\hat{\Omega} = \hat{\sigma}^2 \left[(I_n - \hat{\lambda}W)'(I_n - \hat{\lambda}W) \right]^{-1} \tag{3.19}$$

As noted in Section 3.1.1, applying Newton's method with numerical derivatives to find the optimum produces a numerical estimate of the second derivative of the concentrated log-likelihood at the optimum $\hat{\lambda}$. This numerical estimate of the second derivative can be used in conjunction with other information to produce a variance-covariance matrix estimate.

Note, the SDM model nests the SEM model as a special case. To see this, consider the alternative statement of the SEM model in (3.20). To avoid collinearity problems for row-stochastic W, we assume the matrix X does not contain a constant term and specify this separately. This is necessary to avoid creating a column vector $W\iota_n = \iota_n$ in WX that would duplicate the intercept term.

---

[1] Use of moment matrices requires that we avoid sets of explanatory variables that are poorly scaled or ill-conditioned. In practice, this may not act as a tremendous constraint since even numerically robust computational techniques can be strained by ill-conditioned data sets. In addition, poorly scaled sets of explanatory variables often lead to difficult-to-interpret parameter estimates.

$$y = \alpha \iota_n + X\beta + (I_n - \lambda W)^{-1}\varepsilon$$
$$(I_n - \lambda W)y = \alpha(I_n - \lambda W)\iota_n + (I_n - \lambda W)X\beta + \varepsilon$$
$$y = \lambda Wy + \alpha(I_n - \lambda W)\iota_n + X\beta + WX(-\beta\lambda) + \varepsilon \quad (3.20)$$

The model in (3.20) represents an SDM model where the parameter on the spatial lag of the explanatory variables (WX) has been restricted to equal $-\beta\lambda$. Estimating the more general SDM model ($y = \lambda Wy + X\beta + WX\theta + \varepsilon$) and testing the restriction $\theta = -\beta\lambda$ could lead to rejection of the SEM relative to the SDM.

3.1.3   Estimates for models with two weight matrices

The spatial literature contains a number of models involving two or more weight matrices. Using multiple weight matrices provides a straightforward generalization of the SAR, SDM, and SEM models. For example, Lacombe (2004) uses a two weight matrix SAR model similar to the SDM model shown in (3.21).

$$y = \rho_1 W_1 y + \rho_2 W_2 y + X\beta + W_1 X\gamma + W_2 X\theta + \varepsilon \quad (3.21)$$
$$\varepsilon \sim N(0, \sigma^2 I_n)$$

The weight matrix W_1 was used to capture the effect of neighboring counties within the state, and W_2 captures the effect of neighboring counties in the bordering state. Lacombe (2004) analyzed policies that varied across states, making this model attractive. For a sample of counties that lie on state borders, spatial dependence extends to both counties within the state as well as those across the border in the neighboring state. This SDM variant of Lacombe's model allows for separate influences of the explanatory variables matrix X arising from neighbors within the state versus those in the neighboring state.

The only departure from our discussion of maximum likelihood estimation for this variant of the SDM model involves a bivariate optimization problem over the range of feasible values for ρ_1, ρ_2. Maximizing the (concentrated for $\beta, \gamma, \theta, \sigma^2$) log-likelihood for this variant of the SDM model requires calculating the log-determinant term: $\ln|I_n - \rho_1 W_1 - \rho_2 W_2|$ over a bivariate grid of values for both ρ_1, ρ_2 in the feasible range. These scalar values associated with the bivariate grid would be stored in a matrix rather than a vector. Optimization of the concentrated log-likelihood function over the parameters ρ_1, ρ_2 could repeatedly access this matrix at a very small computational cost.

As another example of specifications involving two weight matrices, the SAC model contains spatial dependence in both the dependent variable and disturbances as shown in (3.22) along with its associated data generating process in (3.23). Unlike the Lacombe model, it is possible to implement this model using the same matrix $W = W_1 = W_2$, but we will have more to say about this later.

$$y = \rho W_1 y + X\beta + u$$
$$u = \lambda W_2 u + \varepsilon \qquad (3.22)$$
$$y = (I_n - \rho W_1)^{-1} X\beta + (I_n - \rho W_1)^{-1}(I_n - \lambda W_2)^{-1}\varepsilon \qquad (3.23)$$
$$\varepsilon \sim N(0, \sigma^2 I_n)$$

The matrices W_1, W_2 can be the same or distinct. Obviously, if the parameter $\rho = 0$, this model collapses to the SEM model, and $\lambda = 0$ yields the SAR model. Normally, the SAC does not contain a separate WX term, so the SAC does not usually nest the SDM. However, one can write an extended SDM that nests the SAC, specifically:

$$y = \rho W y + X\beta + W X\theta + u$$
$$u = \lambda W u + \varepsilon$$

The log-likelihood for the SAC model is shown in (3.24) along with definitions.

$$\ln L = -(n/2)\ln(\pi\sigma^2) + \ln|A| + \ln|B| - \frac{e'e}{2\sigma^2} \qquad (3.24)$$
$$e = B(Ay - X\beta)$$
$$A = I_n - \rho W_1$$
$$B = I_n - \lambda W_2$$

The log-likelihood in (3.24) for the SAC model can also be concentrated with respect to the parameters β, σ^2. Maximizing this likelihood requires computing two log-determinants for the case where $W_1 \neq W_2$, and solving a bivariate optimization problem in the two parameters ρ and λ.

Anselin (1988) raised questions about identification of the SAC model in the case of identical matrices W, but Kelejian and Prucha (2007) provide an argument that the model is identified for this case. Their argument for identification requires that $X\beta$ in the DGP makes a material contribution towards explaining variation in the dependent variable y ($\beta \neq 0$). To see the importance of this, consider (3.25), and note that in the case where $\beta = 0$, a label switching problem exists since $AB = BA$ when A and B are functions of the same weight matrix W. Therefore, the parameters ρ and λ are not identified.

$$y = (I_n - \rho W)^{-1} X\beta + (I_n - \rho W)^{-1}(I_n - \lambda W)^{-1}\varepsilon \qquad (3.25)$$

Although, $\beta \neq 0$ will in principle identify the model, as the noise variance of the disturbances rises, the relative importance of β diminishes. This is shown in (3.26) where the variables are all scaled by σ. This suggests that in low signal-to-noise problems (low variation in the predicted values relative to the noise variance), estimates may show symptoms of this near lack of identification.

$$\sigma^{-1}y = A^{-1}X(\sigma^{-1}\beta) + A^{-1}B^{-1}\sigma^{-1}\varepsilon \qquad (3.26)$$

There is also the SARMA model shown in (3.27), with the corresponding DGP in (3.28).

$$y = \rho W_1 y + X\beta + u$$
$$u = (I_n - \theta W_2)\varepsilon \qquad\qquad\qquad (3.27)$$
$$y = (I_n - \rho W_1)^{-1}X\beta + (I_n - \rho W_1)^{-1}(I_n - \theta W_2)\varepsilon \qquad (3.28)$$

Minor changes would be required to the log-likelihood function for this model as shown in (3.29), where we have replaced the definition $B = (I_n - \lambda W_2)$ from the SAC model with $B = (I_n - \theta W_2)^{-1}$.

$$\ln L = \kappa + \ln|A| + \ln|B| - \frac{e'e}{2\sigma^2} \qquad (3.29)$$
$$e = B(Ay - X\beta)$$
$$A = I_n - \rho W_1$$
$$B = (I_n - \theta W_2)^{-1}$$

Finally, many other models involving multiple weight matrices or combinations of powers of weight matrices have been proposed in the literature such as higher-order spatial AR, MA, and ARMA models (Huang and Anh, 1992). In Chapter 4 we discuss approaches for calculating the determinants that arise in such models.

3.2   Estimates of dispersion for the parameters

So far, the estimation procedures set forth can be used to produce estimates for the spatial dependence parameters ρ and λ using univariate or bivariate maximization of the log-likelihood function concentrated with respect to β and σ^2. Maximum likelihood estimates for the parameters β and σ^2 can be recovered using the maximum likelihood estimates for the dependence parameters $\hat{\rho}$ and $\hat{\lambda}$.

For many purposes, a need exists to conduct inference. Maximum likelihood inference often proceeds using *likelihood ratio* (LR), *Lagrange multiplier* (LM), or *Wald* (W) tests. Asymptotically, these should all yield similar results, although these can differ for finite samples. Often, the choice of one method over the other comes down to computational convenience and other preferences.

Due to the ability to rapidly compute likelihoods, Pace and Barry (1997) propose likelihood ratio tests for hypotheses such as the deletion of a single explanatory variable. To put these likelihood ratio tests in a form similar to t-tests, Pace and LeSage (2003a) discuss use of *signed root deviance* statistics.[2] The signed root deviance applies the sign of the coefficient estimates β to the square root of the deviance statistic (Chen and Jennrich, 1996). These statistics behave similar to t-ratios when the sample is large, and can be used in lieu of t-statistics for hypothesis testing.

Wald inference uses the Hessian (numerical or analytic) or the related information matrix to provide a variance-covariance matrix for the estimated parameters, and thus the familiar t-test. In this case, the Hessian is just the matrix of second-derivatives of the log-likelihood with respect to the parameters. Approaches using either the Hessian (Anselin, 1988, p. 76) or the information matrix (Ord, 1975; Smirnov, 2005) have appeared in the spatial econometrics literature.

An implementation issue is constructing the Hessian or information matrix. We will use the SAR model: $y = \rho W y + X \beta + \varepsilon$ for simplicity in our discussion. Straightforward evaluation of the analytical Hessian or information matrix involves computing a trace term which contains the dense $n \times n$ matrix inverse $(I_n - \rho W)^{-1}$. Chapter 4 provides means of rapidly approximating elements that arise in the Hessian or information matrix. In the following discussion we focus on the Hessian.

Given the ability to rapidly evaluate the log-likelihood function, a purely numerical approach might seem feasible for calculating an estimate of the Hessian. There are some drawbacks to implementing this approach in software for general use. First, practitioners often work with poorly scaled sample data, which makes numerical perturbations used to approximate the derivatives comprising the Hessian difficult. A second point is that univariate optimization takes place using the likelihood concentrated with respect to the parameters β and σ^2, so a numerical approximation to the full Hessian does not arise naturally, as in typical maximum likelihood estimation procedures. This means that computational time must be spent after estimation of the parameters to produce a separate numerical estimate of the full Hessian.

In Section 3.2.1 we discuss ways of marrying the analytic Hessian and numerical Hessian results to take advantage of the strengths of each approach.

---

[2]Deviance is minus twice the log of the likelihood ratio for models fitted by maximum likelihood. The ratio used in these calculations is one involving the likelihood for the model excluding each variable versus that for the model containing all variables.

Namely, most of the analytic Hessian elements do not require much time to compute and have less sensitivity to scaling issues. A numerical approach, however, takes less time and performs well for the single difficult element in the analytic Hessian.

In Chapter 5 Bayesian Markov Chain Monte Carlo (MCMC) estimation methods for spatial regression models are explained, and these can be used to produce estimates of dispersion based on the sample of draws carried out by this sampling-based approach to estimation. Following standard Bayesian regression theory, use of a non-informative prior in these models should result in posterior estimates and inferences that are identical to those from maximum likelihood. Therefore, these estimates of parameter dispersion also provide a valid, but unorthodox, means of conducting maximum likelihood inference.

Also, for large n it often becomes feasible to provide bounded inference. For example, Pace and LeSage (2003a) introduce a lower bound on the likelihood ratio test that allows conservative maximum likelihood inference while avoiding the computationally demanding task of even computing exact maximum likelihood point estimates. They show that this form of *likelihood dominance inference* (Pollack and Wales, 1991) performed almost as well as exact likelihood inference on parameters from a SAR model involving 890,091 observations, where the procedure took less than a minute to compute.

An entirely different approach to the problem of inference in spatial regression models is to rely on an estimation method that is not likelihood based. Examples include the instrumental variable approach of Anselin (1988, p. 81-90), the instrumental variables/generalized moments estimator from Kelejian and Prucha (1998, 1999) or the maximum entropy of Marsh and Mittelhammer (2004). Much of the motivation for using these methods comes from the perceived difficulties of computing estimates from likelihood-based methods, a problem that has been largely resolved. A feature of likelihood-based methods is that the determinant term ensures that resulting dependence parameter estimates are in the interval defined by maximum and minimum eigenvalues of the weight matrix. Some of the alternative estimation methods that avoid using the log-determinant can fail to yield dependence parameter estimates in this interval. In addition, these methods can be sensitive in non-obvious ways to various implementation issues such as the interaction between the choice of instruments and the specification of the model. For these reasons, we focus on likelihood-based techniques.

3.2.1 A mixed analytical-numerical Hessian calculation

For the case of the SAR model, the Hessian we will work with is organized as in (3.30), which we label H. For the case of the SEM model, we would replace the parameter ρ with λ. Of course, some of the derivative expressions change as well.

$$H = \begin{bmatrix} \dfrac{\partial^2 L}{\partial \rho^2} & \dfrac{\partial^2 L}{\partial \rho \partial \beta'} & \dfrac{\partial^2 L}{\partial \rho \partial \sigma^2} \\[2ex] \dfrac{\partial^2 L}{\partial \beta \partial \rho} & \dfrac{\partial^2 L}{\partial \beta \partial \beta'} & \dfrac{\partial^2 L}{\partial \beta \partial \sigma^2} \\[2ex] \dfrac{\partial^2 L}{\partial \sigma^2 \partial \rho} & \dfrac{\partial^2 L}{\partial \sigma^2 \partial \beta'} & \dfrac{\partial^2 L}{\partial (\sigma^2)^2} \end{bmatrix} \tag{3.30}$$

The analytical Hessian which we label $H^{(a)}$ appears in (3.31), where we employ the definitions: $A = (I_n - \rho W)^{-1}$, $B = y'(W + W')y$, $C = y'W'Wy$.

$$H^{(a)} = \begin{bmatrix} -\operatorname{tr}(WAWA) - \dfrac{C}{\sigma^2} & -\dfrac{y'W'X}{\sigma^2} & \dfrac{2C - B + 2y'W'X\beta}{2\sigma^4} \\[2ex] \cdot & -\dfrac{X'X}{\sigma^2} & 0 \\[2ex] \cdot & \cdot & -\dfrac{n}{2\sigma^4} \end{bmatrix} \tag{3.31}$$

For models involving a large number of observations n, the computationally difficult part of evaluating the analytical Hessian in (3.31) involves the term: $-\operatorname{tr}(WAWA) = -\operatorname{tr}(W(I_n - \rho W)^{-1}W(I_n - \rho W)^{-1})$. Done in a computationally straightforward way, this would require calculating the $n \times n$ matrix inverse, $A = (I_n - \rho W)^{-1}$, as well as matrix multiplications involving the n-dimensional spatial weight matrix W. Such an approach would require $O(n^3)$ operations since A is dense for spatially connected problems. The remaining terms involve matrix-vector products, and we note that the spatial weight matrix is often a sparse matrix containing a relatively small number of non-zero elements. As already noted, this allows use of sparse matrix routines that can efficiently carry out the matrix-vector products.

At least three ways exist for handling the term $\operatorname{tr}(W(I_n - \rho W)^{-1}W(I_n - \rho W)^{-1})$. First, one can compute it exactly as in Smirnov (2005). Second, estimating this trace takes little time, and we will examine this in Chapter 4. Third, this term is subsumed in the second derivative of the concentrated log-likelihood with respect to ρ, a quantity that emerges as a byproduct of optimizing the concentrated log-likelihood using Newton's method. We term this latter strategy the *mixed analytical-numerical Hessian*. In this section, we show how this works, and provide an applied illustration demonstrating that this approach is computationally easy to implement and accurate.

To begin, since we rely on univariate optimization of the concentrated log likelihood, this will not produce a full numerical Hessian, but rather a numerical Hessian pertaining only to the parameter ρ (or λ) that arises from the concentrated likelihood labeled L_p in (3.32).

$$\frac{\partial^2 L_p}{\partial \rho^2} \tag{3.32}$$

As noted by Davidson and MacKinnon (2004), we can work with the concentrated likelihood L_p to produce correct values for the parameter ρ (or λ), but we need the full likelihood (L) Hessian H, which can be expressed in terms of the scalar spatial dependence parameter ρ and a vector θ containing the remaining parameters, $\theta = \left(\beta'\ \sigma^2 \right)'$.

$$H = \begin{bmatrix} \dfrac{\partial^2 L}{\partial \rho^2} & \dfrac{\partial^2 L}{\partial \rho \partial \theta'} \\[2ex] \dfrac{\partial^2 L}{\partial \theta \partial \rho} & \dfrac{\partial^2 L}{\partial \theta \partial \theta'} \end{bmatrix} \tag{3.33}$$

It is possible to adjust the empirical concentrated likelihood Hessian so it produces the appropriate element for the full Hessian as illustrated in (3.34).

$$\frac{\partial^2 L}{\partial \rho^2} = \frac{\partial^2 L_p}{\partial \rho^2} + \frac{\partial^2 L}{\partial \rho \partial \theta'} \left(\frac{\partial^2 L}{\partial \theta \partial \theta'} \right)^{-1} \frac{\partial^2 L}{\partial \theta \partial \rho} \tag{3.34}$$

This easily computed expression (3.34) (details to follow) can be substituted into the full Hessian in (3.35).

$$H = \begin{bmatrix} \dfrac{\partial^2 L_p}{\partial \rho^2} + \dfrac{\partial^2 L}{\partial \rho \partial \theta'} \left(\dfrac{\partial^2 L}{\partial \theta \partial \theta'} \right)^{-1} \dfrac{\partial^2 L}{\partial \theta \partial \rho} & \dfrac{\partial^2 L}{\partial \rho \partial \theta'} \\[3ex] \dfrac{\partial^2 L}{\partial \theta \partial \rho} & \dfrac{\partial^2 L}{\partial \theta \partial \theta'} \end{bmatrix} \tag{3.35}$$

Using this approach we can replace the difficult calculation involving $H_{11}^{(a)}$ with the adjusted empirical concentrated likelihood Hessian from (3.34). A key point is that maximum likelihood estimation as set forth in Section 3.1 already yields a vector of the concentrated log-likelihood values as a function of the parameter ρ. Given this vector of concentrated log likelihoods, $\partial^2 L_p / \partial \rho^2$ costs almost nothing to compute.

For the case of the SAR model, this results in the mixed analytical numerical Hessian labelled $H^{(m)}$ in (3.36).

$$H^{(m)} = \begin{bmatrix} \dfrac{\partial^2 L_p}{\partial \rho^2} + Q & -\dfrac{y'W'X}{\sigma^2} & \dfrac{2C - B + 2y'W'X\beta}{2\sigma^4} \\[2ex] \cdot & -\dfrac{X'X}{\sigma^2} & 0 \\[2ex] \cdot & \cdot & -\dfrac{n}{2\sigma^4} \end{bmatrix} \tag{3.36}$$

$$Q = v' \begin{bmatrix} -\dfrac{X'X}{\sigma^2} & 0 \\[2ex] 0 & -\dfrac{n}{2\sigma^4} \end{bmatrix}^{-1} v \tag{3.37}$$

$$v = \begin{bmatrix} -\dfrac{y'W'X}{\sigma^2} & \dfrac{2C - B + 2y'W'X\beta}{2\sigma^4} \end{bmatrix}' \tag{3.38}$$

We note that $\partial^2 L_p/\partial \rho^2$ represents the estimate of the second derivative of the concentrated log likelihood with respect to ρ that arises as a byproduct of optimization. We add this term to the easily calculated quadratic form in Q shown in (3.37) and (3.38). This results in a simple mixed analytical numerical Hessian that can be used for inference regarding the model parameters.

As is well-known, the variance-covariance matrix pertinent to the parameter estimates equals $-H^{-1}$. Given $H^{(m)}$, one could easily simulate the parameter estimates using multivariate normal deviates. This ability to quickly simulate the parameter estimate facilitates finding the distribution of the direct and indirect impacts that we discussed in Chapter 2.

3.2.2   A comparison of Hessian calculations

To compare the various approaches to calculating t-statistics associated with the spatial regression parameters, we used a sample data set from Pace and Barry (1997) containing information for 3,107 US counties on voter participation in the 1980 presidential election. The dependent variable represents voter turnout, those voting as a (logged) *proportion* of those eligible to vote. Explanatory variables included (logged) population over age 18 *Voting Pop*, (logged) population with college degrees *Education*, (logged) population owning homes *Home Owners*, and (logged) median household income *Income*.

The data was fitted using the SDM model, which includes spatial lags of the explanatory variables, labeled *Lag Voting Pop*, *Lag Education*, and so on. Table 3.1 presents the resulting t-statistics calculated using: signed root deviances (SRD), the analytical Hessian (Analytic), Bayesian Markov Chain Monte Carlo (MCMC), the mixture of the empirical and theoretical Hessian (Mixed), and a purely numerical Hessian calculation (Numerical). The results

TABLE 3.1:   A comparison of *t*-statistics calculated using alternative approaches

Variables	SRD	Analytic	MCMC	Mixed	Numerical
Votes/Pop	−29.401	−31.689	−31.486	−31.689	−38.643
Education	7.718	7.752	7.752	7.752	7.922
Home Owners	27.346	29.191	28.977	29.191	29.837
Income	1.896	1.897	1.930	1.897	2.633
Lag Votes/Pop	12.549	12.904	12.961	12.907	13.190
Lag Education	1.570	1.560	1.621	1.560	1.510
Lag Home Owners	−12.114	−12.381	−12.375	−12.382	−12.671
Lag Income	−4.662	−4.661	−4.713	−4.661	−5.038
Intercept	11.603	11.449	11.529	11.453	11.597
ρ	33.709	41.374	41.430	41.427	47.254

in the table demonstrate very similar *t*-statistics from the Analytic, MCMC, and Mixed techniques. The numerical Hessian estimates differ materially from the other Hessian results for some variables such as income, despite the fact that the sample data was well-scaled in this example. The SRD results, which use likelihood ratio inference, match those from the Analytic and Mixed Hessians for the regression parameters, although the SRD regression parameter *t*-statistics appear slightly conservative and the *t*-statistic on ρ is substantially more conservative.

The computational time required was around 0.6 seconds to calculate the analytic terms in the Hessian along with the adjustments from (3.34).

3.3   Omitted variables with spatial dependence

The existence of spatially dependent omitted variables seems a likely occurrence in applied practice. As an example, consider the spatial growth regression literature that analyzes cross-sectional regional income growth as a function of initial period income levels and other explanatory variables describing regional characteristics thought to influence economic growth (Abreu, de Groot, and Florax, 2004; Ertur and Koch, 2007; Ertur, LeGallo and LeSage, 2007; Fingleton, 2001; Fischer and Stirbock, 2006). While regional information on explanatory variables such as human capital may exist, it is likely that sample data information reflecting physical capital and other important determinants of regional economic growth are not readily available. Since physical capital is likely correlated with human capital, and also likely to exhibit spatial dependence, the omitted variables circumstances described in Section 2.2 seem plausible.

In Section 3.3.1, we set forth a statistical test comparing ordinary least-squares (OLS) and SEM estimates that can be used to diagnose misspecification in general, and the potential existence of omitted variables. The motivation for this type of comparison is that theory indicates OLS and SEM estimates should be the same if the true DGP is either OLS, SEM, or any other error model.

A number of authors (Brasington and Hite, 2005; Dubin, 1988; Cressie, 1993, p. 25) have suggested that omitted variables affect spatial regression methods less than least-squares. In Section 3.3.2 we explore this issue by deriving an expression for OLS omitted variable bias in a univariate version of the model. We show that spatial dependence in the explanatory variable exacerbates the usual omitted variables bias produced when incorrectly using OLS to estimate an SEM model in the presence of a spatially dependent omitted variable.

In Section 3.3.3 we explore the conjecture that spatial regression methods suffer less from omitted variables bias. It is shown that the DGP associated with spatially dependent omitted variables matches the SDM DGP. Use of this model in the presence of omitted variables shrinks the bias relative to OLS estimates, which provides a strong econometric motivation for use of the SDM model in applied work. Good theoretical motivations exist for the SDM model as well (Ertur and Koch, 2007).

3.3.1   A Hausman test for OLS and SEM estimates

As already noted in Section 2.2, OLS estimates for the parameters β will be unbiased if the underlying DGP represents the SEM model, but t-statistics from least-squares are biased. As shown in Section 2.2, specification error arising from the presence of omitted variables correlated with the explanatory variable and spatial dependence in the disturbances will lead to a DGP reflecting the SDM model. As shown in Section 3.1.2, the SDM model nests the SEM model as a special case, providing the intuition for this result.

We explore a formal statistical test for equality of the coefficient estimates from OLS and SEM, since passing this test would be a good indication that specification problems (such as omitted variables correlated with the explanatory variables) were not present in the SEM model.

As motivation for the test, we note that if the true DGP is any error model, in a repeated sampling context the average of the error model parameter estimates for β should be equal. This is true even with omitted variables, provided that these are independent of X. To see this, consider the error model DGP in (3.39) where F is some unknown, arbitrary, fixed matrix, and z is an omitted variable that is independent of X. Consider the *generalized least squares* (GLS) estimator in (3.40) based on some arbitrary, fixed variance-covariance matrix G which may bear no relation to a function of F. For any choice of F and G, even in the presence of z, the expected value of the estimates equals β as shown in (3.41).

$$y = X\beta + z + F\varepsilon \tag{3.39}$$
$$\hat{\beta}_G = (X'G^{-1}X)^{-1}X'G^{-1}y \tag{3.40}$$
$$\hat{\beta}_G = (X'G^{-1}X)^{-1}X'G^{-1}X\beta + (X'G^{-1}X)^{-1}X'G^{-1}(z + F\varepsilon)$$
$$E(\hat{\beta}_G) = \beta \tag{3.41}$$

Intuitively, disturbances with a zero expectation whether arising from omitted variables or misspecification (as long as these are orthogonal to the included explanatory variables) do not affect estimates for parameters associated with the explanatory variables.

These theoretical results suggest that a spatial error DGP should result in OLS and SEM parameter estimates that are (on average) equal for the parameters β, despite the presence of some types of model mis-specification. However, the literature contains a number of examples where researchers present estimates from both OLS and SEM that do not seem close in magnitude.

A Hausman test (Hausman, 1978) can be used whenever there are two estimators, one of which is inefficient but consistent (OLS in this case under the maintained hypothesis of the SEM DGP), while the other is efficient (SEM in this case). We set forth a Hausman test for statistically significant differences between OLS and SEM estimates. We argue that this test can be useful in diagnosing the presence of omitted variables that are correlated with variables included in the model. Since this scenario leads to a model specification that should include a spatial lag of the dependent variable, we would expect to see OLS and SEM estimates that are significantly different.

If we let $\gamma = \hat{\beta}_{OLS} - \hat{\beta}_{SEM}$ represent the difference between OLS and SEM estimates, the Hausman test statistic T (under the maintained hypothesis of the SEM DGP) has the simple form in (3.42), where $\hat{\Omega}_O$ represents a consistent estimate of the variance-covariance matrix associated with $\hat{\beta}_{OLS}$ (given a spatial error model DGP). The null hypothesis is that the SEM and OLS estimates are not significantly different. The alternative hypothesis is a significant difference between the two sets of estimates.

$$T = \gamma'(\hat{\Omega}_O - \hat{\Omega}_S)^{-1}\gamma \tag{3.42}$$

Expression (3.43) implies (3.44), and the expectation of the outer product of (3.44) is shown in (3.45). Although the usual OLS estimated variance-covariance matrix $\sigma_o^2(X'X)^{-1}$ is inconsistent for the SEM DGP, Cordy and Griffith (1993) show that (3.45) is a consistent estimator. Under the maintained hypothesis of the SEM DGP, maximum likelihood SEM estimates of $\hat{\sigma}^2$ and λ provide consistent estimates that can be used to replace σ^2 and λ in (3.45) resulting in (3.46).

$$\hat{\beta}_O = \beta + H(I_n - \lambda W)^{-1}\varepsilon \qquad (3.43)$$

$$\hat{\beta}_O - E(\hat{\beta}_O) = H(I_n - \lambda W)^{-1}\varepsilon \qquad (3.44)$$

$$H = (X'X)^{-1}X'$$

$$\Omega_O = \sigma^2 H(I_n - \lambda W)^{-1}(I_n - \lambda W')^{-1}H' \qquad (3.45)$$

$$\hat{\Omega}_O = \hat{\sigma}^2 H(I_n - \hat{\lambda} W)^{-1}(I_n - \hat{\lambda} W')^{-1}H' \qquad (3.46)$$

In (3.42), $\hat{\Omega}_S$ represents a consistent estimate for the variance-covariance associated with $\hat{\beta}_{SEM}$, again under the maintained hypothesis of the spatial error process, where $\hat{\Omega}_S$ is shown in (3.47).

$$\hat{\Omega}_S = \hat{\sigma}^2 (X'(I_n - \hat{\lambda} W)'(I_n - \hat{\lambda} W)X)^{-1} \qquad (3.47)$$

We note that although SEM estimates for β are unbiased, those for the variance-covariance matrix are only consistent due to the dependence on the estimated parameter λ. See Lee (2004) on consistency of spatial regression estimates and Davidson and MacKinnon (2004, p. 341-342) for an excellent discussion of Hausman tests.

The statistic T follows a chi-squared distribution with degrees-of-freedom equal to the number of regression parameters tested. By way of summary, the maximum likelihood estimates for $\hat{\beta}_{SEM}, \hat{\lambda}, \hat{\sigma}^2$ along with $\hat{\beta}_{OLS}$ can be used in conjunction with consistent estimates for $\hat{\Omega}_O$ from (3.46) and $\hat{\Omega}_S$ in (3.47) to calculate the test statistic T. This allows us to test for significant differences between the SEM and OLS coefficient estimates.

If we cannot reject the null hypothesis of equality, this would be an indication that omitted variables do not represent a serious problem or are not correlated with the explanatory variables. If the SEM has a significantly higher likelihood than OLS, but the Hausman test does not find a significant difference between the OLS and SEM estimates, this indicates that the spatial error term in the SEM is capturing the effect of omitted variables, but these are not correlated with the included variables.

The performance of this spatial Hausman test was examined in Pace and LeSage (2008) under controlled conditions using a simulated SEM DGP based on 3,000 observations and varying levels of spatial dependence assigned to the parameter λ. They show that the estimated sizes for this test conformed closely to theoretical sizes.

3.3.2 Omitted variables bias of least-squares

Often explanatory variables used in spatial regression models exhibit dependence, since these reflect regional characteristics. For example, in a housing hedonic pricing model variables such as levels of income, educational attainment, and commuting times to work often exhibit similarity over space, or spatial dependence. Also, housing prices are affected by latent unobservable influences such as architectural quality, attention to landscaping in a neighborhood, convenient access to popular restaurants, walkability, noise, as well as other factors. These latent variables may also exhibit similarity over space. Due to data limitations, these latent variables are likely to be omitted from models. We discuss an expression for the omitted variable bias that arises when OLS estimates are used in circumstances where the included and omitted explanatory variables exhibit spatial dependence and the disturbance process is spatially dependent as in the SEM model. The expression shows that spatial dependence in a single included explanatory variable exacerbates the usual bias that occurs when using OLS to estimate an SEM model in the presence of a spatially dependent omitted variable that is correlated with the included explanatory variable.

We derive an expression for the bias that would arise from using OLS estimates in the presence of spatial dependence in the disturbances, included, and omitted explanatory variables. We work with a vector x representing a single (non-constant) explanatory variable with a mean of zero and following an *iid* normal distribution and let y be the dependent variable. We add an omitted variable to the SEM model and allow for a spatial dependence process to govern this variable as well as the included explanatory variable, leading to the model in (3.48) to (3.51). The vectors ε, and ν represent $n \times 1$ disturbance vectors, and we assume that ε is distributed $N(0, \sigma_\varepsilon^2 I_n)$, ν is distributed $N(0, \sigma_\nu^2 I_n)$, and ε is independent of ν.

$$y = x\beta + u \qquad (3.48)$$
$$u = \lambda W u + \eta \qquad (3.49)$$
$$\eta = x\gamma + \varepsilon \qquad (3.50)$$
$$x = \phi W x + \nu \qquad (3.51)$$

The scalar parameters of the model are: β, λ, ϕ, and γ, and W is an $n \times n$ non-negative symmetric spatial weight matrix with zeros on the diagonal.

Expressions (3.48) and (3.49) are the usual SEM model statements and (3.50) adds an omitted variable, where the strength of dependence (correlation) between the included variable vector x and the omitted variable vector η is controlled by the parameter γ. Finally, (3.51) specifies a spatial autoregressive process to govern the explanatory variable x. We focus on non-negative spatial dependence, by assuming $\lambda, \phi \in [0, 1)$.

Pace and LeSage (2009b) derive theoretical expressions for the bias associated with use of OLS estimates in these circumstances as shown in (3.52) to (3.54).

$$\text{plim}_{n\to\infty} \ \hat{\beta}_o = \beta + T_\gamma(\phi,\lambda)\gamma \tag{3.52}$$

$$T_\gamma(\phi,\lambda) = \frac{\text{tr}[H(\phi)^2 G(\lambda)]}{\text{tr}[H(\phi)^2]} \tag{3.53}$$

$$G(\lambda) = (I_n - \lambda W)^{-1}, \quad H(\phi) = (I_n - \phi W)^{-1} \tag{3.54}$$

As the factor $T_\gamma(\phi,\lambda)$ takes on values greater than unity, this increases the bias in OLS estimates for this model. The magnitude of bias depends on the parameter ϕ representing the strength of spatial dependence in the explanatory variable, the parameter λ reflecting error dependence, and the parameter γ which governs the correlation between the included and omitted variable.

Pace and LeSage (2009b) show that $T_\gamma(\phi,\lambda) > 1$ for $\lambda > 0$ and spatial dependence in the regressor, $\phi > 0$, amplifies these factors. This model encompasses the SEM model as a special case. The asymptotic biases that arise from using least-squares estimates in alternative circumstances such as the presence/absence of omitted variables, and the presence/absence of spatial dependence in the independent variables and disturbances are enumerated below.

1. *Spatial dependence in the disturbances and regressor:* $(\gamma = 0, \lambda, \phi > 0)$, leads to $\text{plim}_{n\to\infty} \ \hat{\beta}_o = \beta$, and there is no asymptotic bias.

2. *Spatial dependence in the regressor in the presence of an omitted variable:* $(\lambda = 0)$, while $(\gamma \neq 0)$, results in $\text{plim}_{n\to\infty} \ \hat{\beta}_o = \beta + \gamma$, representing the standard omitted variable bias.

3. *An omitted variable exists in the presence of spatial dependence in the regressors and disturbances:* $(\gamma \neq 0, \phi, \lambda > 0)$ then $\text{plim}_{n\to\infty} \ \hat{\beta}_o = \beta + T_\gamma(\phi,\lambda)\gamma$, and OLS has omitted variables bias amplified by the spatial dependence in the disturbances and in the regressor.

The first result is well-known, and the second is a minor extension of the conventional omitted variables case for least-squares. The third result shows that spatial dependence in the disturbances (and/or in the regressor) in the presence of omitted variables leads to a magnification of the conventional omitted variables bias. This third result differs from the usual finding that spatial dependence in the disturbances does not lead to bias.

To provide some feel for the magnitude of these biases, we present results from a small Monte Carlo experiment in Table 3.2. We simulated a spatially random set of $1,000$ locations and used these to construct a contiguity-based matrix W. The resulting $1,000 \times 1,000$ symmetric spatial weight matrix W was standardized to be doubly stochastic (have both row and columns sums of unity). The independent variable x was set to an *iid* unit normal vector with zero mean. We set $\beta = 0.75$ and $\gamma = 0.25$ for all trials. Given W and a value for λ and ϕ, we used the DGP to simulate $1,000$ samples of y, and for

Introduction to Spatial Econometrics

each sample we calculated the OLS estimate and recorded the average of the estimates (labeled mean $\hat{\beta}_o$ in Table 3.2). A set of nine combinations of λ and ϕ were used, and the theoretical expectation (labeled $E(\hat{\beta}_o)$ in Table 3.2) was calculated for each of these using expression (3.52).

TABLE 3.2: Omitted variables bias as a function of spatial dependence

Experiment	ϕ	λ	mean $\hat{\beta}_o$	$E(\hat{\beta}_o)$
1	0.0	0.0	0.9990	1.0000
2	0.5	0.0	1.0015	1.0000
3	0.9	0.0	1.0003	1.0000
4	0.0	0.5	1.0173	1.0159
5	0.5	0.5	1.0615	1.0624
6	0.9	0.5	1.1641	1.1639
7	0.0	0.9	1.1088	1.1093
8	0.5	0.9	1.3099	1.3152
9	0.9	0.9	1.9956	2.0035

The table shows the empirical average of the estimates and the expected values for the nine combinations of λ and ϕ. The theoretical and empirical results show close agreement, and the table documents that serious bias can occur when omitted variables combine with spatial dependence in the disturbance process, especially in the presence of spatial dependence in the regressor. For example, OLS estimates yield an empirical average of 1.9956 (which comes close to the theoretical value of 2.0035) when λ and ϕ equal 0.9, even though $\beta = 0.75$ and $\gamma = 0.25$. In this case, $T_\gamma(\phi, \lambda)$ approximately equals 5. If $\beta = -1$ and $\gamma = 0.2$, a $T_\gamma(\phi, \lambda)$ of 5 would mean that an OLS regression would produce an estimate close to 0. Therefore, inflation of the usual omitted variable bias could result in no perceived relation between y and x. A fortiori, the OLS parameter estimate would equal 1 when the true parameter equalled -1, when $\gamma = 0.4$. Therefore, the inflation of omitted variable bias in the presence of spatial dependence can have serious inferential consequences when using OLS.

In addition, Pace and LeSage (2009b) study a more general model that includes spatial dependence in y as well as the disturbances and explanatory variables. Naturally, spatial dependence in y further increases the bias of OLS.

3.3.3 Omitted variables bias for spatial regressions

We consider the conjecture made by a number of authors (Brasington and Hite, 2005; Dubin, 1988; Cressie, 1993, p. 25) that omitted variables affect spatial regression methods less than ordinary least-squares.

We begin by examining the implied DGP for the case of spatial dependence in the omitted variables and disturbances for a given x. This is the DGP associated with the assumptions (3.48) to (3.50). Manipulating these equations yields an equation shown in (3.55) in terms of spatial lags of the dependent and independent variables.

$$y = \lambda W y + x(\beta + \gamma) + W x(-\lambda \beta) + \varepsilon \tag{3.55}$$

$$y = \lambda W y + x\beta + W x\psi + \varepsilon \tag{3.56}$$

We can use the SDM model in (3.56) to produce consistent estimates for the parameters λ and ψ, since this model matches the DGP in the omitted variables circumstances set forth. These consistent estimates would equal the underlying structural parameters of the model in large samples.[3] In other words, for sufficiently large n estimating (3.56) would yield $E(\hat{\beta}) = \beta + \gamma$, $E(\hat{\psi}) = -\lambda \beta$, and $E(\hat{\lambda}) = \lambda$. There is no asymptotic bias in the estimate of λ for the SDM model in (3.56) despite the presence of omitted variables.

There is however asymptotic omitted variable bias in this model's estimates for β, since $E(\hat{\beta}) - \beta = \gamma$. Unlike the results for OLS presented in (3.52), this bias does not depend on x, eliminating the influence of the parameter ϕ that reflects the strength of spatial dependence in the included variable x. Further, the bias does not depend on spatial dependence in the disturbances specified by the parameter λ. Instead, the omitted variable bias is constant and depends only on the strength of relation between the included and omitted explanatory variable reflected by the parameter γ. This is similar to the conventional regression model omitted variable bias result.

These results agree with the earlier observation that omitted variables affect spatial regression methods less than ordinary least-squares. This protection against omitted variables bias is subject to some caveats, since we must produce estimates using a model that matches the implied DGP of the model after taking into account the presence of omitted variables (and the presence of spatial dependence in these and the explanatory variables as well as disturbances). As shown, use of the SEM regression will not contain the spatial lag of the dependent and explanatory variables implied by the presence of omitted variables. Recall from basic regression theory, inclusion of explanatory variables not in the DGP does not lead to bias in the estimates. However,

---

[3]See Kelejian and Prucha (1998), Lee (2004), and Mardia and Marshall (1984) regarding consistency of estimates from spatial regression models.

omitted variables bias arises when variables involved in the DGP are excluded from the model.

On the other hand, the SDM model does match the implied DGP that arises from the presence of omitted variables and spatially dependent explanatory variables. Consider the converse case where we apply the SDM model to produce estimates when the true DGP is that of the SEM and there are no omitted variables. The SDM estimates should still be consistent, but not efficient.

As a somewhat more general approach, Pace and LeSage (2009b) use the SAC DGP and examine the effects of an omitted variable that is correlated with the included variable, x. The presence of an omitted variable also leads to an extended SDM model that includes a spatial lag of the explanatory variables, Wx, and that subsumes the SAC. Consider the case of no omitted variables, where the true DGP is the SAC model. Using the extended SDM model to produce estimates in these circumstances (where the true DGP is the SAC model) results in inefficient, but consistent extended SDM model estimates for the explanatory variable. Note, efficiency of the estimates is often not the main concern for large spatial samples. Now consider the converse case where the true DGP is the extended SDM model, but we estimate the SAC model. The estimates for the explanatory variable coefficients will be biased due to an incorrect exclusion of the spatially lagged explanatory variables (WX) from the model. In other words, when the true DGP is associated with the extended SDM model where explanatory variables from neighboring regions are important, use of the SAC model will produce biased estimates that suffer from the omitted variables problems of the type we have considered.

By way of conclusion, we examined the impact of omitted variables on least-squares and various spatial regression model estimates when the DGP reflects spatial dependence in: the dependent variable, the independent variable and the disturbances. We find that the conventional omitted variables bias is amplified when OLS estimation procedures are used for these models. Use of certain spatial regression models such as the SDM in conjunction with consistent estimators will produce estimates that do not suffer from the amplified bias. These results provide a strong motivation for use of the SDM model specification in applied work where omitted variable problems seem likely.

3.4   An applied example

To provide a simple illustration, we rely on a relationship between regional *total factor productivity* (tfp) as the dependent variable y and regional knowledge stocks as the single explanatory variable. As illustrated in Chapter 1, the tfp dependent variable can be constructed using the residuals from a log-linear

Cobb-Douglas production function regression with constant returns to scale imposed. The dependent variable used here was constructed using an empirical estimate of the relative shares of labor and the assumption of constant returns to scale.

The dependent variable (total factor productivity) represents what is sometimes referred to as the Solow residual, as motivated in Chapter 1. Taking this view, we can plausibly rely on a single explanatory variable vector A representing the regional stock of knowledge, resulting in the model in (3.57), where we use a in (3.58) to represent $\ln A$.

$$y = \alpha \iota_n + \beta \ln A + \varepsilon \qquad (3.57)$$
$$y = \alpha \iota_n + \beta a + \varepsilon \qquad (3.58)$$

The variable A was constructed using the stock of regional patents appropriately discounted as a proxy for the regional stock of knowledge. LeSage, Fischer and Scherngell (2007) provide a detailed description of the sample data which covers 198 European Union regions from the 15 pre-2004 EU member states. The model relates regional knowledge stocks to regional total factor productivity to explore whether knowledge stocks impact the efficiency with which regions use their physical factors of production.

Although we use the regional stock of patents as an empirical proxy for technology, these are unlikely to capture the true technology available to regions. This is because knowledge produced by innovative firms is only partly appropriated due to the public good nature of knowledge which spills over to other firms within the region and in nearby regions. We might posit the existence of unmeasured knowledge a^* that is excluded from the model but correlated with the included variable a. It is well-known that regional patents exhibit spatial dependence (Parent and LeSage, 2008; Autant-Bernard, 2001), so as already motivated this would lead to an SDM model:

$$y = \alpha_0 \iota_n + \rho W y + \alpha_1 a + \alpha_2 W a + \varepsilon \qquad (3.59)$$

The SDM model in (3.59) subsumes the spatial error model SEM as a special case when the parameter restriction: $\alpha_2 = -\rho \alpha_1$. The SEM model would arise if there were no correlation between measured and unmeasured knowledge stocks, a and a^*, and when the restriction $\alpha_2 = -\rho \alpha_1$ is true.[4] In Chapter 6 we apply a simple likelihood-ratio test of the SEM versus SDM model to test the restriction $\alpha_2 = -\rho \alpha_1$ for this model and sample data.

3.4.1 Coefficient estimates

Recall that we showed how spatially dependent omitted variables will lead to the presence of spatial lags of the explanatory variables in Section 2.2.

---

[4] Anselin (1988) labels this the "common factor restriction."

Estimates from the SEM and SDM along with t-statistics are presented in Table 3.3.

It is frequently the case that applied studies compare estimates such as those from an SEM model to those from models containing a spatial lag of the dependent variable such as SAR or SDM. This is not a valid comparison as the SEM does not provide for spillovers. The SDM summary impact estimates based on partial derivatives are reported in Table 3.4, and will be discussed shortly.

TABLE 3.3: SEM and SDM model estimates

Parameters	SEM model estimates		SDM model estimates	
	Coefficient	t-statistic	Coefficient	t-statistic
α_0	2.5068	17.28	0.5684	3.10
α_1	0.1238	6.02	0.1112	5.33
α_2			−0.0160	−0.48
ρ	0.6450	8.97	0.6469	9.11

Many studies misinterpret the coefficient α_2 on the spatial lag of the knowledge capital variable $(W \cdot a)$ as a test for the existence of spatial spillovers. Since this coefficient is not significantly different from zero, they would erroneously conclude that there are no spatial spillovers associated with knowledge capital.

3.4.2 Cumulative effects estimates

Inference regarding the SDM model direct and indirect (spillover) impacts would be based on the summary measures of direct and indirect impacts for the SDM model. The matrix expression reflecting the own- and cross-partial derivatives for this model takes the form:

$$S_r(W) = V(W)(I_n\alpha_1 + W\alpha_2)$$
$$V(W) = (I_n - \rho W)^{-1} = I_n + \rho W + \rho^2 W^2 + \rho^3 W^3 + \dots$$

Table 3.4 reports effects estimates that were produced by simulating parameters using the maximum likelihood multivariate normal parameter distribution and the mixed analytical Hessian described in Section 3.2.1. A series of 2,000 simulated draws were used. The reported means, standard deviations and t-statistics were constructed from the simulation output.

If we consider the direct impacts, we see that these are close to the SDM model coefficient estimates associated with the variable a reported in Table 3.3. The difference between the coefficient estimate of 0.1112 and the

TABLE 3.4: Cumulative effects scalar summary estimates

	Mean effects	Std deviation	*t*-statistic
direct effect	0.1201	0.0243	4.95
indirect effect	0.1718	0.0806	2.13
total effect	0.2919	0.1117	2.61

direct effect estimate of 0.1201 equal to 0.0089 represents feedback effects that arise as a result of impacts passing through neighboring regions and back to the region itself. The discrepancy is positive since the impact estimate exceeds the coefficient estimate, reflecting some positive feedback. Since the difference between the SDM coefficient and the direct impact estimate is very small, we would conclude that feedback effects are small and not likely of economic significance.

In contrast to the similarity of the direct impact estimates and the SDM coefficient α_1, there are large discrepancies between the spatial lag coefficient α_2 from the SDM model and the indirect impact estimates. For example, the indirect impact is 0.1718, and significantly different from zero using the *t*-statistic. The SDM coefficient estimate associated with the spatial lag variable $W \cdot a$ reported in Table 3.3 is -0.0160, and not significant based on the *t*-statistic. If we incorrectly view the SDM coefficient α_2 on the spatial lag of knowledge stocks $(W \cdot a)$ as reflecting the indirect impact, this would lead to an inference that the knowledge capital variable $W \cdot a$ exerts a negative and insignificant indirect impact on total factor productivity. However, the true impact estimate points to a positive and significant indirect impact (spillover) arising from changes in the variable a.

It is also the case that treating the sum of the SDM coefficient estimates from the variables a and $W \cdot a$ as total impact estimates would lead to erroneous results. The total impact of knowledge stocks on total factor productivity is a positive 0.2919 that is significant, whereas the total impact suggested by summing up the SDM coefficients would equal less than half this magnitude. These differences will depend on the size of indirect impacts which cannot be correctly inferred from the SDM coefficients. In cases where the indirect impacts were zero, and the direct impact estimates are close to the SDM estimates on the non-spatially lagged variables, the total impact could be correctly inferred. Of course, one would not know if the indirect impacts were small or insignificant without calculating the scalar summary impact measures presented in Table 3.4.

We can interpret the total impact estimates as elasticities since the model is specified using logged levels of total factor productivity and knowledge stocks. Based on the positive 0.2919 estimate for the total impact of knowledge stocks, we would conclude that a 10 percent increase in regional knowledge would result in a 2.9 percent increase in total factor productivity. Around 2/5 of this impact comes from the direct effect magnitude of 0.1201, and 3/5 from the indirect or spatial spillover impact based on its scalar impact estimate of 0.1718.

3.4.3   Spatial partitioning of the impact estimates

We can spatially partition these impacts to illustrate the nature of their influence as we move from immediate to higher-order neighbors. This might be of interest in applications where the spatial extent of the spillovers is an object of inference.

These are presented for the SDM model in Table 3.5, which shows the mean, standard deviation and a t-statistic for the *marginal* effects associated with matrices W of orders 0 to 9. Direct effects for W^1 will equal zero and the indirect effects for W^0 equal zero as discussed in Chapter 2. Of course, if we cumulated the marginal effects in the table over all orders of W until empirical convergence of the infinite series, these would equal the cumulative effects reported in Table 3.4.

TABLE 3.5:   Marginal spatial partitioning of impacts

	Direct effects	Standard deviation	t-statistic
W^0	0.1113	0.0205	5.4191
W^1	0.0000	0.0000	—
W^2	0.0046	0.0013	3.5185
W^3	0.0016	0.0007	2.3739
W^4	0.0010	0.0005	2.0147
W^5	0.0006	0.0003	1.6643
W^6	0.0004	0.0002	1.4208
W^7	0.0002	0.0002	1.2285
W^8	0.0001	0.0001	1.0761
W^9	0.0001	0.0001	0.9516
	Indirect effects	Standard deviation	t-statistic
W^0	0.0000	0.0000	—
W^1	0.0622	0.0188	3.3085
W^2	0.0353	0.0131	2.6985
W^3	0.0243	0.0105	2.3098
W^4	0.0160	0.0083	1.9220
W^5	0.0107	0.0066	1.6283
W^6	0.0072	0.0052	1.3978
W^7	0.0049	0.0040	1.2151
W^8	0.0033	0.0031	1.0672
W^9	0.0023	0.0024	0.9455

From the table we see that direct and indirect effects exhibit the expected decay with higher order W matrices. If we use a t-statistic value of 2 as a measure of when the effects are no longer statistically different from zero, we see that the spatial extent of the spillovers from regional knowledge stocks is around W^4. For our matrix W based on 7 nearest neighbors, the matrix W^2 contains 18 non-zero elements representing second-order neighbors (for the average region in our sample). The matrix W^3 contains (an average) 30.8 third-order neighbors and W^4 has 45 fourth-order neighbors. This suggests that spatial spillover effects emanating from a single region exert an impact on a large proportion of the 198 regions in our sample. However, we note that the size of the spillover effects is not likely to be economically meaningful for higher-order neighboring regions.

Using our elasticity interpretation, we can infer that a relatively large increase of 10 percent in knowledge stocks would have indirect or spatial spillover effects corresponding to a 0.6 percent increase in first-order neighboring region factor productivity, 0.35 percent increase in second-order neighbors factor productivity, 0.24 for third-order neighbors, and so on.

The other notable feature of Table 3.5 is the small amount of feedback effect shown in the marginal direct effects, and the relatively quick decay with orders of W.

3.4.4 A comparison of impacts from different models

It is interesting to compare the SDM model estimates and scalar summary of effects with those from the SAR and SAC models. The coefficient estimates are presented in Table 3.6. Given the lack of significance of the spatial lag variable $W \cdot a$ in the SDM model, we would expect to see estimates from the SAR and SDM models that are quite similar, as shown in Table 3.6. The SAC model resulted in an insignificant estimate for the spatial dependence parameter λ associated with the disturbances. This also produces estimates similar to those from the SAR and SDM models.

TABLE 3.6: SAR and SAC model estimates

Parameters	SAR model estimates		SAC model estimates	
	Coefficient	t-statistic	Coefficient	t-statistic
α_0	0.5649	3.10	0.5625	2.11
α_1	0.1057	5.93	0.1144	5.09
α_2				
ρ	0.6279	10.12	0.6289	6.27
λ			-0.0051	-0.02
σ^2	0.1479		0.1509	
Log-Likelihood	-29.30		-30.65	

Effects estimates for the SAR and SAC model have the same analytical form since they are based on the matrix expressions from Section 2.7.

$$S_r(W) = V(W)I_n\beta_r$$
$$V(W) = (I_n - \rho W)^{-1} = I_n + \rho W + \rho^2 W^2 + \rho^3 W^3 + \dots$$

The difference between impacts from these two models and those for the SDM model is the additional term $W\theta$ that appears in the case of the SDM model. Since the distribution of θ is centered near zero according to the point estimate and associated t-statistic, we would expect similar impact estimates from the SAR, SDM and SAC models in this particular illustration.

TABLE 3.7: A comparison of cumulative impacts from SAR, SAC and SDM

	SAR effects	Std deviation	t-statistic
direct effect	0.1145	0.0207	5.53
indirect effect	0.1746	0.0620	2.81
total effect	0.2891	0.0827	3.49
	SDM effects	Std deviation	t-statistic
direct effect	0.1201	0.0243	4.95
indirect effect	0.1718	0.0806	2.13
total effect	0.2919	0.1117	2.61
	SAC effects	Std deviation	t-statistic
direct effect	0.1199	0.0241	4.98
indirect effect	0.1206	0.0741	1.62
total effect	0.2405	0.0982	2.44

We note that invalid comparisons of point estimates from different spatial regression model specifications has lead practitioners to conclude that changing the model specifications will lead to very different inferences. This may also have lead to excessive focus in the spatial econometrics literature on procedures for comparative testing of alternative model specifications, a subject we take up in Chapter 6. However, using the correct partial derivative interpretation of the parameters from various models results in less divergence in the inferences from different model specifications. This result is related to the partial derivative interpretation of the impact from changes to the variables from different model specifications which represents a valid basis for these comparisons.

This is not meant to imply that model specification is not important. For example, use of an SEM model would lead to omission of the important spatial spillover (indirect effects) found here. In addition, the SAC effects estimates lead to an inference that the indirect spillover impacts are not significantly different from zero based on the t-statistic reported in Table 3.7.

3.5   Chapter summary

In Section 3.1 we set forth computationally efficient approaches to maximum likelihood estimation of the basic family of spatial regression models. The most challenging part of maximum likelihood estimation is computing the log determinant term that appears in the log-likelihood function, and Chapter 4 will provide details regarding this. In addition to point estimates there is also a need to provide a variance-covariance matrix estimate that can be used for inference. Section 3.2 discussed various strategies and set forth a mixed approach that uses numerical Hessian results to modify a single computationally challenging term from the analytical Hessian.

The public domain *Spatial Econometrics Toolbox* (LeSage, 2007) and *Spatial Statistics Toolbox* (Pace, 2007) provide code examples written in the MATLAB language that implement most of the methods discussed in this text. This should allow the interested reader to examine detailed examples that implement the ideas presented here.

Modeling spatial relationships often results in omitted latent influences that are spatial in nature. For example, hedonic home price regressions usually rely on individual house characteristics that may exclude important neighborhood variables that reflect accessibility, school quality, amenities, etc. In Section 3.3, we examined the nature of bias that will arise from omitted variables in both least-squares and spatial regression estimates. An interesting feature of omitted variables in spatial regression models is that they will lead to data generating processes that include spatial lags of the explanatory variables, providing a powerful motivation for use of the spatial Durbin model.

An applied illustration was provided in Section 3.4 to reinforce the ideas set forth in this chapter. A comparison of maximum likelihood estimates from a family of spatial regression models along with an interpretation of the parameters was provided. The simple one-variable model was based on regional variation in factor productivity for a sample of 198 European Union regions. The focus of this applied illustration was on the role of regional knowledge stocks in explaining variation in regional total factor productivity.

Chapter 4

Log-determinants and Spatial Weights

Many spatial applications involve large data sets. For example, the US Census provides data on blocks ($n = 8, 205, 582$), block groups ($n = 208, 790$), census tracts ($n = 65, 443$), and other geographies. In the case of the Census data, each of these observations represents a region. If spatial dependence is material and each region affects every other region, this leads to $n \times n$ dependence relations. Since some applications involve elaborate models, computational aspects of spatial econometrics have been an active area of research for some time (Ord, 1975; Martin, 1993; Pace and Barry, 1997; Griffith, 2000; Smirnov and Anselin, 2001; LeSage and Pace, 2007).

This chapter addresses theoretical and numerical issues that arise when fitting models to spatial data using likelihood-based techniques. Likelihood-based techniques involve the determinant of the variance-covariance matrix which measures the degree of dependence among observations. This chapter addresses both exact and approximate calculation of the log-determinant and bounds for the dependence parameter.

In addition, the chapter deals with calculation of other quantities used in spatial estimation and inference such as the diagonal of the variance-covariance matrix and the derivative of the log-determinant.

The chapter addresses efficient computation of the estimated spatial effects and shows a general approach to obtaining closed-form solutions to many single parameter spatial models. Finally, the chapter discusses aspects of quickly constructing spatial weight matrices.

4.1 Determinants and transformations

Statistical applications often involve transformations of the dependent variable. Unless these transformations are handled properly, statistical procedures have the potential to produce pathological behavior. We use a simple example to illustrate the type of problem that can arise. Suppose the interest is in a least-squares fit of the relationship in (4.1) involving transformation of the dependent variable by the scalar T.

$$Ty = X\beta + \varepsilon \tag{4.1}$$

A least-squares fit that allowed a setting for $T = 0$ would yield a vector of zeros on the left hand side (LHS) resulting in a perfect fit when $\beta = 0$. Therefore, selecting T to minimize errors in (4.1) leads to a pathological solution.

As a slightly more complex example, suppose that $y = \begin{bmatrix} u & v & w \end{bmatrix}'$ and that $X = \iota_3$. Let T be a 3×3 matrix shown in (4.2).

$$T = \begin{bmatrix} 1 & c & c \\ c & 1 & c \\ c & c & 1 \end{bmatrix} \tag{4.2}$$

When $c = 1$, $Ty = \begin{bmatrix} u+v+w & u+v+w & u+v+w \end{bmatrix}'$ and least squares could set $\beta = u + v + w$ to perfectly explain y. One can devise similar pathological examples in cases involving more observations.

Something about the transformation T acts to reduce variability of Ty which can be exploited by a statistical procedure attempting to maximize goodness-of-fit. One common characteristic of both of these pathological examples is that the determinant of T equals zero (T is singular).

Determinants and the role they play in transformations can be considered at a more basic level outside the context of statistical applications. To demonstrate this, we examine some basic geometry. The unit square shown as the solid line segments in Figure 4.1 has positive coordinates as in S of (4.3). The area bounded by the solid line segments is 1. Suppose we transform the coordinates of the unit square S in (4.3), multiplying by the transformation matrix T in (4.4) as shown in (4.5).

$$S = \begin{bmatrix} 0 & 1 & 0 & 1 \\ 0 & 0 & 1 & 1 \end{bmatrix} \tag{4.3}$$

$$T = \begin{bmatrix} 1 & c \\ c & 1 \end{bmatrix} \tag{4.4}$$

$$S_T = TS \tag{4.5}$$

If $c = 0.0$, the coordinates remain the same leaving us with a unit square after transformation. If $c = 0.9$, the new coordinates appear in (4.6).

$$S_T = \begin{bmatrix} 0 & 1 & 0.9 & 1.9 \\ 0 & 0.9 & 1 & 1.9 \end{bmatrix} \tag{4.6}$$

Figure 4.1 shows the original unit square ($c = 0$) as well as two transformed unit squares based on $c = 0.5$ and $c = 0.9$. These transformations stretch the coordinates of the unit square to produce parallelograms.

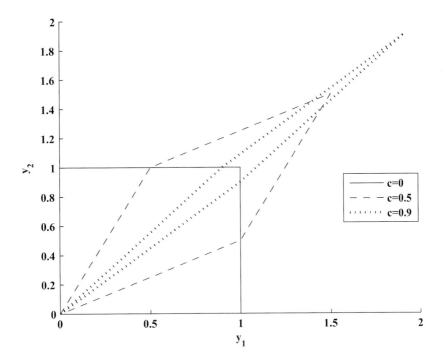

FIGURE 4.1:   Bivariate (y_1, y_2) transformation

These parallelograms seem to have less area than the original unit square which had an area of 1. In fact, the absolute value of the determinant of the transformation matrix T equals the area of the resulting parallelograms. For the case of the 2×2 matrix T, the determinant of this matrix, $|T|$, equals $T_{11}T_{22} - T_{12}T_{21}$. For this particular T, $|T| = 1 - c^2$, so values of $c = 0, 0.5, 0.9$ lead to determinants of T that equal $1, 0.75, 0.19$. This implies that the area of the parallelograms decreases under these transformations where c is positive. If $c = 1.1$, the determinant would equal -0.21, but the absolute value of 0.21 would be the area of the resulting parallelogram. More importantly, if $c = 1$, the determinant equals 0 and thus the transformation collapses the unit square.

In this two-dimensional case, the absolute value of the determinant of T measures the area of the parallelogram (transformed unit square). In three dimensions, a 3×3 transformation matrix T can transform the unit cube into a parallelepiped. In this case, the absolute value of the determinant of T would measure the volume of the parallelepiped formed by transforming the unit cube. In n dimensions, the $n \times n$ matrix T can transform the unit hypercube to yield an n-parallelotope. As before, the absolute value of the determinant of T would measure the n-dimensional volume.

Returning to statistical issues, examine the point clouds in Figure 4.2 from a transformed normal bivariate distribution based on varying the level of correlation between y_1 and y_2, which we denote $\rho = 0, 0.33, 0.66, 0.99$. The bivariate distribution before the transformation was *proper* (the volume under the bivariate density equals 1). These point clouds show a pattern similar to that of the parallelograms with higher correlations corresponding to more severe transformations.

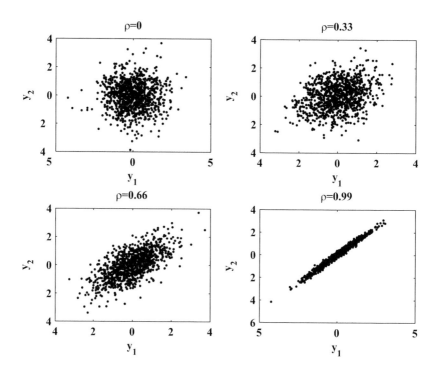

FIGURE 4.2: Bivariate (y_1, y_2) normal point clouds by correlation (ρ)

It is visually evident that the point clouds associated with the transformations exhibit shrinkage relative to the original point cloud. To counteract the shrinkage aspect of transformations, distributions must include an adjustment to preserve the volume of unity under the joint probability density. As a result, continuous multivariate densities such as the multivariate normal in (4.7) include the determinant of the variance-covariance matrix as the adjustment.

$$N(u) = (2\pi)^{-n/2}|\Omega|^{-1/2} \exp\left(\frac{1}{2}(u-\mu)'\Omega^{-1}(u-\mu)\right) \qquad (4.7)$$

The multivariate t distribution (shown below), Wishart, and other distributions also contain a determinant term.

$$t(u) = \Gamma\left(\frac{n+p}{2}\right)\Gamma\left(\frac{n}{2}\right)^{-1}(n\pi)^{-\frac{p}{2}}|\Omega|^{-\frac{1}{2}}\left(1 + \frac{(u-\mu)'\Omega^{-1}(u-\mu)}{n}\right)^{-\frac{n+p}{2}}$$

4.2   Basic determinant computation

The product of the diagonal elements of a triangular or a diagonal matrix yields the determinant. Most algorithms for calculating the determinant exploit this by reducing the matrix under consideration to a diagonal or a triangular matrix. There are various approaches that can be used to create diagonal or triangular matrices. For example, one can compute the determinant of a matrix using eigenvalues which reduce the matrix under consideration to a diagonal matrix. Alternatively, various forms of Gaussian elimination such as the LU or Cholesky decompositions reduce the matrix under consideration to a triangular matrix. These matrix decomposition techniques also use the multiplicative property of determinants, $|CD| = |C||D|$ to reduce the general problem to a product involving simpler problems.

The matrix decomposition approach to calculating determinants can be illustrated with simple examples. We begin with the Gaussian elimination approach which relies on elementary operations such as adding a multiple of one row to another to achieve a triangular form. These row operations do not change the determinant, and therefore the final triangular matrix has the same determinant as the initial matrix of interest. We begin with A in (4.8) and add the first row of A times a to the second row of A, which zeros out the element below the diagonal yielding A^1 in (4.9). The diagonal elements of the resulting triangular matrix A^1 are known as *pivots*, and the product of the pivots equals the determinant as shown in (4.10) (Strang, 1976).

$$A = \begin{bmatrix} 1 & -a \\ -a & 1 \end{bmatrix} \qquad (4.8)$$

$$A^1 = \begin{bmatrix} 1 & -a \\ 0 & 1-a^2 \end{bmatrix} \qquad (4.9)$$

$$|A| = |A^1| = 1 - a^2 \qquad (4.10)$$

A zero pivot would yield a zero determinant and a singular matrix. Therefore, non-singular matrices yield non-zero pivots and positive definite matrices yield strictly positive pivots. Obviously, A is non-singular when $\text{abs}(a) \neq 1$ and A is positive definite when $\text{abs}(a) < 1$.

Note, the determinant of the first sub-matrix of A or A_{11} equals 1. Not coincidentally, this is the first pivot of A^1. In fact, the pivots from the triangular matrix produced by row operations yield the sequence of determinants for sub-matrices of increasing size. This is a valuable feature that is useful in examining spatial systems composed of subsets of the observations such as used in local spatial autoregressions (Pace and LeSage, 2003a).

Most computer routines actually yield a unit lower triangular matrix L containing ones on the diagonal, an upper triangular matrix U and a *permutation* matrix P so that $PA = LU$. The result of PA is to reorder the rows of A. This can be used to produce $|A| = |L||U|/|P|$, which represents an LU decomposition with permutations or reordering of rows for numerical accuracy. Fortunately, when $A = I_n - \rho W$ where W is row-stochastic and $\text{abs}(\rho) < 1$, A is *strictly diagonally dominant*. This means that the diagonal element (which equals 1) strictly exceeds the sum of the other elements in the row (which equals ρ since W is row-stochastic). Strictly diagonally dominant matrices are invertible (non-singular). Moreover, strictly diagonally dominant matrices do not require reordering rows of A, allowing us to set $P = I_n$ (Golub and Van Loan, 1996, Theorem 3.4.3). For the case where $P = I_n$, L is unit lower triangular having a determinant of 1, so $|A| = |U|$ (Strang, 1976, p. 21).

For symmetric positive definite matrices, an algorithmic variant yields the Cholesky decomposition so that $A = R'R$ where R is triangular. Therefore $|A| = |R||R'| = |R|^2$, since the determinant of a matrix and its transpose are identical. The Cholesky decomposition is almost twice as fast as the LU decomposition since it takes account of symmetry. In addition, symmetric positive definite matrices do not require reordering or permutations for numerical accuracy which reduces the computational cost.

For non-symmetric A interest often lies in $B = A'A$. In this case, a non-singular A implies a symmetric positive definite B. Therefore, it sometimes pays to find the determinant of the symmetric B rather than work directly with A. In this case, non-singular A can have negative (but not zero) elements on the diagonal of U. The determinant of A equals the product of the pivots (some of which may be positive and others negative) and thus can be positive or negative. This creates a problem when calculating the log of the determinant for B. However, this problem can be resolved by taking the absolute value of the determinant which implies finding the absolute value of the pivots before using logs as shown in (4.11) to (4.13).

$$|B| = |A'||A| = \text{abs}(|A|)^2 \tag{4.11}$$

$$\text{abs}(|A|) = \prod_{i=1}^{n} \text{abs}(U_{ii}) \tag{4.12}$$

$$\ln|B| = 2\ln\text{abs}(|A|) = 2\sum_{i=1}^{n} \ln(\text{abs}(U_{ii})) \tag{4.13}$$

For symmetric real matrices (or those that can be transformed to exhibit symmetry), one can obtain the spectral decomposition $B = V\Lambda_B V'$. This decomposition was applied to our 2×2 example to produce the results shown in (4.17) and (4.18). The diagonal matrix Λ_B contains the n real eigenvalues, and V in (4.17) is a matrix comprised of n orthogonal eigenvectors (each with n elements). By construction, $VV' = I_n$ and this sets up (4.19). Since $|V| = |V'|$, it follows that $|V| = |V'| = 1$ and this allows us to find the determinant using the product of the diagonal elements of $I_2 - a\Lambda_B$ as shown in (4.21). This is equal to the usual 2×2 determinant expression which can be seen from (4.22).

$$A = I_2 - aB \tag{4.14}$$

$$B = \begin{bmatrix} 0 & 1 \\ 1 & 0 \end{bmatrix} \tag{4.15}$$

$$B = V\Lambda_B V' \tag{4.16}$$

$$V = \begin{bmatrix} -0.7071 & -0.7071 \\ -0.7071 & 0.7071 \end{bmatrix} \tag{4.17}$$

$$\Lambda_B = \begin{bmatrix} -1 & 0 \\ 0 & 1 \end{bmatrix} \tag{4.18}$$

$$A = V(I_2 - a\Lambda_B)V' \tag{4.19}$$

$$|A| = |V||I_2 - a\Lambda_B||V'| \tag{4.20}$$

$$|A| = (1 - a\Lambda_{11})(1 - a\Lambda_{22}) \tag{4.21}$$

$$|A| = (1 + a)(1 - a) = 1 - a^2 \tag{4.22}$$

An advantage of the eigenvalue approach is that the eigenvalues of B only need to be computed once. Subsequent updating of the determinant of A for differing values of a requires very little additional computation. However, the eigenvalue approach does not scale well to larger problems.

In terms of computation, all of these techniques require order of n cubed ($O(n^3)$) calculations for general matrices. However, the Cholesky requires $n^3/3$, the LU requires $2n^3/3$. There are many methods for calculating eigenvalues which require a multiple of the LU and Cholesky counts since the methods have some iterative component.

In practice, it becomes numerically challenging to compute the actual determinant using the product of all pivots, since small pivots can lead to a very small numerically inaccurate determinant. Instead of calculating the determinant *per se*, we can find the log-determinant using the sum of the logged pivots which will produce a more accurate numerical solution. This naturally requires positive pivots such as those from a positive definite matrix.

Some rules pertaining to determinants and related quantities of interest in spatial econometrics are enumerated below. We will discuss their application in circumstances specific to various spatial econometric modeling situations in Section 4.3.

$$|CD| = |C||D| \tag{4.23}$$

$$|C'| = |C| \tag{4.24}$$

$$|G| = \prod_{i=1}^{n} G_{ii} \quad \text{diagonal or triangular } G \tag{4.25}$$

$$\ln|I_n - \rho W| = -\sum_{i=1}^{\infty} \frac{\rho^i \operatorname{tr}(W^i)}{i} \tag{4.26}$$

$$|e^C| = e^{\operatorname{tr}(C)} \tag{4.27}$$

$$|C^a| = |C|^a \tag{4.28}$$

$$|C \otimes D| = |C|^{dim(D)}|D|^{dim(C)} \tag{4.29}$$

$$\operatorname{tr}(C \otimes D) = \operatorname{tr}(C)\operatorname{tr}(D) \tag{4.30}$$

$$(C \otimes D)^m = C^m \otimes D^m \tag{4.31}$$

$$(A \otimes B)(C \otimes D) = (AC \otimes BD) \tag{4.32}$$

$$\operatorname{tr}(A^j) = \sum_{i=1}^{n} \lambda_i^j \tag{4.33}$$

$$\operatorname{diag}(AB) = (A \odot B')\iota_n \tag{4.34}$$

$$\operatorname{tr}(AB) = \iota_n'(A \odot B')\iota_n = \iota_n'(A' \odot B)\iota_n \tag{4.35}$$

In the above rules $\otimes$ represents the Kronecker product and $\odot$ represents elementwise or Hadamard multiplication.

4.3   Determinants of spatial systems

Spatial models using weight matrices have additional structure and features which greatly aid computation of determinants, equation solutions, and other quantities of interest for spatial modeling.

If each observation depends on some, but not all other $n-1$ observations, the weight matrix will have a number of zeros and is therefore a *sparse* matrix. In particular, the common contiguity weight matrix will have an average of approximately six neighbors for each observation for spatially random data on a plane. An implication is that a contiguity-based weight matrix will have approximately $6n$ non-zeros and $n^2 - 6n$ zeros. In terms of the proportion of non-zero elements this equals $6/n$, so this matrix becomes increasingly sparse with increasing n. One can generalize this to the m nearest neighbor problem which would have a proportion of non-zeros equal to m/n. Sparsity also extends to other dependence structures such as those used in geostatistics (Barry and Pace, 1997).

Sparsity provides tremendous storage and computational advantages. In the 2000 US Census there were $65,443$ tracts, $208,790$ block groups, and $8,205,582$ blocks. A dense weight matrix would require 31.90, 324.80, and $501,659.33$ gigabytes of storage respectively for these sample sizes. In contrast, a sparse contiguity matrix would require less than 0.01, 0.03, and 1.10 gigabytes of storage for these data samples. In terms of computation, some sparse matrix techniques require only linear in n calculations, whereas dense matrix techniques often require calculations that are cubic in n.

In addition, weight matrices have known properties, which can often be used to simplify required calculations (Bavaud, 1998; Martellosio, 2006). One example of this is the row-stochastic matrix W, which results in $\rho \in [0, 1)$ as a sufficient condition for non-singular $Z = I_n - \rho W$. An example where zeros on the main diagonal of W were used can be found in LeSage and Pace (2007). They show that zeros on the diagonal of W lead to a zero log-determinant for the matrix exponential spatial specification which we discuss in Chapter 9. This arises since $\ln |e^{\alpha W}| = 0$, where $e^{\alpha W}$ plays the role of the transformation matrix T in (4.1) in the case of the matrix exponential spatial specification.

To illustrate how known properties of spatial systems can be exploited to computational advantage, we use a sample of $3,107$ counties in the continental United States. For our illustrations we use a matrix W based on 6 nearest neighbors constructed using Euclidean distance between points of the projected coordinates.[1] Figure 4.3 shows the *graph* (as in graph theory) where edges (line segments) between nodes (points) indicate nodes that are neighbors. If only one edge is needed, the nodes are first-order neighbors. If t edges are needed to transverse between nodes, the nodes are t-order neighbors.

Orderings of rows and columns greatly affect decomposition times for sparse spatial weight matrices. Ordering rows and columns corresponds to the operation $W_P = PWP'$ where P is an $n \times n$ permutation matrix. Permuta-

---

[1] The earth is a three dimensional sphere, but maps are two dimensional. Accordingly, maps can only approximate the surface of a sphere with error. *Map projections* attempt to make useful two dimensional approximations of the sphere (Snyder and Voxland, 1989). We used a transverse Mercator projection of the latitude and longitude coordinates to arrive at new locational coordinates such as shown in Figure 4.3.

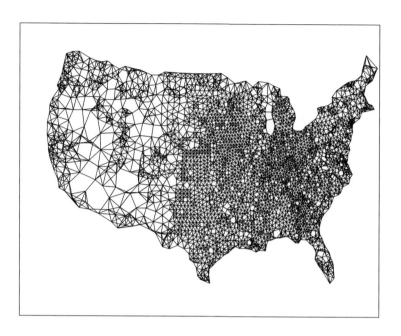

FIGURE 4.3:   Graph of W based on six nearest neighbors for US Counties

tion matrices have a number of convenient properties such as $P^{-1} = P'$ and $|P| = 1$. Using these, we can show that reordering elements of W will not affect the log-determinant calculation. This follows from: $|P(I_n - \rho W)P'| = |P||I_n - \rho W||P'|$, which equals $|I_n - \rho W|$ (as well as $|I_n - \rho PWP'|$).

The simplest ordering is geographic. For example, we ordered the rows and the columns of the matrix W so that the most northern tract is in the row and column 1 position, and the most southern tract is in row and column n. This simple ordering often concentrates non-zero elements closer to the diagonal than in the original ordering. Intuitively, this makes the permuted system more like a *band matrix* which has non-zeros concentrated in fixed bands around the diagonal. The reverse Cuthill-McKee ordering provides a more systematic way of reducing the *bandwidth* of matrices ($\max(\text{abs}(i-j))$ for non-zero elements) .

Other sophisticated ordering algorithms such as minimum degree and nested dissection can provide computational benefits (Golub and Van Loan, 1996). The approximate minimum degree ordering is designed to aid Gaussian elimination. Although its workings are not as straightforward as the bandwidth reducing orderings, it often results in the lowest *fill-in* that arises from additional non-zero elements introduced in L and U. The fill-in occurs in Gaussian elimination as non-zero elements are introduced during elementary row operations to eliminate elements in the earlier rows that have to be eliminated later. A good ordering results in low fill-in.

We use a sample of $62,226$ US Census tracts from the year 2000 to illustrate how alternative orderings impact computational time required for operations such as the Cholesky and LU matrix decompositions. Table 4.1 shows the time in seconds required to calculate permutations as well as Cholesky and LU matrix decompositions for $I_n - \rho W$. The approximate minimum degree ordering resulted in the lowest fill-in based on the percentage of non-zeros, and the fastest computational time. In Chapter 3, we discussed computing the log-determinant over a grid of values for the parameter ρ. From the table, we can infer the time necessary to calculate 100 determinants for a grid of ρ values and interpolating these to produce a finer grid. This would require less than one minute for the approximate minimum degree ordering. In contrast, it was not feasible to calculate even a single (log) determinant for our $62,226 \times 62,226$ matrix when the sample exhibited a random ordering, since the calculation required more that 12 gigabytes of computer memory.

TABLE 4.1: Times in seconds for different orderings

Operation	Geographic ordering	Cuthill-McKee ordering	Minimum Degree ordering
Permutation Time	0.058	0.058	0.061
Cholesky Time	1.586	1.147	0.115
LU Time	6.856	8.429	0.316
% non-zeros in U	0.201	0.209	0.022

By way of conclusion, software capable of producing these orderings of the sample data is a requirement when computing log-determinants via LU or Cholesky decompositions.

4.3.1 Scalings and similarity transformations

The permutation transformation $Z_P = P(I_n - \rho W)P'$ represents one type of *similarity transformation*. Given a matrix A, the matrix $A_1 = CAC^{-1}$ is *similar to* A, which means it will have the same eigenvalues, and therefore the same determinant. Given a spatial transformation $Z_B = I_n - \rho RB$ where B is a symmetric binary weight or adjacency matrix and R is a diagonal matrix containing the inverse of the row sums of B, RB is a non-symmetric, row-stochastic weight matrix. In general, non-symmetric matrices have complex eigenvalues. However, in this case RB has real eigenvalues that are the same as the symmetric (but not row-stochastic) matrix $R^{\frac{1}{2}}BR^{\frac{1}{2}}$. Consider the similarity transformation $R^{-\frac{1}{2}}Z_BR^{\frac{1}{2}}$ which produces the symmetric matrix $I_n - \rho R^{\frac{1}{2}}BR^{\frac{1}{2}}$. From a statistical perspective, using the row-stochastic RB may yield better results. However, from a numerical analysis perspective using the similar, but symmetric matrix $R^{\frac{1}{2}}BR^{\frac{1}{2}}$ will usually perform better (Ord, 1975). Therefore, the best strategy in many cases is to use the row-stochastic W for calculations related to the statistical portion of the estimation problem, but work with a similar symmetric matrix when calculating the log-determinant. Given a table of determinants or the eigenvalues, similarity transformations represent low-cost computational operations.

Computation is often simpler when W has a maximum eigenvalue of 1, which is the case for row-stochastic matrices or matrices that are *similar to* row-stochastic matrices. Consider a candidate weight matrix of interest W_a that does not have a maximum eigenvalue $\max(\lambda_a)$ equal to 1. This can be transformed using $W_b = W_a \max(\lambda_a)^{-1}$ to have a maximum eigenvalue of 1. An implication is that any weight matrix W can be scaled to have a maximum eigenvalue of 1. We note that this facilitates interpretation of the powers of W, since these would also have a maximum eigenvalue of 1.

Symmetric *doubly stochastic* weight matrices are those that have rows and columns that sum to 1 and exhibit symmetry. This means the maximum eigenvalue equals 1 and all eigenvalues are real. Transforming a matrix W_t to doubly stochastic form involves an iterative process: 1) calculating the diagonal matrix of row sums R_t for the symmetric weight matrix W_t, 2) calculating $W_{t+1} = R_t^{-\frac{1}{2}}W_tR_t^{-\frac{1}{2}}$, and, 3) repeating steps 1) and 2) until convergence. The resulting doubly stochastic weight matrix is not *similar* to the initial weight matrix W_t.

4.3.2 Determinant domain

Which values of ρ lead to non-singular $Z = I_n - \rho W$? For symmetric matrices, the compact open interval for $\rho \in (\lambda_{min}^{-1}, \lambda_{max}^{-1})$ will lead to a symmetric positive definite Z. In the case of symmetric matrices similar to row-stochastic matrices where $\lambda_{max} = 1$, the interval for ρ becomes $(\lambda_{min}^{-1}, 1)$.

The situation becomes more difficult when W has complex eigenvalues. Assume that W is scaled to be row-stochastic so that $W\iota_n = \iota_n$. If W is not similar to a symmetric matrix, it may have complex eigenvalues. If a real matrix has complex eigenvalues, these come in complex conjugate pairs (Bernstein, 2005, p. 131). Let λ represent the n by 1 vector of eigenvalues of W. The determinant of $(I_n - \rho W)$ equals,

$$|I_n - \rho W| = \prod_{i=1}^{n}(1 - \rho\lambda_i) = \left[\prod_{i=3}^{n}(1 - \rho\lambda_i)\right](1 - \rho\lambda_1)(1 - \rho\lambda_2) \qquad (4.36)$$

where, without loss of generality, one of the complex conjugate pairs of eigenvalues appears in λ_1 and the other in λ_2. If the product $(1 - \rho\lambda_1)(1 - \rho\lambda_2)$ equals 0, this would lead to a zero determinant which would imply singular $(I_n - \rho W)$ and a singular variance-covariance matrix.

What value of ρ could lead to a singular $(I_n - \rho W)$? To focus on the complex conjugate nature of λ_1 and λ_2, we express these as $\lambda_1 = r + jc$ and $\lambda_2 = r - jc$, where r is the real part of λ_1, λ_2, jc is the complex part of λ_1, λ_2, and j is the square root of -1, so that $j^2 = -1$. We assume $c \neq 0$, since a value of 0 would lead to a real number representation. Equations (4.37)–(4.40) form the complex quadratic equation given complex conjugate pairs of the eigenvalues and set the complex quadratic equation to 0 to find values of ρ associated with a singularity.

$$0 = (1 - \rho\lambda_1)(1 - \rho\lambda_2) \tag{4.37}$$
$$0 = (1 - \rho r - \rho jc)(1 - \rho r + \rho jc) \tag{4.38}$$
$$0 = 1 - 2\rho r + \rho^2 r^2 - j^2 \rho^2 c^2 \tag{4.39}$$
$$0 = 1 - 2\rho r + \rho^2 (r^2 + c^2) \tag{4.40}$$

Rewriting (4.40) using the discriminant $d = b^2 - 4ac$ from the quadratic formula $ax^2 + bx + c = 0$ we find that $d < 0$.

$$d = 4 \left[r^2 - r^2 - c^2 \right] \tag{4.41}$$
$$d = -4c^2 \tag{4.42}$$

Since c^2 is always positive (we ruled out $c = 0$ by assumption), the discriminant d is negative so the quadratic equation will yield two complex roots. This means that a real ρ can never result in a product of the function of two complex conjugate eigenvalues equaling 0. In other words, complex conjugate eigenvalues do not affect whether $I_n - \rho W$ is singular. Only purely real eigenvalues can affect the singularity of $I_n - \rho W$.

Consequently, for W with complex eigenvalues, the interval of ρ which guarantees non-singular $I_n - \rho W$ is $(r_s^{-1}, 1)$ where r_s equals the most negative purely real eigenvalue of W. Fortunately, sparse eigenvalue routines such as "eigs" in Matlab can be used to rapidly find the eigenvalue with the smallest real part.

4.3.3 Special cases

In the following sections, we discuss issues related to calculating log determinants for special cases that arise in applied practice.

4.3.3.1 Naturally triangular systems

In some cases, such as with temporal or spatiotemporal data where a unidirectional order such as time implies that no observation depends upon a future observation, we can arrange observations so W is triangular with zeros on the main diagonal. In this case, $I_n - \rho W$ has a determinant of 1 and a log-determinant of 0. For example, Pace et al. (2000) exploit this type of situation to estimate a spatiotemporal model involving real estate data.

4.3.3.2   Regular locational grid

Data from remote sensing appears either locally or globally on a regular locational grid. For example, a satellite may record the value of a variable every three square meters, collecting this information over a wide area. Use of regular locational grids is also popular in theoretical and Monte Carlo work. Typically, the plot of non-zeros of $I_n - \rho W$ in this case has a band structure as shown in Figure 4.4, where interior points in the grid depend on 20 other observations.

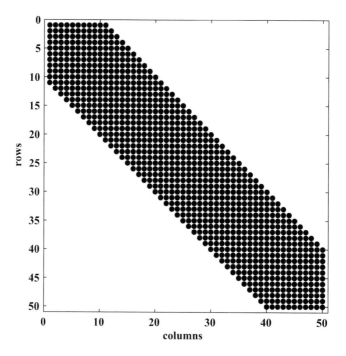

FIGURE 4.4:   Plot of non-zeros of $I_n - 0.8W$ based on a regular locational grid

Given the large number of observations produced by remote sensing, calculating the log-determinant would seem difficult, but regular locational grids prove advantageous in Gaussian elimination. Figure 4.5 shows the first 100 pivots from Gaussian elimination applied to a regular locational grid of 25,000 observations, where each interior observation has 20 neighbors. The figure shows pivots that reach an asymptotic value after about 30 observations, taking the same value for all remaining observations. This allows us to compute pivots for the first 100 observations and extrapolate these for remaining observations, leading to the same answer that would be obtained by calculating all the pivots. Despite the large number of observations, regular locational systems usually require only simple calculations to find the log-determinant.

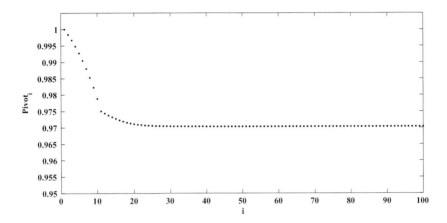

FIGURE 4.5:   Plot of pivots of $I_n - 0.8W$ based on a regular locational grid

4.3.3.3   Closest neighbors

Pace and Zou (2000) examined the case of spatial dependence based on only a single closest neighbor. Let B represent the spatial weight matrix where $B_{ij} = 1$ when observation j is the closest neighbor to observation i. These relations are not necessarily symmetric. For example, if i is on the edge of town while j is located in a subdivision, j may be the closest neighbor to i, but i may not be the closest neighbor to j. Given irregular point data only some observations will be closest neighbors to each other. In fact, for spatially random data on a square, $3\pi(8\pi + 33^{1/2})^{-1}$ proportion of the data (roughly 31 percent) are closest neighbors (Epstein et al., 1997, p. 6).

Pace and Zou (2000) show that the log-determinant, $\ln|I_n - \rho B|$ equals $n_s \ln(1 - \rho^2)$, where n_s equals the number of symmetric pairs of elements in the binary weight matrix B. The number of symmetric pairs equals $0.5 \operatorname{tr}(B^2)$, and for random data distributed on a square, $E(n_s) = 3n\pi(8\pi + 33^{1/2})^{-1}$.

4.3.3.4   Matrix exponential

LeSage and Pace (2007) explored the matrix exponential specification $Z = e^{\alpha W}$. Conveniently, $\ln|e^{\alpha W}| = \ln(e^{\alpha \operatorname{tr}(W)})$, and since $\operatorname{tr}(W) = 0$ for typical matrices W that have zeros on the diagonal, the determinant for this model is quite simple. The matrix exponential spatial specification often produces empirical estimates and inferences that are similar to models based on $Z = I_n - \rho W$, but has computational advantages due to the simple determinant. Chapter 9 discusses the benefits of matrix exponential models in detail.

4.3.3.5   Fractional transformations

The log-determinant of $(I_n - \rho W)^\delta$ is $\delta \ln|I_n - \rho W|$ and this is real provided $|I_n - \rho W| > 0$. Chapter 9 discusses this idea in the context of spatial fractional differencing models.

4.3.3.6   Polynomial

Often a desire exists to work with $\mathrm{AR}(p)$ or $\mathrm{MA}(q)$ processes that use matrices $R(\rho)$, $A(\phi)$. The vectors ρ and ϕ are $p \times 1$ and $q \times 1$ real parameter vectors with the integers $p, q > 1$.

$$R(\rho) = I_n - \rho_1 W - \cdots - \rho_p W^p \tag{4.43}$$
$$A(\phi) = I_n + \phi_1 W + \cdots + \phi_q W^q \tag{4.44}$$

If we attempt to find $\ln|R(\rho)|$ or $\ln|A(\phi)|$ directly using expressions (4.43) or (4.44), this may become difficult. This occurs because W^j in (4.43) and (4.44) becomes progressively less sparse as j rises.

One can factor polynomials such as $R(\rho)$ and $A(\phi)$ that have real coefficients into a product of linear and quadratic polynomials with real coefficients. The quadratic polynomials can be factored into two linear polynomials with two roots. The two roots must both be real or a complex conjugate pair. That is, a polynomial can be factored into a product of linear polynomials with real roots, or complex conjugate roots. The roots for $R(\rho)$ appear as the $p \times 1$ vector $\lambda(\rho)$ and the roots for $A(\phi)$ appear as the $q \times 1$ vector $\lambda(\phi)$. Some of these roots may have identical values.

$$R(\rho) = (I_n + \lambda(\rho)_1 W) \cdots (I_n + \lambda(\rho)_p W) \tag{4.45}$$
$$A(\phi) = (I_n + \lambda(\phi)_1 W) \cdots (I_n + \lambda(\phi)_q W) \tag{4.46}$$

Computationally, the factoring process has nothing to do with W. The same process works for a scalar variable such as x. Specifically, given the coefficients of (4.43) or (4.44), one can form equivalent equations (have the same factorization terms of the powers of x (use x in place of W in (4.43) or (4.44)). The equivalent equations are polynomials and the roots of the polynomials yield $\lambda(\rho)_i$ or $\lambda(\phi)_j$ where $i = 1, \ldots, p$, $j = 1, \ldots, q$.

Given $\lambda(\rho)_i$ or $\lambda(\phi)_j$, the overall log-determinants are the sum of the component log-determinants.

$$\ln|R(\rho)| = \ln|I_n + \lambda(\rho)_1 W| + \cdots + \ln|I_n + \lambda(\rho)_p W| \qquad (4.47)$$
$$\ln|A(\phi)| = \ln|I_n + \lambda(\phi)_1 W| + \cdots + \ln|I_n + \lambda(\phi)_q W| \qquad (4.48)$$

As a simple example, consider the spatial error components (Kelejian and Robinson, 1995) specification with symmetric W,

$$\Omega(\theta) = I_n + \theta W^2 \qquad (4.49)$$

where θ is a scalar parameter. One can factor the spatial error component specification and find the log-determinants of each part.

$$\Omega(\theta) = (I_n + (-\theta)^{0.5} W)(I_n - (-\theta)^{0.5} W) \qquad (4.50)$$
$$\ln|\Omega(\theta)| = \ln|I_n + (-\theta)^{0.5} W| + \ln|I_n - (-\theta)^{0.5} W| \qquad (4.51)$$

If $\theta > 0$, $(-\theta)^{0.5}$ is complex. However, logarithms and determinants are defined for complex arguments and this poses no problem. Consider the example in (4.52)–(4.59) where the sum of the individual log-determinants of the complex factors equal the log-determinant of $\Omega = \ln|I_3 + 0.5W^2|$. In the example, G_1 and G_2 are the result of applying Gaussian elimination to A_1 and A_2. As demonstrated below, Gaussian elimination works for complex numbers.

$$W = \begin{bmatrix} 0 & 0.5 & 0.5 \\ 0.5 & 0 & 0.5 \\ 0.5 & 0.5 & 0 \end{bmatrix} \qquad (4.52)$$

$$\Omega = I_3 + 0.5W^2 \qquad (4.53)$$

$$\Omega = A_1 A_2 = (I_3 - (0.707i)W)(I_3 + (0.707i)W) \qquad (4.54)$$

$$G_1 = \begin{bmatrix} 1.0000 & 0 - 0.3536i & 0 - 0.3536i \\ 0 & 1.1250 & 0.1250 - 0.3536i \\ 0 & 0 & 1.2222 + 0.0786i \end{bmatrix} \qquad (4.55)$$

$$G_2 = \begin{bmatrix} 1.0000 & 0 + 0.3536i & 0 + 0.3536i \\ 0 & 1.1250 & 0.1250 + 0.3536i \\ 0 & 0 & 1.2222 - 0.0786i \end{bmatrix} \qquad (4.56)$$

$$\ln|G_1| = 0 + 0.1178 + 0.2027 + 0.0642i = 0.3205 + 0.0642i \qquad (4.57)$$
$$\ln|G_2| = 0 + 0.1178 + 0.2027 - 0.0642i = 0.3205 - 0.0642i \qquad (4.58)$$
$$\ln|\Omega| = \ln|G_1| + \ln|G_2| = 0.6410 \qquad (4.59)$$

To summarize, the coefficients of the AR or MA in (4.43) or (4.44) can be factored into products of simple linear factors as in (4.45) or (4.46). The factorization process works with the simpler scalar polynomial and requires almost no time to compute. These factors take the form $I_n + \lambda W$ where λ may be real or complex.

One can still form a table of log-determinants and interpolate over this to accelerate calculations. However, now the table relating log-determinants to arguments is three-dimensional. Specifically, the table contains a real part argument, complex part argument, and the log-determinant. Using this technique we can rely on a single table to store all log-determinants for different factors in the polynomial. Consequently, the log-determinant of polynomial functions of weight matrices has almost the same computational cost as the log-determinant of the linear function $I_n - \rho W$.

We discussed SARMA models in Chapter 3 and we will discuss fitting multiple weight matrices in Chapter 5. The determinant table approach presented above could allow higher-order SARMA models to be fitted with large data sets. This approach could be used in conjunction with MCMC to deal with potential local optima that can arise in these models.

4.3.3.7 Kronecker products

Spatial simultaneous equations, spatial vector autoregressions, and origin-destination flow data may require finding $\ln|I_{nm} - A_{nm}|$ where $A_{nm} = (\Lambda \otimes W)$ and Λ is a $m \times m$ matrix. Fortunately, the structure of Kronecker products greatly facilitates calculation of the log-determinant. Using the Taylor series expansion of the log-determinant produces (4.60) and (4.61).

$$\ln|I_{nm} - A_{nm}| = -\sum_{i=1}^{\infty} \frac{\text{tr}(A_{nm}^i)}{i} \qquad (4.60)$$

$$\ln|I_{nm} - A_{nm}| = -\sum_{i=1}^{\infty} \frac{\text{tr}(\Lambda^i)\,\text{tr}(W^i)}{i} \qquad (4.61)$$

The use of a trace estimator discussed in Section 4.4.1.2 enables efficient calculation of traces for the powers of W. In the typical case where m is substantially less than n, exact computation of $\text{tr}(\Lambda^i)$ requires little time. Therefore, what appears to be an $nm \times nm$ problem reduces to separate $n \times n$ and $m \times m$ problems.

A related point is that eigenvalues of A_{nm} equal the nm cross-products between the eigenvalues of W and Λ, which facilitate calculating the log-determinant. The computational demands of calculating eigenvalues limits this technique to problems involving small or moderate n.

Models used for origin-destination flow data discussed in Chapter 8 provide another example where the log-determinant involves Kronecker products of matrices. LeSage and Pace (2008) consider a model involving spatial dependence among $N = n^2$ flows, resulting in the log-determinant expression in (4.62).

$$B = I_N - \rho_1(I_n \otimes W) - \rho_2(W \otimes I_n) - \rho_3(W \otimes W) \qquad (4.62)$$
$$B = I_N - A \qquad (4.63)$$

They proceed by finding $tr(A^j)$ where A appears in (4.63). While finding the trace of a $n^2 \times n^2$ matrix seems difficult, in reality the log-determinant of B simply involves weighted traces of the powers of W. To see this, consider that any power of A greater than 1 will result in products and cross-products involving the components of A. The cross-product $(I_n \otimes W)$ and $(W \otimes W)$ can be expressed as $(W \otimes W^2)$ using the Kronecker mixed product rule. This expression has a trace of $tr(W)\, tr(W^2) = 0$, since $tr(W) = 0$, and this is a scalar. Because each product and cross-product has a form involving traces of powers of W, a pre-computed table of these traces facilitates reweighting these to arrive at an approximation of the overall log-determinant. For any combination of ρ_1, ρ_2, and ρ_3, the products and cross-products are also scalars, and these represent coefficients associated with the traces. Multiplying the pre-computed traces by these coefficients yields an estimate of the log-determinant. Chapter 8 discusses spatial modeling of flow data in detail.

4.3.3.8 Multiple and parameterized W

We have already discussed models where $Z = I_n - \rho_1 W_1 - \rho_2 W_2$, and W_1 and W_2 are not functionally related (Lacombe, 2004). More complicated models can be constructed that rely on functions of functionally unrelated multiple weight matrices. If both W_1 and W_2 are row-stochastic matrices, $\rho_1 + \rho_2 < 1$, and $\rho_1, \rho_2 \geq 0$, Z is strictly diagonally dominant and thus non-singular. The log-determinant as a function of these parameters is smooth when $\rho_1 + \rho_2 < 1$ and $\rho_1, \rho_2 \geq 0$, since the first and second derivatives are continuous. This allows tabulation and interpolation of the log-determinants to produce a finer grid over values of ρ_1, ρ_2. However, in this case of two log-determinants, the table now has two arguments (ρ_1 and ρ_2) which will require more time to compute as well as more storage space. A natural extension of this applies to three or more weight matrices. However, the logistical, computational and storage difficulties increase with additional weight matrices and parameters.

In many cases, the individual weight matrices exhibit some natural order or smoothness conditions. For example, consider $Z = I_n - \rho_1 W_1 - \rho_2 W_2 - \rho_3 W_3$, where W_1, W_2, W_3 represent first-, second- and third-nearest neighbor matrices. In this case, one might impose a monotonic restriction on ρ_1, ρ_2, and ρ_3, using the restriction: $\rho_i = \rho\gamma^i(\gamma + \gamma^2 + \gamma^3)^{-1}$. This converts a multiple parameter problem to a simpler two parameter problem involving only ρ and γ, which can easily be tabulated. Possible functions that could serve as smoothing restrictions include polynomials (in the spirit of Almon distributed lags), geometric decay, exponential decay, and so on.

4.3.3.9 Local W

Suppose we wish to examine a more local system. For example, given a cluster of homes at the city center we begin adding observations on progressively more distant homes. How do these new observations affect the dependence as measured by the log-determinant? Suppose $Z = I_3 - 0.8W$ where W is the same as in (4.52). In numerical terms, Z appears in (4.65) and Gaussian elimination of Z, labeled G, appears in (4.66).

$$Z = I_3 - 0.8W \tag{4.64}$$

$$Z = \begin{bmatrix} 1.0000 & -0.4000 & -0.4000 \\ -0.4000 & 1.0000 & -0.4000 \\ -0.4000 & -0.4000 & 1.0000 \end{bmatrix} \tag{4.65}$$

$$G = \begin{bmatrix} 1.0000 & -0.4000 & -0.4000 \\ 0 & 0.8400 & -0.5600 \\ 0 & 0 & 0.4667 \end{bmatrix} \tag{4.66}$$

The determinant of Z_{11} is 1 and this matches the first pivot of G. The determinant of the first two rows and columns of Z is $1 - (-0.4)^2 = 0.84$, and this matches the product of the first and second pivots of G. Finally, using the formula for the determinant of Z yields a value of 0.3920, and this matches the product of all three pivots of G. Consequently, a bonus arising from Gaussian elimination is the sequence of log-determinants that result from adding additional observations to the system. Pace and LeSage (2003a) use this feature of log-determinants to estimate a sequence of SDM local estimates around each observation in the sample.

4.4   Monte Carlo approximation of the log-determinant

The log-determinant equals the trace of the matrix logarithm as shown in (4.67). In turn the matrix logarithm has a simple infinite series expansion in terms of the powers of W as shown in (4.68). Since the trace operation is linear $(\text{tr}(A + B) = \text{tr}(A) + \text{tr}(B))$, the log-determinant is a weighted series of traces of the powers of W as shown in (4.69).

$$\ln|I_n - \rho W| = \text{tr}\left(\ln(I_n - \rho W)\right) \tag{4.67}$$

$$\ln(I_n - \rho W) = -\sum_{i=1}^{\infty} \frac{\rho^i W^i}{i} \tag{4.68}$$

$$\ln|I_n - \rho W| = -\sum_{i=1}^{\infty} \frac{\rho^i \, \text{tr}(W^i)}{i} \tag{4.69}$$

Since the logarithm is defined over the complex plane (except for $\ln(0)$), (4.69) still holds for non-symmetric W that may have complex eigenvalues or for complex ρ.

One can partition the infinite series into a finite, lower order series and a remainder denoted by R composed of higher-order infinite expressions such as (4.70) and (4.71). Martin (1993) initially proposed this approach to dealing with the log-determinant.

$$\ln|I_n - \rho W| = -\sum_{i=1}^{o} \frac{\rho^i \, \text{tr}(W^i)}{i} - \sum_{i=o+1}^{\infty} \frac{\rho^i \, \text{tr}(W^i)}{i} \tag{4.70}$$

$$\ln|I_n - \rho W| = -\sum_{i=1}^{o} \frac{\rho^i \, \text{tr}(W^i)}{i} - R \tag{4.71}$$

If R is small, one can approximate the log-determinant through a finite, lower-order series as in (4.72).

$$\ln|I_n - \rho W| \approx -\sum_{i=1}^{o} \frac{\rho^i \, \text{tr}(W^i)}{i} \tag{4.72}$$

From a computational perspective, forming W^j and then taking the trace is costly and inefficient having an operational count of up to $O(n^3)$ when W^j is dense.

Fortunately, other methods exist for estimating traces. For example, let u represent a 2×1 vector of independent unit normals and A a 2×2 matrix as in (4.73) and (4.74). The expectation of the quadratic form $u'Au$ equals $\text{tr}(A)$ since u_i^2 follows a χ^2 distribution with one degree of freedom. Of course, this has an expectation equal to 1, whereas $E(u_i u_j) = 0$ for $i \neq j$ as in (4.75) to (4.78).

$$A = \begin{bmatrix} a & b \\ c & d \end{bmatrix}, \quad u = \begin{bmatrix} u_1 \\ u_2 \end{bmatrix} \tag{4.73}$$

$$u'Au = u_1^2 a + u_2^2 d + u_1 u_2 (b + c) \tag{4.74}$$

$$E(u'Au) = E(u_1^2)a + E(u_2^2)d + E(u_1 u_2)(b + c) \tag{4.75}$$

$$E(u_i^2) = 1 \tag{4.76}$$

$$E(u_i u_j) = 0 \quad i \neq j \tag{4.77}$$

$$E(u'Au) = a + d = \text{tr}(A) \tag{4.78}$$

Let $\widetilde{T}_{(j)}^{(i)}$ represent the estimate of $\text{tr}(W^i)$ using the jth n by 1 random unit normal vector $u_{(j)}$ as shown in (4.79). For convenience, we will term $u_{(j)}$ the jth seed, and Girard (1989) proposed normal seeds in estimating traces.[2]

$$\widetilde{T}_{(j)}^{(i)} = u_{(j)}' W^i u_{(j)} \tag{4.79}$$

Given the estimated $\text{tr}(W^i)$ which equals $\widetilde{T}_{(j)}^{(i)}$, one could use this to form an estimate of the log-determinant based on the estimated trace.[3]

$$\ln |I_n - \rho W|_{(j)} \approx - \sum_{i=1}^{o} \frac{\rho^i \widetilde{T}_{(j)}^{(i)}}{i} \tag{4.80}$$

Calculating $\ln |I_n - \rho W|_{(j)}$ across m independent $u_{(j)}$ and averaging the results can improve the accuracy of the log-determinant estimate. Each of the $\ln |I_n - \rho W|_{(j)}$ are independent, so the variance of the averaged estimate is $1/m$ times the variance of a single estimate $\ln |I_n - \rho W|_{(j)}$.

$$\ln |I_n - \rho W| \approx \left(\frac{1}{m} \right) \sum_{j=1}^{m} \ln |I_n - \rho W|_{(j)} \tag{4.81}$$

Also, calculating m log-determinant estimates $\ln |I_n - \rho W|_{(j)}$ allows estimating the variance of the log-determinant, and therefore confidence intervals associated with the averaged log-determinant estimate.

From a computational perspective, the algorithm begins by picking the seed $u_{(j)}$, initializing the loop by setting $z(1)$ to $u_{(j)}$, and by calculating a matrix-vector product $z(t + 1) = W z(t)$ as well as $\widetilde{T}_{(j)}^{(i)} = z(1)' z(t + 1)$ over a loop

---

[2]Barry and Pace (1999) rediscovered the normal seed without the benefit of the work of Girard (1989) and gave different proofs of the performance of normal seeds in estimating the trace.

[3]Barry and Pace (1999) actually used $\widehat{T}_{(j)} = [n(u_{(j)}' u_{(j)})^{-1}]\widetilde{T}_{(j)}$. This has slightly lower variability than $\widetilde{T}_{(j)}$, but $\widetilde{T}_{(j)}$ has the advantage of simplicity.

from $t = 1, \ldots, o$. This can be repeated m times to improve the precision of the trace estimates and thus the log-determinant estimates by averaging over $\widetilde{T}_{(j)}^{(i)}$ for $j = 1, \ldots, m$ to yield $\widetilde{T}^{(i)}$. In addition, having the $\widetilde{T}_{(j)}^{(i)}$ allows easy calculation of confidence intervals for the estimated log-determinant. The algorithm has computational complexity $O(nmo)$ and therefore is linear in n.

Given the estimated $\widetilde{T}^{(i)}$, an outstanding advantage of the algorithm is that computation of $\ln |I_n - \rho W|$ for any ρ requires almost no time. Let a represent an $o \times 1$ vector so that $a_i = -\widetilde{T}^{(i)}/i$ and let $b = \left[\rho, \quad \rho^2, \ldots, \rho^o \right]'$. The log-determinant estimate for a particular ρ is just $a'b$, a simple dot product between two vectors of length o where o might be 100. Therefore, updating the estimate of the log-determinant for a new value of ρ is virtually costless.

A few refinements boost the computational speed and accuracy. First, one can employ symmetry to reduce the work by half. For symmetric W, let $v = W^i u$ and therefore $v'v = u'W^{2i}u$ which estimates the trace of (W^{2i}). Second, the lower-order exact moments are either known or easily computed, and using these can materially reduce the approximation error. For example, $\text{tr}(W) = 0$ by construction whereas $\text{tr}(W^{p+q}) = \iota_n'((W^p)' \odot W^q)\iota_n$. Therefore, $\text{tr}(W^2)$ for a symmetric matrix is the sum of squares of all the elements in W. It does not take long to compute these exact traces. For example, using a contiguity-based W where $n = 1,024,000$, it takes 0.43 seconds to compute W^2 and 6.3 seconds to find $\text{tr}(W^4)$. The computational requirements of the exact traces in general rise at a faster than linear rate, but computing the lower order exact traces is quite feasible for sparse W. Third, one can improve the choice of seeds $u_{(j)}$ by rejecting "bad" seeds where "bad" in this context means that the estimated moments differ significantly from the known lower order exact moments (Zhang et al., 2008).

This raises the issue of seed choice. Alternatives to the normal seed proposed by Girard (1989) include using n independent draws of -1 and 1 with equal probability as a seed (Hutchinson, 1990). Also, a seed where the kth element of $u_{(j)}$ equals 1 and the other elements equal 0 will yield the diagonal element of W_{kk}^i. Selecting k randomly over $[1, n]$ for m samples and averaging these samples yields an estimate of $n^{-1} \text{tr}(W^i)$. For all of these approaches, the use of m independent realizations naturally facilitates parallel processing. All seed choices work well for moderate m, but we have found that the normal seed performs better for very small m (including $m = 1$).

Ideally, approximations should provide a means to assess accuracy which might be measured in terms of the impact on variables of interest in applied problems. One approach to this is to consider how independent estimates of the log-determinant affect the estimated parameter ρ in the autoregressive model. We take this approach in Section 4.4.1. Since one can rapidly solve for the spatial dependence parameter estimate given the log-determinant function, we could use m independent estimates of the log-determinant. This would lead to m estimates for ρ, and variation in these estimates would serve as a guide to the approximation accuracy. If the variation is small relative to the standard error of the parameter estimate, this indicates that the approximation error does not materially degrade parameter estimation.

4.4.1 Sensitivity of ρ estimates to approximation

Useful approximations contain errors that do not materially affect the results. In Section 4.4.1.1, we explore the dispersion in the estimated value of ρ that arises from a log-determinant approximation using a Monte Carlo experiment. We show that the approximation is very accurate in that it does not materially affect the estimate or inference regarding the parameter ρ. Section 4.4.1.2 and Section 4.4.1.3 provide detailed Monte Carlo experiments that explore the nature of the approximation error that leads to accuracy of the Monte Carlo log-determinant approximation of Barry and Pace (1999) in applied practice.

4.4.1.1 Error in $\tilde{\rho}$ using individual MC log-determinant estimates

To assess the accuracy of the Monte Carlo log-determinant estimator, we conducted an experiment with sample sizes n ranging from $1,000$ to $1,024,000$. For each sample size we generated a random set of points, calculated a contiguity weight matrix W_n, and simulated the dependent variable using $y_n = (I_n - 0.75W_n)^{-1}(X\beta + \varepsilon)$. The matrix X contained an intercept as well as a vector of standard normal random deviates, and β was set to ι_2. An *iid* normal vector was used for ε with a standard deviation of 0.25. For each y_n we calculated 100 separate estimated log-determinants using the Barry and Pace (1999) Monte Carlo approximation.

These estimated log-determinants did not involve averaging across multiple estimated log-determinants as is typically done in application of the Barry and Pace (1999) approach. The reported results were based on a single normal vector (normal seed) used in the Monte Carlo log-determinant estimator. Using a single vector to estimate the log-determinant is seemingly quite extreme, since Barry and Pace (1999) recommend use of 30 to 50 such vectors in their statistical approximation. To gauge performance of the individual log-determinant estimates, we also averaged over all 100 log-determinant estimates to produce a more accurate estimate.

Table 4.2 shows the results from using an average of all log-determinant estimates versus 100 separate estimates of the log-determinant where each estimate corresponds to a different normal seed. The estimated log-determinants used four exact lowest-order moments and 100 total moments.

Rather surprisingly, the estimated $\tilde{\rho}$ was not sensitive to numerical approximation error present in the individual MC log-determinant estimates. Across 100 trials of the individual MC log-determinant estimates, the estimated dependence parameter varied over a very small range that declined with n. The average estimated dependence parameter was very close to the actual value of 0.75. For example, when $n = 1,024,000$, the average estimated ρ across the 100 separate log-determinants was 0.749404, and the range between the largest and smallest estimate was only 0.000052. Using an average of the 100 log-determinants produces the same estimate for ρ to six decimal places. It took only 23.7 seconds, on average, to estimate the log-determinant of the $1,024,000 \times 1,024,000$ matrix.

TABLE 4.2: Estimates of ρ based on aggregate vs. individual MC log-determinant estimates

n	$\tilde{\rho}_{\overline{\ln|z|_i}}$	mean $\tilde{\rho}_{\ln|z|_i}$	range $\tilde{\rho}_{\ln|z|_i}$	Seconds per ln-det
1,000	0.762549	0.762550	0.002119	0.005281
2,000	0.752098	0.752099	0.001359	0.007593
4,000	0.751474	0.751475	0.000873	0.013504
8,000	0.752152	0.752152	0.000748	0.032123
16,000	0.747034	0.747034	0.000485	0.070086
32,000	0.748673	0.748673	0.000339	0.197200
64,000	0.750238	0.750238	0.000224	0.453033
128,000	0.750622	0.750622	0.000147	1.153523
256,000	0.749766	0.749766	0.000122	3.917425
512,000	0.750144	0.750144	0.000068	10.498189
1,024,000	0.749404	0.749404	0.000052	23.710256

4.4.1.2 Trace estimation accuracy

The accuracy of the Monte Carlo estimates for the log-determinant raise the question of why it works so well. Either the procedure estimates traces extremely well, the shape of the log-likelihood is not sensitive to small variations in the log-determinant, or some combination of both factors is at work. In this section, we investigate the accuracy of the trace estimates with an experiment. We estimate $n^{-1}\operatorname{tr}(W^2)$ which we denote $\tilde{tr}$ using a contiguity weight matrix and compare these estimates to the exact trace $tr = n^{-1}\operatorname{tr}(W^2)$. Table 4.3 shows the results from this experiment for varying n. As expected, the Monte Carlo estimate of the trace appears more or less unbiased having small average errors that vary in sign across n. In addition, the standard deviation and ranges of the estimated traces are small. For sample sizes of 16,000 and above, we see a difference of less than 0.01 between the largest and smallest estimated traces across the 100 trials. Again, these results are for single trials or iterations. Aggregating the different trials or iterations would further improve the performance.

These results conform to the theoretical investigations of Girard (1989, p. 5) as well as Barry and Pace (1999, p. 52) where $\sigma^2(\widetilde{T}_{(j)}^{(i)}) = \kappa n^{-1}$ and κ is a constant.

TABLE 4.3:   Individual MC trace estimates of $n^{-1}\operatorname{tr}(W^2)$ across n

n	tr	average $\tilde{tr} - tr$	s.d. $\tilde{tr}$	range $\tilde{tr}$
1,000	0.166809	−0.000017	0.009841	0.046782
2,000	0.165991	0.000311	0.006742	0.034160
4,000	0.165782	−0.000296	0.004303	0.023378
8,000	0.165640	−0.000213	0.003138	0.018299
16,000	0.165505	0.000078	0.001957	0.009831
32,000	0.165420	0.000278	0.001690	0.009949
64,000	0.165391	0.000083	0.001050	0.005029
128,000	0.165395	0.000093	0.000799	0.003943
256,000	0.165377	0.000011	0.000634	0.003241
512,000	0.165359	−0.000023	0.000393	0.001820
1,024,000	0.165353	−0.000032	0.000277	0.001319

4.4.1.3   Theoretical analysis of sensitivity of $\tilde{\rho}$ to log-determinant error

The autoregressive model: $y = X\beta + \rho W y + \varepsilon$ has a concentrated or profile likelihood $(L_p(\rho))$ which involves a scalar spatial dependence parameter ρ, as shown in (4.82).

$$L_p(\rho) = C + \ln|I_n - \rho W| - \frac{n}{2}\ln(e(\rho)'e(\rho)) \tag{4.82}$$

Recall the Taylor series expansion of the log-determinant shown in (4.83).

$$\ln|I_n - \rho W| = -\sum_{j=1}^{\infty}\rho^j \operatorname{tr}(W^j)/j \tag{4.83}$$

Substituting (4.83) into (4.82) leads to (4.84) which emphasizes the role of $\operatorname{tr}(W^j)$ in the profile likelihood.

$$L_p(\rho) = C - \sum_{j=1}^{\infty}\rho^j \operatorname{tr}(W^j)/j - \frac{n}{2}\ln(e(\rho)'e(\rho)) \tag{4.84}$$

To examine the sensitivity of the system to an approximation error, we introduce a scalar parameter δ_j to model proportional error in estimation of $\operatorname{tr}(W^j)$, the jth moment. This proportional error could be positive or negative. Consider a new approximate profile likelihood $L_p(\rho|\delta_j)$ for a given δ_j. Substitution of (4.84) in the first equation, along with the notation $G(\rho) = -\rho^j \operatorname{tr}(W^j)/j$ allows rewriting $L_p(\rho|\delta_j)$ in (4.86) as the original profile likelihood plus an approximation error.

$$L_p(\rho|\delta_j) = C - \delta_j \rho^j \, \text{tr}(W^j)/j - \sum_{j=1}^{\infty} \rho^j \, \text{tr}(W^j)/j - \frac{n}{2} \ln(e(\rho)'e(\rho))$$

$$L_p(\rho|\delta_j) = -\delta_j \rho^j \, \text{tr}(W^j)/j + L_p(\rho) \tag{4.85}$$

$$L_p(\rho|\delta_j) = \delta_j G(\rho) + L_p(\rho) \tag{4.86}$$

We form the score function, $S_p(\rho|\delta_j)$ for a given δ_j.

$$S_p(\rho|\delta_j) = \frac{dL_p(\rho|\delta_j)}{d\rho} = \delta_j \frac{dG(\rho)}{d\rho} + \frac{dL_p(\rho)}{d\rho} \tag{4.87}$$

The score function set to 0 is the first-order condition for an optimum and this forms an implicit function F.

$$F = S_p(\rho|\delta_j) = 0 \tag{4.88}$$

Given unimodality of the profile likelihood for this problem, a unique value of ρ solves the implicit equation $F = 0$, but the solution depends upon the error δ_j. We can examine the sensitivity of the dependence parameter ρ with respect to the error δ_j using the implicit function theorem for two variables (which allows for total instead of partial derivatives) as shown in (4.89).

$$\frac{d\rho}{d\delta_j} = -\frac{dF}{d\delta_j}\bigg/\frac{dF}{d\rho} \tag{4.89}$$

Taking the derivative of F with respect to δ_j yields (4.90).

$$\frac{dF}{d\delta_j} = \frac{dG(\rho)}{d\rho} \tag{4.90}$$

The derivative of F with respect to ρ yields (4.91).

$$\frac{dF}{d\rho} = \delta_j \frac{d^2G(\rho)}{d\rho^2} + \frac{d^2L_p(\rho)}{d\rho^2} \tag{4.91}$$

Using (4.89), (4.90), and (4.91) yields (4.92).

$$\frac{d\rho}{d\delta_j} = -\left[\delta_j \frac{d^2G(\rho)}{d\rho^2} + \frac{d^2L_p(\rho)}{d\rho^2}\right]^{-1} \frac{dG(\rho)}{d\rho} \tag{4.92}$$

We now consider simplifying this expression, noting that the second derivative of $L_p(\rho)$ with respect to ρ is the Hessian $H(\rho)$ (Davidson and MacKinnon, 1993, p. 267-269).

$$\frac{d^2L_p(\rho)}{d\rho^2} = \frac{dS_p(\rho)}{d\rho} = H(\rho) \tag{4.93}$$

Given an estimated $\tilde{\rho}$, the estimated variance of $\tilde{\rho}$ is $\tilde{\sigma}^2(\tilde{\rho})$. Equation (4.94) expresses the well-known relation between the Hessian and the variance of a parameter estimate for a univariate function such as our concentrated log likelihood.

$$\tilde{\sigma}^2(\tilde{\rho}) = -H(\tilde{\rho})^{-1} \tag{4.94}$$

Using (4.94) provides an additional simplification.

$$\frac{d^2 L_p(\tilde{\rho})}{d\tilde{\rho}^2} = H(\tilde{\rho}) = -\tilde{\sigma}^{-2}(\tilde{\rho}) \tag{4.95}$$

Taking (4.95) and (4.92) in conjunction with $dG(\rho)/d\rho = -\rho^{j-1}\,\mathrm{tr}(W^j)$, evaluated around the point of $\delta_j = 0$, leads to (4.96).

$$\left.\frac{d\tilde{\rho}}{d\delta_j}\right|_{(\delta_j=0)} = -\tilde{\sigma}^2(\tilde{\rho})\rho^{j-1}\,\mathrm{tr}(W^j) \tag{4.96}$$

The fact that $d\tilde{\rho}/d\delta_j < 0$ for positive ρ makes intuitive sense. Positive δ_j results in a more severe log-determinant penalty, and this should depress $\tilde{\rho}$. Turning to the relative magnitudes of these variables, the estimated variance $(\tilde{\sigma}^2(\tilde{\rho}))$ of $\tilde{\rho}$ decreases with n, while $\mathrm{tr}(W^j)$ rises with n. This suggests the product of these two variables should not greatly vary with n.

Putting this in differential form in (4.97), we note that from the starting point of $\delta_j = 0$, $d\delta_j$ equals the proportional error in the jth moment, which could be negative or positive. In other words, going to $\delta_j^{(1)}$ from $\delta_j^{(0)} = 0$ means that $d\delta_j$, the change in the variable, equals $\delta_j^{(1)}$.

$$d\tilde{\rho} = \frac{d\tilde{\rho}}{d\delta_j}\delta_j \tag{4.97}$$

However, as was demonstrated, δ_j, the proportional error in the jth moment tends to decrease with n, so the overall numerical error associated with $\tilde{\rho}$ tends to decrease with n.

To examine the relative magnitudes of these terms in applied practice, we performed a simple Monte Carlo experiment. A set of $2,500$ random points were generated and a contiguity weight matrix W was used to simulate: $y = (I_n - 0.75W)^{-1}(X\beta + \varepsilon)$, where X contains an intercept column plus a standard normal vector. The parameter β was set equal to ι_2, and an *iid* normal ε having a standard deviation of 0.25 was used. Exact traces for orders 1 to 100 were calculated using a set of $1,000$ instances of y, with the exception of $\mathrm{tr}(W^6)$. For this single exception, we added a proportional error equal to 0.01, and calculated the theoretical as well as empirical change in $\tilde{\rho}$ in response to a one percent error in $\mathrm{tr}(W^6)$.

On average, the empirical change was $-0.629453 \cdot 10^{-5}$ while the predicted change was $-0.630461 \cdot 10^{-5}$. The difference between empirical and theoretical errors was not statistically significant. Given that $\delta_j = 0.01$, $d\tilde{\rho}/d\delta_j$ approximately equals -0.00063 for this case. Consequently, $\tilde{\rho}$ is not very sensitive to proportional error in one of the traces. A similar experiment using other values of n demonstrated that this derivative does not vary materially with n.

To summarize, the sensitivity of $\tilde{\rho}$ to approximation errors for the log-determinant in a single trace depends on: the variance associated with the estimate of ρ, true value of the trace, true value of ρ, and the proportional error in estimating the trace. The combination of the first three factors should have a value which does not vary greatly with n. This is because the trace rises linearly with n (all else equal) and the variance declines linearly with n. However, the proportional error in estimating the trace declines with n. Specifically, $\sigma^2(\tilde{T}_{(j)}^{(i)}) = \kappa n^{-1}$ where κ is a constant (Barry and Pace, 1999, p. 52). Consequently, the sensitivity of $\tilde{\rho}$ to log-determinant error decreases with sample size.

4.5 Chebyshev approximation

Taylor series approximations provide good results around the point of expansion ($\rho = 0$ in the previous section). In contrast, Chebyshev approximations minimize errors over a range, attempting to minimize the maximum error over the entire interval. Pace and LeSage (2004) applied the Chebyshev approximation technique to the log-determinant problem.

The basic idea is to approximate the matrix function $\ln(I_n - \rho W)$ using the Chebyshev polynomials as well as Chebyshev coefficients, and use the trace of these to produce an estimate of the log-determinant. Following Press et al. (1996), let c_j represent the Chebyshev coefficients (4.98) associated with the function $\ln(1 - \rho x)$, where x is real and lies on $[-1, 1]$. We furthermore assume W is symmetric with a maximum eigenvalue of 1 and we restrict ρ to $(-1, 1)$. The coefficients in (4.98) depend on the evaluation points x in (4.99) as well as the specific scalar function $f(x)$ under consideration. In this case, the function of interest is $\ln(1 - \rho x)$ as shown in (4.100), and we note that the desired matrix function inherits the same coefficients.

$$c_j(\rho) = \frac{2}{q+1} \sum_{k=1}^{q+1} f(x_k) \cos\left(\frac{\pi(j-1)\left(k-\frac{1}{2}\right)}{q+1}\right) \tag{4.98}$$

$$x_k = \cos\left(\frac{\pi\left(k-\frac{1}{2}\right)}{q+1}\right) \tag{4.99}$$

$$f(x) = \ln(1 - \rho x) \tag{4.100}$$

Given the Chebyshev polynomials in (4.102)–(4.107) and the coefficients in

(4.98), the approximation of the matrix logarithm appears in (4.101).[4] Even though this is a matrix function, it uses the coefficients from the scalar function $f(x) = \ln(1 - \rho x)$ and, in fact, this is part of the definition of matrix functions.

$$\ln(1 - \rho W) \approx \sum_{k=1}^{q+1} c_k T_{k-1}(W) - \frac{1}{2} c_1 I_n \tag{4.101}$$

$$T_0(W) = I_n \tag{4.102}$$

$$T_1(W) = W \tag{4.103}$$

$$T_2(W) = 2W^2 - I_n \tag{4.104}$$

$$T_3(W) = 4W^3 - 3W \tag{4.105}$$

$$T_4(W) = 8W^4 - 8W^2 + I_n \tag{4.106}$$

$$T_{n+1}(W) = 2W T_n(W) - T_{n-1}(W) \quad n \geq 1 \tag{4.107}$$

Taking the trace of the matrix logarithm yields the log-determinant (4.108) and this leads to (4.109).

$$\ln|I_n - \rho W| = \text{tr}(\ln(1 - \rho W)) \tag{4.108}$$

$$\approx \sum_{j=1}^{q+1} c_j \, \text{tr}(T_{j-1}(W)) - \frac{n}{2} c_1 \tag{4.109}$$

As Figure 4.6 illustrates, a low-order (quintic in this case) Chebyshev approximation to the log-determinant can closely tract the exact log-determinant. The figure was constructed using a $1,024,000$ by $1,024,000$ contiguity-based W.

To provide an idea about the accuracy of the Chebyshev log-determinant approximation, we conducted an experiment where we set n equal to $10,000$, generated a random set of points, calculated a contiguity weight matrix W, and simulated the dependent variable using $y = (I_n - 0.75W)^{-1}(X\beta + \varepsilon)$. The matrix X contains an intercept column of ones and a random unit normal vector, with β set to ι_2, and ε is *iid* normal with a standard deviation of 0.25. We generated $1,000$ trials of y, estimated the model via maximum likelihood using the exact log-determinant as well as the Chebyshev log-determinant approximation. The difference in $\tilde{\rho}$ between the two estimates exhibited a mean absolute error of 0.0008630 for the quadratic approximation and 0.000002 for the quintic approximation.

---

[4]Equation (3) in Pace and LeSage (2004) left out I_n from the last term in (4.101). Thanks to Janet Walde for pointing this out.

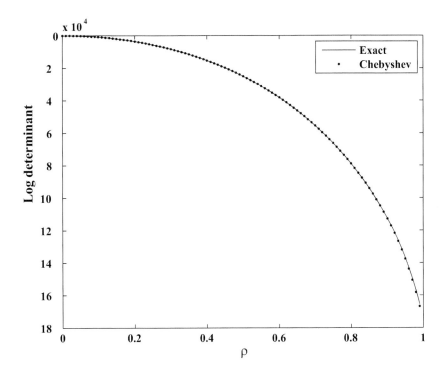

FIGURE 4.6: Exact and fifth order Chebyshev log-determinants

For the problem involving a $1,024,000 \times 1,024,000$ matrix, it took less than 0.5 second to compute $\mathrm{tr}(W^2)$ via $\iota'_n(W' \odot W)\iota_n$ used in the quadratic approximation compared to 7.62 minutes for the exact log-determinant. It required 23.3 seconds to calculate all traces needed for the quintic approximation.

The approximation has a great advantage in more complicated circumstances such as those involving multiple weight matrices (Pace and LeSage, 2002). Consider the case of a linear combination of component weight matrices so that $W = \alpha_1 W_1 + \alpha_2 W_2$ and for simplicity, W_1, W_2 are symmetric and doubly stochastic. If $\alpha_1 + \alpha_2 = 1$, α_1 and $\alpha_2 \geq 0$, then W is symmetric and doubly stochastic and has real eigenvalues with a maximum eigenvalue equal to 1. The overall $\mathrm{tr}(W^2)$ is a quadratic form of the individual traces of the cross-products formed by W_1 and W_2. Using the relation that $\mathrm{tr}(W_i W_j) = \iota'_n(W'_i \odot W_j)\iota_n$ accelerates computation of the traces. Given that traces of the cross-products have been pre-computed prior to estimation, updating an approximation to the log-determinant requires a multiplication of a 1×2 vector times a 2×2 matrix followed by multiplication of a 2×1 vector which is almost instantaneous.

$$W = \alpha_1 W_1 + \alpha_2 W_2 \qquad (4.110)$$

$$\alpha = \begin{bmatrix} \alpha_1 & \alpha_2 \end{bmatrix}' \qquad (4.111)$$

$$1 = \alpha' \iota_2, \quad \alpha_1, \alpha_2 \geq 0 \qquad (4.112)$$

$$\mathrm{tr}(W^2) = \alpha' \begin{bmatrix} \mathrm{tr}(W_1^2) & \mathrm{tr}(W_1 W_2) \\ \mathrm{tr}(W_1 W_2) & \mathrm{tr}(W_2^2) \end{bmatrix} \alpha \qquad (4.113)$$

Therefore, the quadratic Chebyshev approximation can greatly aid in working with systems like those in (4.110) through (4.113). Note, $\mathrm{tr}(W) = 0$ for any combination of W_1 and W_2. See Pace and LeSage (2002) for an example involving an additive system containing a large number of weight matrices.

4.6   Extrapolation

Figure 4.7 shows a sequence of log-determinants of $I_n - \rho W_n$ as n goes from 1 to $10,000$. Specifically, we took $10,000$ randomly located points, formed W, and ordered the rows and columns of W from North to South. That is the first row and column would correspond to the most northern location and $I_1 - \rho W_1$ is just 1. The first two rows and columns would refer to the spatial system comprised of the two most northerly observations, and so on. Obviously, the initial observations have a paucity of spatial relations with other observations, but this quickly changes as n becomes larger. This is an example of an *increasing domain* ordering and the initial observations exhibit more of an *edge effect* since neighboring observations reflect points lying near the edge of the map (Cressie, 1993).

The approximate linearity of Figure 4.7 suggest the potential of taking a sequence of log-determinants and extrapolating these. Measuring the slope of the log-determinant curve at locations sufficiently far apart can serve as approximately independent samples, and these can be used to conduct inference regarding unobserved future slopes. Pace and LeSage (2009a) follow this approach to find the log-determinant of a $3,954,400$ by $3,954,400$ weight matrix associated with Census block locations.

4.7   Determinant bounds

Sometimes a bound on the log-determinant can greatly reduce mathematical complexity of a problem, allow assessing the quality of an approximation,

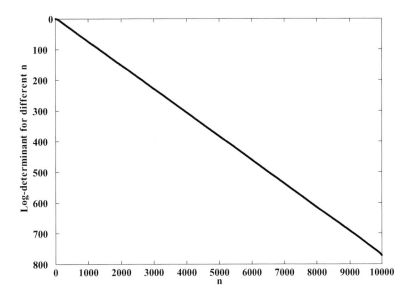

FIGURE 4.7: Log-determinants for different n associated with increasing domain ordering

or bound possible parameter estimates (Pace and LeSage, 2002, 2003a, 2004). Pace and LeSage (2002) provided a simple quadratic bound shown in (4.114) for symmetric W with maximum eigenvalue equal to 1. They derived this using the definition of the Taylor series for the log-determinant in conjunction with the fact that the maximum trace is $\text{tr}(W^2)$ and higher-order traces must be positive.

$$(\rho + \ln(1 - \rho))\,\text{tr}(W^2) < \ln|I_n - \rho W| < -\frac{\rho^2}{2}\,\text{tr}(W^2) \qquad (4.114)$$

Another set of bounds can be constructed using the fact that triangular block systems have log-determinants that can be calculated using the sum of log-determinants from each of the main blocks. This suggests that for a general $I_n - \rho W$ in (4.115), it would simplify the problem if U contained more zeros. This would allow partitioning the problem into two smaller, more tractable problems.

$$I_n - \rho W = I_n - \rho S - \rho U \tag{4.115}$$

$$S = \begin{bmatrix} A & B \\ 0 & D \end{bmatrix} \tag{4.116}$$

$$U = \begin{bmatrix} 0 & 0 \\ C & 0 \end{bmatrix} \tag{4.117}$$

The basic idea is that reducing the number of non-zero elements in U would lead to a positive bound on the log-determinant. In addition, subtracting ρ times the elements in U from the main diagonal produces a lower bound on the log-determinant. Let $r = U\iota_n$ and $R_{ii} = r_i$ for $i = 1, \ldots, n$. In this case, one can bound the log-determinant as in (4.118).

$$\ln |I_n - \rho S| \geq \ln |I_n - \rho W| \geq \ln |(I_n - \rho R) - \rho S| \tag{4.118}$$

Skillful reordering of the observations can reduce the number and magnitude of the elements of U. Pace and LeSage (2009c) use this approach to bound the log-determinant of a $3,954,400 \times 3,954,400$ matrix. Since almost the same estimate of dependence was obtained using the lower and upper log-determinant bounds, these proved as informative as the actual log-determinant. An advantage of partitioning is the possibility of using parallel processing as well as reduced memory requirements.

Bounds may not yield a precise answer, but may yield a range of answers. Pace and LeSage (2003a) invoked likelihood dominance ideas to demonstrate that bounds could permit qualitative inferences concerning parameter estimates.

Various types of bounds can be combined with approximations such as the Monte Carlo log-determinant estimator or the Chebyshev approximation to provide bounds on the approximations. For example, Barry and Pace (1999) bound the remainder term in the Taylor series approximation. For W with real eigenvalues and a maximum eigenvalue of 1, $\mathrm{tr}(W^{2j+1}) < \mathrm{tr}(W^{2j})$ and $\mathrm{tr}(W^{2(j+1)}) < \mathrm{tr}(W^{2j})$. Consequently, the last even order estimated trace sets an upper bound on the remaining omitted traces. Using this sets up lower and upper bounds on the omitted terms.

4.8   Inverses and other functions

Although the focus of this chapter has been on computing log-determinants, other functions involving the spatial weight matrix W deserve attention. First, $(I_n - \rho W)^{-1}$ often appears in various contexts. In the vast majority of those contexts, it appears in conjunction with a matrix or vector. For example,

$v = (I_n - \rho W)^{-1}\varepsilon$ where ε is a $n \times 1$ vector. In this case, it would be poor computational practice in terms of speed, memory requirements, and accuracy to compute $(I_n - \rho W)^{-1}$ and then multiply it by the vector ε. It would be better to solve the equation $(I_n - \rho W)v = \varepsilon$ for v.

One can do this using iterative techniques such as conjugate gradients or with the LU or Cholesky decompositions. For example, suppose $I_n - \rho W = LU$. In this case, $LUv = \varepsilon$ which is identical to $Lz = \varepsilon$ where $z = Uv$. Solving the triangular system $Lz = \varepsilon$ for z, and then solving a second triangular system $Uv = z$ for v yields the desired solution for v. Although it seems like more work, it actually performs much better than the brute force approach, especially for sparse matrices. This is because $(I_n - \rho W)^{-1}$ is dense for a spatially connected system requiring n^2 storage locations. In contrast, the sparse matrix has a much smaller storage footprint. The brute force approach is guaranteed to take $O(n^3)$ operations while solving the equation could take as few as $O(n)$ operations.

As an example, for $n = 10,000$, forming $(I_n - 0.75W)^{-1}$ and multiplying it by a $10,000 \times 1$ vector ε takes 86.0 seconds, while solving the equation takes 0.071 seconds. It may seem that having the inverse available would save time when dealing with a new vector ε. Given the inverse, it takes 0.69 seconds to multiply the inverse and vector, and this is slower than solving the system again. However, for a given L and U, solving the system again with a new ε takes 0.0056 seconds. From a memory perspective, it requires around 1.6 gigabytes of memory to hold the $10,000 \times 10,000$ matrix whereas it takes 4.62 megabytes for L and U. In addition, for symmetric W the Cholesky decomposition would take around half the memory and time required by the LU decomposition.

Chebyshev and Taylor series approximations can also aid in solving equations. The function to be approximated is $f(x) = 1/(1 - \rho x)$. Given the coefficients $c_0, \ldots, c_q$ of the series associated with the powers of W, the problem in (4.119) reduces to (4.121) in which ε is multiplied by a power of W. Forming powers of W times the vector ε in the efficient manner described in (4.122) allows us to form $u_{(1)} = W\varepsilon$, then $Wu_{(1)} = W^2\varepsilon$. Therefore, calculation of the q powers of the matrix times a vector just involves a set of q matrix-vector operations.

$$v = (I_n - \rho W)^{-1}\varepsilon \qquad (4.119)$$

$$v \approx \left[c_0 I_n + c_1 W + c_2 W^2 +, \ldots, + c_q W^q\right]\varepsilon \qquad (4.120)$$

$$v \approx c_0\varepsilon + c_1 W\varepsilon +, \ldots, + c_q W^q\varepsilon \qquad (4.121)$$

$$W^j\varepsilon = W(W^{j-1}\varepsilon) \qquad (4.122)$$

To provide an indication of the performance of this approach, we simulated a spatial system with $n = 10,000$, $\rho = 0.75$, and a vector of *iid* unit normals for ε. The correlation between the sixth-degree Chebyshev approximation of

v versus the exact version of v was equal to 0.999986. The time and storage requirements of this approximation approach were *de minimus*.

Another problem which often arises is finding the diagonal of the inverse, $\text{diag}((I_n - \rho W)^{-1})$, as shown in (4.123). In practice, computing $(I_n - \rho W)^{-1}$ and finding the diagonal breaks down due to memory problems for data sets such as the US Census tracts.

$$d = \text{diag}((I_n - \rho W)^{-1}) \tag{4.123}$$
$$d \approx c_0 \, \text{diag}(I_n) + c_1 \, \text{diag}(W) +, \ldots, + c_q \, \text{diag}(W^q)$$
$$d \approx c_0 \iota_n + c_1 0_n + c_2 \, \text{diag}(W^2) +, \ldots, + c_q \, \text{diag}(W^q)$$

The exact diagonals of the powers of W can be efficiently computed via (4.124)

$$\text{diag}(W_i W_j) = (W_i \odot W_j') \iota_n \tag{4.124}$$

The diagonal of a matrix A can also be estimated via Monte Carlo as shown in (4.125)–(4.130).

$$A = \begin{bmatrix} a & b \\ c & d \end{bmatrix}, \quad u = \begin{bmatrix} u_1 \\ u_2 \end{bmatrix} \tag{4.125}$$
$$Au = \begin{bmatrix} u_1 a + u_2 b & u_1 c + u_2 d \end{bmatrix}' \tag{4.126}$$
$$u \odot Au = \begin{bmatrix} u_1^2 a + u_1 u_2 b & u_2 u_1 c + u_2^2 d \end{bmatrix}' \tag{4.127}$$
$$E(u_i^2) = 1 \tag{4.128}$$
$$E(u_i u_j) = 0 \quad i \neq j \tag{4.129}$$
$$E(u \odot Au) = \begin{bmatrix} a & d \end{bmatrix}' = \text{diag}(A) \tag{4.130}$$

To do this efficiently, let v be an $n \times m$ matrix of *iid* unit normal deviates, where m is the number of vectors used in the approximation procedure. We begin by setting the initial values $v_{(0)} = v$. Monte Carlo estimates of the diagonals can be efficiently computed via (4.131)–(4.132).

$$v_{(j)} = W v_{(j-1)} \tag{4.131}$$
$$\text{diag}(W^j) \approx (v \odot v_{(j)}) \frac{\iota_m}{m} \tag{4.132}$$

The term $v \odot v_{(j)}$ is an $n \times m$ matrix where each column is an independent estimate of the diagonal. Post multiplying this by ι_m sums across the m independent estimates and dividing by m converts this into an average of the m estimates. Relative to estimating the trace, m needs to be much larger for an accurate estimate the diagonal. Selection of m depends on the size of the diagonal elements and their variation. If memory is a problem, estimation

could proceed by setting m to a smaller value and then repeating the calculation. An average over all estimates would provide the approximate diagonal.
As with the trace estimator, using exact diagonals for the lower order powers
of W can boost accuracy at a low computational cost.

To demonstrate the performance of approaches based on using diagonals of
W^j, we used the same scenario as in the equation solution problem where $n = 10,000$, $\rho = 0.75$, and ε was a vector of *iid* unit normals. For this problem, the
correlation between the sixth-degree Chebyshev approximation of the diagonal
for $(I_n - 0.75W)^{-1}$ versus the exact version of (4.123) was equal to 0.99958.
This correlation rises to 0.9999994 for a tenth-degree approximation.

Another matrix function often encountered is $\mathrm{tr}(W(I_n - \rho W)^{-1})$, which is
the derivative of the log-determinant with respect to ρ shown in (4.133). The
scalar function to be approximated is $f(x) = x/(1 - \rho x)$.

$$h = \mathrm{tr}(W(I_n - \rho W)^{-1}) \tag{4.133}$$
$$h \approx c_0 \,\mathrm{tr}(I_n) + c_1 \,\mathrm{tr}(W) + c_2 \,\mathrm{tr}(W^2) +, \ldots, + c_q \,\mathrm{tr}(W^q) \tag{4.134}$$

We used the same scenario as above to demonstrate the performance of
the matrix function in (4.133). For this problem, the exact trace of $W(I_n - 0.75W)^{-1}$ was 2282.04 and the quintic Chebyshev approximation in (4.134)
was equal to 2281.67, a difference of 0.37. However, the exact trace of
$W(I_n - 0.749W)^{-1}$ is 2274.82 which is a difference of 7.23 from the exact trace
associated with $\rho = 0.75$. Similarly, the exact trace for $W(I_n - 0.751W)^{-1}$
is 2289.31, which differs by 7.26 relative to the trace for $\rho = 0.75$. These
differences in the exact traces that arise from a very small change of 0.001 in
ρ are many times the small approximation error equal to 0.37, suggesting the
estimate is close to the exact trace.

Similar approaches could be employed for other matrix functions of interest,
and to obtain selected elements other than the diagonal. A major advantage
of these approximations over exact computation (besides the storage requirements and time) is the ability to pre-compute quantities such as $\mathrm{tr}(W^j)$ or
$\mathrm{diag}(W^j)$ which can be updated to produce new approximations at very little
cost. In addition, some of these quantities appear in multiple functions. For
example, pre-computing $\mathrm{diag}(W^j)$ allows computation of $\mathrm{diag}((I_n - \rho W)^{-1})$
and because the trace is the sum of the diagonal, having the diagonal facil-
itates approximation of the *derivative* of the log-determinant as well as the
log-determinant. As already noted, given these diagonals or traces, reweighting these to approximate a function is almost costless.

4.9 Expressions for interpretation of spatial models

In Chapter 2, we showed that the impact on the expected value of the dependent variable arising from changes in the rth non-constant explanatory variable was a function of the multiplier matrix $S_r(W)$ in (4.135), where we use α^* to represent the intercept coefficient.

$$E(y) = \sum_{r=1}^{p} S_r(W)x_r + \alpha^* \iota_n \qquad (4.135)$$

The matrix $S_r(W)$ for the SAR, SDM, and extended SDM models are shown in (4.136), (4.138), and (4.140). By extended SDM model we mean: $y = \rho W y + X\beta + W X\theta + W^2 X\gamma + \alpha^* \iota_n + \dots$ where X is an $n \times p$ matrix containing n observations on p non-constant explanatory variables. The overall number of independent variables in these models equals k, and $k = op + 1$ where o is 1 in the case of SAR, 2 in the case of SDM, and 3 for the extended version of the SDM.

$$S_r(W) = (I_n - \rho W)^{-1}\beta_r \qquad (4.136)$$
$$= I_n\beta_r + \rho W\beta_r + \rho^2 W^2\beta_r + \dots \qquad (4.137)$$
$$S_r(W) = (I_n - \rho W)^{-1}(I_n\beta_r + W\theta_r) \qquad (4.138)$$
$$= [I_n\beta_r + W\theta_r] + \rho[W\beta_r + W^2\theta_r] + \dots \qquad (4.139)$$
$$S_r(W) = (I_n - \rho W)^{-1}(I_n\beta_r + W\theta_r + W^2\gamma_r + \dots) \qquad (4.140)$$

We proposed summary measures of these impacts which involve $S_r(W)$.

$$\bar{M}(r)_{direct} = n^{-1}tr(S_r(W)) \qquad (4.141)$$
$$\bar{M}(r)_{total} = n^{-1}\iota_n' S_r(W)\iota_n \qquad (4.142)$$
$$\bar{M}(r)_{indirect} = \bar{M}(r)_{total} - \bar{M}(r)_{direct} \qquad (4.143)$$

As noted in Chapter 2, it is computationally inefficient to directly calculate summary measures of these impacts using the definition of $S_r(W)$, since this would involve $n \times n$ dense matrices. The main challenge in this case is calculating $tr(S_r(W))$. As discussed in this chapter, forming estimates of the trace does not require much computational effort.

Let T represent an $o \times (q + 1)$ matrix containing the average diagonal elements of the powers of W. We show the case of $o = 2$ in (4.144).

$$T = \begin{bmatrix} 1 & 0 & n^{-1}tr(W^2) & n^{-1}tr(W^3) & \dots & n^{-1}tr(W^q) \\ 0 & n^{-1}tr(W^2) & n^{-1}tr(W^3) & n^{-1}tr(W^4) & \dots & n^{-1}tr(W^{q+1}) \end{bmatrix} \qquad (4.144)$$

Let G represent a diagonal $(q+1) \times (q+1)$ matrix as shown in (4.146) that contains powers of ρ, and a $p \times o$ matrix P populated with parameters as in (4.147).

$$g = \begin{bmatrix} 1 & \rho & \rho^2 & \cdots & \rho^q \end{bmatrix} \tag{4.145}$$

$$G_{ii} = g_i \quad i = 1, \ldots, q+1 \tag{4.146}$$

$$P = \begin{bmatrix} \beta_1 & \theta_1 \\ \beta_2 & \theta_2 \\ \vdots & \vdots \\ \beta_p & \theta_p \end{bmatrix} = \begin{bmatrix} \beta & \theta \end{bmatrix} \tag{4.147}$$

We can produce p-vectors containing the cumulative scalar summary impact measures for each of the $r = 1, \ldots, p$ non-constant explanatory variables as in (4.148) to (4.150) using $a = \iota_{q+1}$.

$$\bar{M}_{direct} = PTGa \tag{4.148}$$

$$\bar{M}_{total} = (\beta + \theta)ga \tag{4.149}$$

$$\bar{M}_{indirect} = \bar{M}_{total} - \bar{M}_{direct} \tag{4.150}$$

There may also be interest in calculating *spatially partitioned impact estimates* that show how the effects decay as we move to higher order neighboring regions. We note that the vector $a = \iota_{q+1}$ acts to cumulate the effects over all orders 0 to q, using the global multiplier matrix $(I_n - \rho W)^{-1}$. If we replace this $(q+1) \times 1$ vector with a vector containing a 1 in the first element and zeros for the remaining elements, the effects would take the form: $\bar{M}_{direct} = \beta$, $\bar{M}_{total} = \beta + \theta$, and $\bar{M}_{indirect} = \theta$. These impact estimates represent only impacts aggregated over the zero-order neighbors, represented by the first term from the global multiplier, $\rho^0 W^0 = I_n$. This produces a spatially partitioned effect that represents only the first bracketed term in (4.139). If we set the first two elements of the vector a to values of 1 with zeros for the remaining elements, this would result in effects estimates cumulated over the first two bracketed terms in (4.139), reflecting the zero-order neighbors plus the first-order neighbors. Therefore, setting the first t elements in a to values of one with the remaining elements set to zero allows us to produce spatially partitioned versions of the direct, indirect, and total impact estimates. These would represent a cumulation of the effects out to the $(t-1)$th order neighbors based on the first t terms in the expansion (4.139).

4.10 Closed-form solutions for single parameter spatial models

We discuss closed-form solutions that can be applied to some spatial models that involve only a single dependence parameter. The basic idea is that the first-order conditions for maximizing the concentrated log likelihood lead to a polynomial involving the single dependence parameter. This problem has a closed-form solution in the same sense that limited information maximum likelihood (LIML) has a closed-form solution. By this we mean that one can express the optimizing value for the parameter of interest in terms of eigenvalues of a small matrix.

We begin with the SAR or SDM model where we set up a quadratic form (quadratic polynomial) for the sum-of-squared error term in (4.151) to (4.157).

$$Y = \begin{bmatrix} y & Wy \end{bmatrix} \tag{4.151}$$

$$M(X) = I_n - X(X'X)^{-1}X' \tag{4.152}$$

$$u(\rho) = \begin{bmatrix} 1 & -\rho \end{bmatrix}' \tag{4.153}$$

$$e(\rho) = M(X)YG_1 u(\rho) \tag{4.154}$$

$$G_1 = I_2 \tag{4.155}$$

$$Q = G_1 Y' M(X) Y G_1 \tag{4.156}$$

$$S(\rho) = e(\rho)'e(\rho) = u(\rho)'Qu(\rho) \tag{4.157}$$

For this specification, $G_1 = I_2$, so it does not play a role. However, for other specifications G_1 could become something other than an identity matrix. Equations (4.158) to (4.162) lay out the polynomial approximation of the log-determinant.

$$\tau = \begin{bmatrix} \operatorname{tr}(W) & \operatorname{tr}(W^2) & \ldots & \operatorname{tr}(W^p) \end{bmatrix}' \tag{4.158}$$

$$G_2 = \begin{bmatrix} -1 & 0 & \ldots & 0 \\ 0 & -1/2 & \ldots & 0 \\ 0 & 0 & \ddots & 0 \\ 0 & 0 & \ldots & -1/p \end{bmatrix} \tag{4.159}$$

$$c = G_2\tau \tag{4.160}$$

$$v(\rho) = \begin{bmatrix} \rho & \rho^2 & \rho^3 & \ldots & \rho^p \end{bmatrix}' \tag{4.161}$$

$$\ln|I_n - \rho W| \approx c'v(\rho) \tag{4.162}$$

We restate the concentrated log likelihood in (4.163).

$$L_p(\rho) = c'v(\rho) - \frac{n}{2}\ln(u(\rho)'Qu(\rho)) \tag{4.163}$$

Given the simple log-likelihood in (4.163), the gradient in (4.164) is equally simple.

$$\frac{dL_p(\rho)}{d\rho} = \frac{d\ln|I_n - \rho W|}{d\rho} - \frac{n}{2}S(\rho)^{-1}\frac{dS(\rho)}{d\rho} \qquad (4.164)$$

Setting this to zero, and simplifying yields the first order condition in (4.165).

$$\frac{d\ln|I_n - \rho W|}{d\rho}S(\rho) - \frac{n}{2}\frac{dS(\rho)}{d\rho} = 0 \qquad (4.165)$$

Multiplying (4.165) by n^{-1} results in the mathematically equivalent condition in (4.166), that has better numerical properties for large n.

$$\frac{1}{n}\frac{d\ln|I_n - \rho W|}{d\rho}S(\rho) - \frac{1}{2}\frac{dS(\rho)}{d\rho} = 0 \qquad (4.166)$$

The positive definite quadratic form $S(\rho)$ is a polynomial of degree 2 in ρ whose coefficients are the sum along the antidiagonals of Q. The derivative of the log-determinant is a polynomial of order $p - 1$ which is multiplied by the quadratic polynomial $S(\rho)$ yielding a polynomial of degree $p + 1$. Of course, the derivative of $S(\rho)$ wrt ρ is a polynomial of degree 1 (linear). Consequently, the expression in (4.165) is a polynomial of degree $p+1$, and has $p+1$ solutions or zeros.

These $p + 1$ solutions can be found by calculating eigenvalues of the $p + 1$ by $p + 1$ companion matrix (Horn and Johnson, 1993, p. 146-147). So this problem has a closed-form in the same sense as limited information maximum likelihood (LIML), expressing the answer in terms of the eigenvalues of a small matrix.[5] We note that calculating the eigenvalues in this case requires $O((p + 1)^3)$ operations and does not depend upon n, making this approach suitable for large spatial applications. Another advantage is that one can test possible solutions to make sure the proposed solution represents a global optimum. In addition, the second derivative of the concentrated log likelihood with respect to the dependence parameter is available and this facilitates calculating the Hessian as discussed in Chapter 3.

One can express other specifications in this format as well. LeSage and Pace (2007) provide more information on this approach in the context of a similar closed-form solution for the matrix exponential spatial specification that we discuss in Chapter 9.

---

[5] See Anderson, T.W. and H. Rubin (1949) for more on LIML. Other methods also exist for finding the roots of polynomials. See Press et al. (1996, p. 362-372) for a review of these.

4.11   Forming spatial weights

Forming a matrix W based on contiguity or nearest neighbors in terms of a Euclidean distance function or other distance *metric* seems intuitive. However, straightforward calculation usually involves $O(n^2)$ operations requiring substantial time. Fortunately, there are more elegant ways to form W which usually require $O(n\ln(n))$ operations for locations on a plane.

Computational geometry provides tools for many of the tasks relevant for specifying weight matrices (Goodman and O'Rourke, 1997). For example, points or sites in the *Voronoi* diagram shown in Figure 4.8 represent the location of observations, with edges around each site represented by the Voronoi polygon for that site. The edges of the Voronoi polygons provide boundaries for regions surrounding the respective sites. These regions have the property that the interior points are closer to their site than to any other site. This type of approach has been used in the context of developing retail trade areas (Dong, 2008), site selection, facilities planning, and so on. Voronoi polygons represent a generalization for irregular location data of the traditional regular polygons from central place theory.

Segments connecting sites between contiguous Voronoi polygons form the legs of *Delaunay* triangles. A number of computationally elegant means exist to compute Voronoi diagrams and Delaunay triangles in $O(n\ln(n))$ operations for planar data. Since the legs of the Delaunay triangles connect the sites, these form a way of specifying contiguity. Therefore, given the Delaunay triangles associated with a set of sites, one uses this information to specify W. On average, there are six neighbors to each site and so there will be close to $6n$ triangles (defined as 3 points). Given the triangles, it requires $O(n)$ operations to form a sparse weight matrix W.

The matrix W derived using Delaunay triangles can be used to find the m nearest neighbors for each site. Since W represents the neighbors, W^2 represents neighbors to neighbors, and so on. Neighboring relations from various orders form a good candidate set for the nearest neighbors. For example, let $Z = W + W^2 + W^3 + W^4$. The non-zero elements of Z_i represent neighbors up to the fourth order for each observation i. Having identified the candidate observations, we can use their locational coordinates to calculate the distance (for some chosen metric) from observation i to the candidate neighbors. Sorting this short set of candidate neighbor distances for the m nearest neighbors takes little time and the size of the candidate neighbor set is not related to n for moderate to large values of n. Doing this n times yields a set of $n \times m$ matrices containing observation indices to the m nearest neighbors for each observation. Given the $n \times m$ matrix of observation indices, it takes $O(n)$ calculations to populate the m nearest neighbor W matrix.

It is sometimes useful to employ the $n \times m$ matrix of observation indices to define a first nearest neighbor matrix $W_{(1)}$, a second nearest neighbor matrix

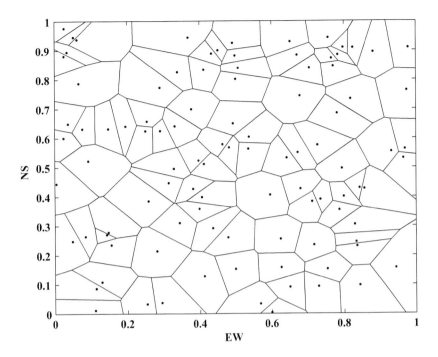

FIGURE 4.8: Voronoi diagram of 100 spatially random points

$W_{(2)}$, and so forth. One can use these various individual neighbor matrices to form spatial weight matrices consisting of a weighted set of m nearest neighbors. Possible weighting schemes include geometric decline with order, exponential decline with order, or smooth polynomial weights in a fashion similar to Almon distributed lags.

The Monte Carlo log-determinant estimator and many other matrix functions of interest rely heavily on sparse matrix-vector multiplication. Usually these operations require specialized routines for general matrix-vector products. However, sparse matrix-vector products can be easily implemented in almost any programming language. We illustrate this using m nearest neighbor weight matrices, where we let D represent an $n \times m$ matrix of indices to the m neighbors for each of the n observations. Let D_j, the jth column of D, represent the $n \times 1$ vector of indices associated with the jth neighbor. The first element of D_2 would contain the index of the first observation's second nearest neighbor. The tenth element of D_5 would contain the tenth observation's fifth nearest neighbor, and so on. For an $n \times q$ matrix V, $W_j V = V(D_j)$. If the nearest neighbors had weights w_j, the overall $WV = w_1 V(D_1) + w_2 V(D_2) + \cdot + w_m V(D_m)$.

A very interesting feature is that for many programming languages the time required to index a matrix is the same as for a vector, since indexing only requires altering a row label that pertains to all columns. To illustrate this, we indexed a $1,000,000$ by 50 matrix and a $1,000,000$ by 1 vector. In Matlab it took 0.051 seconds to index the matrix and 0.043 seconds to index the vector.

We note that sparse matrix-vector products needed to compute the log-determinant estimator and other matrix functions require only indexing and addition, two of the fastest operations possible on digital computers. This means that nearest neighbor calculations can be implemented in a variety of computing languages such as FORTRAN or C, and various statistical software environments.

For multivariate and spatiotemporal weight matrices, it may be necessary to compute the n^2 possible distances and sort these to find the m nearest neighbors. Fortunately, partial sorting algorithms exist (see fortran2000.com for some examples) that require some function of m operations for each sort rather than a function of n. This greatly accelerates formation of multivariate and spatiotemporal weight matrices. Often there exist ways to reduce the nearest neighbor candidate list for each observation. For example, in the case of a spatiotemporal weight matrix one could require neighbors to be no more than four years old, which would materially reduce the time required to form spatiotemporal weight matrices spanning a 20 year period.

4.12   Chapter summary

Determinants and other functions of the $n \times n$ spatial weight matrix often arise in spatial econometrics. Brute force approaches to dealing with determinants and matrix functions limit the possible sample size or the ability to perform more involved estimation or testing. This chapter dealt with these computational issues. Section 4.1 introduced the need for determinants when working with dependent variable transformations of the type that arise in spatial regression models. Section 4.2 described basic approaches to calculating determinants. Special features of spatial systems such as sparsity were discussed in Section 4.3 along with ways to take advantage of sparsity by ordering rows and columns of W when using direct calculations for the log-determinant.

Section 4.4 introduced a Monte Carlo approximation to the log-determinant that takes little time and memory and showed that this approach leads to small errors in estimation of the dependence parameter, ρ. Similarly, Section 4.5 introduced a Chebyshev approximation that also performed well. Section 4.6 discussed another approximation approach that treats the sample as arising from an increasing domain process and extrapolates the log-determinant. Section 4.7 described a few approaches to bounding log-determinants that may be of both theoretical and practical interest.

Section 4.8 extended the log-determinant approximation techniques to other commonly encountered tasks that arise in spatial modeling. These included solving systems of equations, finding the diagonal of the variance-covariance matrix (without computing the entire variance-covariance matrix), and calculating the derivative of the log-determinant. Section 4.9 applied some of these techniques to computing the summary measures of impacts (partial derivatives) that are needed for interpretation of spatial econometric models. Section 4.10 showed a method that can be used to produce closed-form solutions for a number of spatial models involving a single dependence parameter. Finally, efficient approaches to forming spatial weight matrices were discussed in Section 4.11.

Chapter 5

Bayesian Spatial Econometric Models

This chapter describes application of Bayesian methodology to spatial econometric modeling and estimation. Bayesian methodology has existed for a long time, but recent approaches to estimation of these models have led to a revival of interest in these methods. The estimation approach known as Markov Chain Monte Carlo (MCMC) decomposes complicated estimation problems into simpler problems that rely on the conditional distributions for each parameter in the model (Gelfand and Smith, 1990). This innovation makes application of the Bayesian methodology far easier than past approaches that relied on analytical solution of the posterior distribution. A result of this is that an extensible toolkit for solving spatial econometric estimation problems can be developed at both a theoretical and applied level.

We begin by introducing Bayesian econometric methodology in Section 5.1. This is followed in Section 5.2 by a conventional Bayesian approach to estimating the SAR model, which is shown to require numerical integration over the spatial dependence parameter ρ. The historical need to carry out numerical integration of the *posterior distribution* with respect to parameters in Bayesian models made conventional Bayesian methodology relatively difficult for models with a large number of parameters.

Section 5.3 introduces *Markov Chain Monte Carlo* (MCMC) estimation approaches for the family of spatial econometric models discussed in Chapter 2. Bayesian methodology focuses on distributions involving data and parameters, which has the effect of structuring estimation problems in such a way as to produce a posterior distribution that can be decomposed into a sequence of *conditional distributions*. These "conditionals" characterize the distribution of a single parameter given all other parameters in the model, and they are extremely useful from both a theoretical and applied perspective. From an applied perspective, conditional distributions are required for Markov Chain Monte Carlo (MCMC) estimation. This method of estimation became popular when Gelfand and Smith (1990) demonstrated that MCMC sampling from the sequence of complete conditional distributions for all parameters in a model produces a set of estimates that converge in the limit to the true (joint) posterior distribution of the parameters. Therefore, if we can decompose the posterior distribution into a set of conditional distributions for each parameter in the model, drawing samples from these will provide us with valid

Bayesian parameter estimates. This approach to estimation is the subject of Sections 5.3 and 5.4 with an applied illustration provided in Section 5.5.

One benefit of Bayesian methodology used in conjunction with MCMC estimation is that an extensible toolkit can be devised for dealing with a wide array of spatial econometric problems. As a concrete example of this, Section 5.6 extends the conventional spatial regression models from Chapter 2 to allow for a heteroscedastic disturbance structure in place of the conventional assumption of homoscedastic disturbances. Other uses for the MCMC method are also provided in this section.

Additional material on the MCMC approach to modeling and estimation is provided in a number of the remaining chapters.

5.1   Bayesian methodology

An important aspect of Bayesian methodology is the focus on distributions for the data as well as parameters. Bayes' rule involves combining the data distribution embodied in the likelihood function with prior distributions for the parameters assigned by the practitioner, to produce a posterior distribution for the parameters. The posterior distribution forms the basis for all inference, since it contains all relevant information regarding the estimation problem. Relevant information includes both sample data information coming from the likelihood, as well as prior or subjective information embodied in the distributions assigned to the parameters.

As noted in Chapter 1, econometric models focus on: 1) estimation and inference about parameters, 2) model specification and model comparisons, and 3) model prediction or out-of-sample forecasting. The Bayesian approach to estimation arises from some basic axioms of probability.[1] For two random variables A and B we have that the *joint probability* $p(A, B)$ can be expressed in terms of *conditional probability* $P(A|B)$ or $P(B|A)$ and the *marginal probability* $P(B)$ (or $P(A)$) as shown in (5.1) and (5.2).

$$p(A, B) = p(A|B)p(B) \qquad (5.1)$$
$$p(A, B) = p(B|A)p(A) \qquad (5.2)$$

Setting these two expressions equal and rearranging gives rise to *Bayes' Rule*:

$$p(B|A) = \frac{p(A|B)p(B)}{p(A)} \qquad (5.3)$$

---

[1] See Koop (2003) for a more complete introduction to Bayesian econometrics. Much of this introductory material mirrors that presented in the introduction of Koop (2003).

For our purposes, we let $\mathcal{D} = \{y, X, W\}$ represent model data and $\theta = B$ denote model parameters so that:

$$p(\theta|\mathcal{D}) = \frac{p(\mathcal{D}|\theta)p(\theta)}{p(\mathcal{D})} \qquad (5.4)$$

A point to note is that Bayesian modeling assumes the parameters θ have a *prior distribution* $p(\theta)$ that reflects previous knowledge as well as uncertainty we have prior to observing the data. This distribution is used to provide a formalized probabilistic statement that overtly specifies both prior information and uncertainty. If we know very little from prior experience, then this distribution should represent a vague probabilistic statement, whereas a great deal of previous experience would lead to a very narrow distribution centered on parameter values gained from previous experience.

The left-hand side of (5.4) represents the post-data inference for θ and is called the *posterior distribution* of θ. This represents an update of the *prior distribution* for the parameter θ after conditioning on the sample data. The right-hand side of (5.4) indicates that Bayesian parameter inference represents a compromise between prior information embodied in $p(\theta)$ and new information provided by the sample data and model represented by the likelihood $p(\mathcal{D}|\theta)$. The Bayesian approach relies on conditional probability to provide a formal structure of rules that allows us to learn (update our prior knowledge) about some unknown quantity such as θ using the model and data.

The nature of the compromise between prior and sample data takes the following form. If a spatial sample of data exhibits wide variation across regions and we are only able to collect a small sample of regions, then the posterior distribution would place more emphasis on prior experience (the distribution $p(\theta)$) than the distribution implied by the small sample of observable data and model contained in the likelihood, $p(\mathcal{D}|\theta)$. Conversely, if our prior experience (information) was very limited and a great deal of sample data resulted in a likelihood that indicated values of θ distributed tightly around a particular value, then the posterior distribution would place more emphasis on the model and sample data information, embodied in the likelihood. There is no explicit process such as this in classical statistics that allows for a tradeoff between prior information and sample data information, since inference is based entirely on the model and sample data, represented by the likelihood function.

This distinction becomes unimportant for inference based on maximum likelihood and Bayesian estimation procedures when we have large data samples and very little prior information. In this situation, both methods rely almost entirely on the sample data information to provide inferences regarding the parameter θ.

Econometric modeling issue 1) is estimation and inference about parameters. We note that all Bayesian inference (sometimes called learning) about $\theta|\mathcal{D}$ is based on the *posterior density* $p(\theta|\mathcal{D})$ for the parameters θ given the data $\mathcal{D}$. We can simplify (5.4) by ignoring the data distribution $p(\mathcal{D})$, since this distribution doesn't involve the parameters θ.

$$p(\theta|\mathcal{D}) \propto p(\mathcal{D}|\theta)p(\theta) \qquad (5.5)$$

Expression (5.5) indicates that the *posterior distribution* of the parameter vector θ is a (proportional) product of the term $p(\mathcal{D}|\theta)$ representing the model likelihood and the term $p(\theta)$ representing the *prior distribution* for the parameters θ. Since non-Bayesians do not subscribe to the notion that parameters have *prior* distributions, they cannot express prior beliefs or uncertainty about the parameters using prior distributions.

Econometric modeling issue 2) is model specification and comparison. Given a set $i = 1, \ldots, m$ of Bayesian models, each would be represented by a likelihood function and prior distribution as in (5.6).

$$p(\theta^i|\mathcal{D}, M_i) = \frac{p(\mathcal{D}|\theta^i, M_i)p(\theta^i|M_i)}{p(\mathcal{D}|M_i)} \qquad (5.6)$$

Treating the posterior distributions in this case as conditional on the model specification M_i, we can apply Bayes' rule to expand terms like $p(\mathcal{D}|M_i)$ in a fashion similar to (5.3). This leads to a set of unconditional posterior model probabilities:

$$p(M_i|\mathcal{D}) = \frac{p(\mathcal{D}|M_i)p(M_i)}{p(\mathcal{D})} \qquad (5.7)$$

These serve as the basis for inference about different models, given the sample data. The term $p(\mathcal{D}|M_i)$ that appears on the right-hand-side of expression (5.7) is called the *marginal likelihood*, and we can solve for this key quantity needed for model comparison finding:

$$p(\mathcal{D}|M_i) = \int p(\mathcal{D}|\theta^i, M_i)p(\theta^i|M_i)d\theta^i \qquad (5.8)$$

An important point is the unconditional nature of the model probabilities, which do not depend on the posterior mean parameter values alone, but the entire posterior distribution over which we integrate. Maximum likelihood approaches to model comparison rely on the likelihoods of two models evaluated at the mean values of the parameter estimates. This means that model comparison inferences depend on scalar values of the parameter estimates used to evaluate the likelihood. In contrast, Bayesian model comparison uses the entire posterior distribution of values for the parameters.

Expression (5.8) makes it clear that the theory behind Bayesian model comparison is quite simple and follows directly from formal probability axioms of statistics. However, implementation may be hindered by the need to integrate over the parameter vector θ. We will discuss several approaches to dealing with the integration problem in Chapter 6 which is devoted to model comparison.

The Bayesian theory for econometric issue 3) requires prediction of an out-of-sample set of observations that we denote y^* based on sample data observations in $\mathcal{D}$. Here again, we rely on rules of probability to arrive at the *posterior distribution* of the prediction sample given the sample data $\mathcal{D}$, and model parameters θ. This also involves expressing a joint distribution in terms of a conditional and marginal.

$$p(y^*|\mathcal{D}) = \int p(y^*, \theta|\mathcal{D})d\theta = \int p(y^*|\mathcal{D}, \theta)p(\theta|\mathcal{D})d\theta \qquad (5.9)$$

5.2 Conventional Bayesian treatment of the SAR model

As noted in the previous section, Bayesian methods require analysis of the posterior distribution of the model parameters. This can be carried out using analytical or numerical methods. In Section 5.2.1 we show that analytical approaches are possible in spatial econometric modeling. However, numerical methods for univariate or bivariate integration are required to produce posterior inferences in the family of spatial econometric models discussed in Chapter 2, which places limits on both theoretical and applied work. Conditioning provides a way to avoid these problems, and is ideally suited to match theoretical and MCMC estimation approaches to spatial econometric modeling.

5.2.1 Analytical approaches to the Bayesian method

As already indicated, one aspect of the Bayesian method is the introduction of prior information in the modeling process. Investigators specify their prior beliefs using distributions, which are combined with the data distribution to produce the posterior distribution used for inference. The requirement that prior beliefs be revealed as part of solving the estimation problem is viewed by some to be a disadvantage of the Bayesian method. We show that in a spatial econometric setting where data samples are typically large, prior information will tend to play a minor role in determining the character of the posterior distribution. The fundamental Bayesian identity works to create a matrix-weighted average of sample and prior information in the posterior, but the weights are strongly influenced by the quantity of sample data information available relative to prior information. When large samples are available they provide a simplification that can facilitate analytical evaluation of the posterior distribution.

Using the spatial autoregressive model (SAR) from the family of models in Chapter 2, we can demonstrate the combination of prior and sample information. The likelihood for the SAR model: $y = \rho W y + X\beta + \varepsilon$, can be written as in (5.10), where we rely on $A = (I_n - \rho W)$ for notational convenience and $|A|$ denotes the determinant of this matrix.

$$p(\mathcal{D}|\beta, \sigma, \rho) = (2\pi\sigma^2)^{-\frac{n}{2}} |A| \exp\left(-\frac{1}{2\sigma^2}(Ay - X\beta)'(Ay - X\beta)\right) \quad (5.10)$$

As noted, we are required to specify prior distributions for the parameters in the model, which will be combined with the likelihood to produce the posterior distribution. We might rely on what is known as a *normal-inverse gamma prior* (NIG) distribution for the parameters β and σ^2. This form of prior makes the normal prior distribution for β conditional on an inverse gamma distribution for the parameter σ^2. Since the parameter ρ plays such an important role in this model, and is often a subject of inference, we might specify a uniform prior over the feasible range for this parameter, $(1/\lambda_{min}, 1/\lambda_{max})$, where $\lambda_{min}, \lambda_{max}$ represent the minimum and maximum eigenvalues of the spatial weight matrix. As noted in Chapters 2 and 4, *row-stochastic* spatial weight matrices W typically used in applications of these models lead to ρ in the interval $(\lambda_{min}^{-1}, 1)$.

A formal statement of the Bayesian SAR model is shown in (5.11), where we assume an $n \times k$ explanatory variables matrix X. We have added a *normal-inverse gamma* (NIG) prior for β and σ, with the prior distributions indicated using π. This prior specifies that β given σ is distributed multivariate normal $N(c, \sigma^2 T)$, and the marginal distribution for σ takes the form of an inverse gamma density denoted $IG(a, b)$ in (5.12).

$$y = \rho W y + X\beta + \varepsilon \quad (5.11)$$
$$\varepsilon \sim N(0, \sigma^2 I_n)$$
$$\pi(\beta, \sigma^2) \sim NIG(c, T, a, b)$$
$$= \pi(\beta|\sigma^2)\pi(\sigma^2)$$
$$= N(c, \sigma^2 T)IG(a, b) \quad (5.12)$$
$$= \frac{b^a}{(2\pi)^{k/2}|T|^{1/2}\Gamma(a)}(\sigma^2)^{-(a+(k/2)+1)}$$
$$\times \exp[-\{(\beta - c)'T^{-1}(\beta - c) + 2b\}/(2\sigma^2)]$$
$$\pi(\sigma^2) = \frac{b^a}{\Gamma(a)}(\sigma^2)^{-(a+1)}\exp(-b/\sigma^2) \quad (5.13)$$
$$\sigma^2 > 0, \ a, b > 0$$
$$\pi(\rho) \sim U(\lambda_{min}^{-1}, \lambda_{max}^{-1})$$

Note that we have parameterized the inverse-gamma distribution in (5.13), where $\Gamma(\cdot)$ represents the standard gamma function, $\Gamma(a) = \int_0^\infty t^{a-1}e^{-t}dt$. The parameters used to specify our prior beliefs are those from the NIG prior, c, T, a, b, and might also include prior parameters for ρ if we relied on a different type of informative prior for this model parameter.

In cases where we have a great deal of prior uncertainty regarding the parameters β, we can set $c = 0$, and assign a very large prior variance for β with zero covariance between parameters in the vector β. This might be accomplished by setting $T = I_k \cdot 10^{10}$, or some other large magnitude. This is known as a *diffuse* or *uninformative* prior. An uninformative prior can be set for the parameter σ^2 by assigning values of $a = b = 0$. We will argue shortly that this approach is intuitively reasonable when dealing with large spatial data sets. For ρ, we assign a prior that indicates all outcomes within the feasible range $(\lambda_{\min}^{-1}, \lambda_{\max}^{-1})$ are equally probable. An additional point to note is that we assume independence between the prior assigned to β and σ^2 and that for ρ. Prior independence does not imply independence in the posterior distributions of the model parameters, it simply reflects prior beliefs which can be inconsistent with posterior outcomes.

From Bayes' Theorem we know that the posterior distribution for the model parameters takes the form:

$$p(\beta, \sigma^2, \rho | \mathcal{D}) = \frac{p(\mathcal{D}|\beta, \sigma^2, \rho)\pi(\beta, \sigma^2)\pi(\rho)}{p(\mathcal{D})} \qquad (5.14)$$

By multiplying the expression for the likelihood and prior we can determine the form of the posterior up to a constant term, $p(\mathcal{D})$, that does not involve the model parameters. An identity in (5.15) that has been labeled *completing the square* is useful in arriving at the result shown in (5.16).

$$(Ay - X\beta)'(Ay - X\beta) + (\beta - c)'T^{-1}(\beta - c) + 2b$$
$$\equiv (\beta - c^*)'(T^*)^{-1}(\beta - c^*) + 2b^* \qquad (5.15)$$

$$p(\beta, \sigma^2, \rho | \mathcal{D}) \propto (\sigma^2)^{a^* + (k/2) + 1} |A|$$
$$\times \exp\{-\frac{1}{2\sigma^2}[2b^* + (\beta - c^*)'(T^*)^{-1}(\beta - c^*)]\} \qquad (5.16)$$
$$c^* = (X'X + T^{-1})^{-1}(X'Ay + T^{-1}c)$$
$$T^* = (X'X + T^{-1})^{-1}$$
$$a^* = a + n/2$$
$$b^* = b + \left(c'T^{-1}c + y'A'Ay - (c^*)'(T^*)^{-1}c^*\right)/2$$
$$A = I_n - \rho W$$

One thing to note concerning this result is that the usual case for non-spatial regression models where the NIG prior serves as a *conjugate* prior distribution does not hold here (Zellner, 1971). A general definition of conjugate prior distributions are those that result in computationally tractable posterior distributions, and this term is often used to refer to situations where the posterior distribution takes the same form as the prior distribution except that model parameters are updated. In the case of a non-spatial regression model an $NIG(c, T, a, b)$ prior placed on the parameters β and σ would result in an $NIG(c^*, T^*, a^*, b^*)$ posterior distribution for these parameters, where we use the star superscript to denote parameter updating. This would be the result if $\rho = 0$, so that $A = I_n$ in expression (5.16).

Because of the informative prior distributions used in this model, matters become complicated in two ways. First, there is the need to specify or assign values for the parameters of the NIG prior distribution. Second, the posterior distribution in (5.16) is difficult to analyze because it requires integration of σ^2 as well as ρ to arrive at a posterior expression for β. Similarly, it requires that we integrate with respect to β and ρ to find the posterior for σ^2. We should ask the question — is the prior information likely to exert an impact on our inferences regarding β? If not, we can greatly simplify the posterior in (5.16), by relying on uninformative priors.

We note that the term: $y'A'Ay - (c^*)'(T^*)^{-1}c^*$ would equal the sum of squared residuals from the SAR model if we knew the posterior means for the parameters ρ and c^*. This implies that the ratio b^*/a^* would be approximately equal to the residual sum of squares for the case where $a, b \to 0$ and $T^{-1} \to 0$. These NIG prior parameter settings would result in uninformative prior distributions assigned to the parameters β and σ^2.

5.2.2   Analytical solution of the Bayesian spatial model

It is worthwhile to pursue solution of the Bayesian SAR model in the context of simplifications offered by a non-informative prior (Hepple, 1995a,b). We replace the NIG prior from the previous section with an uninformative prior based on $a, b = 0$ and $T^{-1} = 0$, and assume independence between the prior assigned to ρ and that for β and σ, i.e., $\pi(\beta, \sigma, \rho) = \pi(\beta, \sigma)\pi(\rho)$. Using this type of prior and applying Bayes' Theorem that combines the likelihood and prior leads to the simplified posterior distribution shown in (5.17). This result is consistent with our earlier observation relating the term: $y'A'Ay - (c^*)'(T^*)^{-1}c^*$ and the residual sum of squares.

$$p(\beta, \sigma, \rho|\mathcal{D}) \propto p(\mathcal{D}|\beta, \sigma, \rho) \cdot \pi(\beta, \sigma) \cdot \pi(\rho) \tag{5.17}$$

$$\propto \sigma^{-n-1} |A| \exp\left(-\frac{1}{2\sigma^2}(Ay - X\beta)'(Ay - X\beta)\right) \pi(\rho)$$

With this simpler posterior, we can treat σ as a nuisance parameter and analytically integrate this out of the expression in (5.17). This can be accomplished using properties of the inverse gamma distribution, leading to:

$$p(\beta, \rho | \mathcal{D}) \propto |A| \, \{(Ay - X\beta)'(Ay - X\beta)\}^{n/2} \pi(\rho) \qquad (5.18)$$
$$= |A| \, \{(n-k)s^2 + (\beta - c^*)'X'X(\beta - c^*)\}^{-n/2} \pi(\rho)$$
$$c^* = (X'X)^{-1}X'Ay$$
$$s^2 = (Ay - Xc^*)'(Ay - Xc^*)/(n-k)$$

Conditional on ρ, the expression in (5.18) represents a multivariate Student-t distribution that we can integrate with respect to β, leaving us with the marginal posterior distribution for ρ, shown in (5.19).

$$p(\rho | \mathcal{D}) \propto |A| \, (s^2)^{-(n-k)/2} \pi(\rho) \qquad (5.19)$$

There is no analytical solution for the posterior expectation or variance of ρ, which we would be interested in for purposes of inference. However, simple univariate numerical integration methods would allow us to find this expectation as well as the posterior variance of ρ. The integrals required are shown in (5.20).

$$E(\rho | \mathcal{D}) = \rho^* = \frac{\int \rho \cdot p(\rho | \mathcal{D}) d\rho}{\int p(\rho | \mathcal{D}) d\rho}$$
$$\mathrm{var}(\rho | \mathcal{D}) = \frac{\int [\rho - \rho^*]^2 \cdot p(\rho | \mathcal{D}) d\rho}{\int p(\rho | \mathcal{D}) d\rho} \qquad (5.20)$$

Referring to expression (5.19), we see that the integration in (5.20) would involve evaluating the $n \times n$ determinant: $|A| = |I_n - \rho W|$, over the domain of support values for ρ. This can be accomplished efficiently using either the direct sparse matrix approach of Pace and Barry (1997) or the Monte Carlo estimator for the log determinant of Barry and Pace (1999) discussed in Chapter 4.

There is still the problem of determining the feasible range for ρ. We can take alternative approaches here depending on prior information available. In the case where no prior information is available, we could rely on an interval based on the minimum and maximum eigenvalues of the $n \times n$ matrix W, which determine the theoretical feasible range for ρ. In problems involving large spatial samples, the matrix W is sparse, containing a large number of zero elements. For frequently used row-stochastic matrices W, $\lambda_{max} = 1$, so we need only compute λ_{min}, which can be found using sparse matrix algorithms.[2] A second approach would be to use a prior to impose a restriction to the interval $(-1, 1)$. This imposition reflects prior knowledge that most applications of spatial regression models report estimates for the parameter ρ within this range of values. It may also express the prior sentiment that values

---

[2] See LeSage (1999) for a discussion of how to accomplish this using MATLAB software.

of the parameter ρ less than -1 would likely be indicative of problems with the weight matrix or model specification and of little interest. This approach has the advantage of eliminating the need to compute the minimum eigenvalue of the potentially large $n \times n$ matrix W. LeSage and Parent (2007) introduce a beta prior distribution for ρ, which we denote $\mathcal{B}(\alpha, \alpha)$. This alternative to the uniform prior distribution is defined on the interval $(-1, 1)$ and centered on zero. We will discuss and demonstrate use of this prior in Section 5.4. A third approach would be to argue that negative spatial dependence is of little interest in a particular problem, so the prior on ρ could be used to restrict values of ρ to the interval $[0, 1)$.

In all of these cases, we can work with the log of the expression in (5.19), and construct a vector associated with a grid of q values for ρ in the relevant interval that takes the form in (5.21). Here, we assume the prior for ρ from (5.19) is uniform over the range $(1/\lambda_{\min}, 1/\lambda_{\max})$. Therefore, $\ln(\pi(\rho))$ does not vary with ρ and is constant. The constant κ contains this and other constant terms.

$$
\begin{pmatrix} \ln\ p(\rho_1|y) \\ \ln\ p(\rho_2|y) \\ \vdots \\ \ln\ p(\rho_q|y) \end{pmatrix} = \kappa + \begin{pmatrix} \ln|I_n - \rho_1 W| \\ \ln|I_n - \rho_2 W| \\ \vdots \\ \ln|I_n - \rho_q W| \end{pmatrix} - \left(\frac{n-k}{2}\right) \begin{pmatrix} \ln(s^2(\rho_1)) \\ \ln(s^2(\rho_2)) \\ \vdots \\ \ln(s^2(\rho_q)) \end{pmatrix} \quad (5.21)
$$

We draw on the vectorization scheme for the grid of q values for ρ from Pace and Barry (1997) described in Chapter 3, to produce the following:

$$
\begin{aligned}
s^2(\rho_i) &= e_o' e_o - 2\rho_i e_d' e_o + \rho_i^2 e_d' e_d \\
e &= e_o - \rho e_d \\
e_o &= y - X c_o \\
e_d &= W y - X c_d \\
c_o &= (X'X)^{-1} X' y \\
c_d &= (X'X)^{-1} X' W y
\end{aligned} \quad (5.22)
$$

This vector allows univariate numerical integration using a simple method such as Simpson's rule. Note the way in which computational advances that improve maximum likelihood estimation can also be used in Bayesian approaches to estimation. This is an excellent example of cross-fertilization that arises from computational advances in maximum likelihood and Bayesian methods. We will see that this same approach can be used in the context of MCMC estimation to even greater advantage.

Despite this simplicity, there is still the point that we need to carry out the integration twice to obtain the mean and variance for the parameter ρ. In the case of the SAR model, the posterior mean for β takes a form: $E(\beta|y, X, W) =$

$c^* = (X'X)^{-1}X'(I_n - \rho^* W)y$, which does not involve numerical integration. However, we note this is not true of other members of the family of spatial econometric models introduced in Chapter 2. It is also the case that univariate integration would be needed to obtain posterior variances for β, and the same would be true for the parameter σ^2 in this model.

5.3 MCMC estimation of Bayesian spatial models

As already motivated the posterior distribution for the SAR model requires univariate numerical integration to obtain the posterior mean and variance for the parameter ρ, as well as other parameters in the model. This section is devoted to an alternative methodology known as Markov Chain Monte Carlo (MCMC), which has become very popular in econometrics and mathematical statistics. Section 5.3.1 provides basic background for this approach to estimation, and introduces the method in the context of our basic family of spatial econometric models. The power and generality of this approach is demonstrated with extensions of the basic spatial autoregressive model to the case of heteroscedastic disturbances in Section 5.6.1.

Additional illustrations of the flexibility and power of this approach to estimation are provided in Chapter 10, where the topic of spatial dependence for models involving censored and binary dependent variables (spatial Tobit and probit) are discussed. We also introduce Bayesian MCMC estimation of origin-destination flow models in Chapter 8, and for the matrix exponential spatial specification in Chapter 9. Due to the extensible nature of Bayesian methods in conjunction with MCMC, estimation of these models that deal with a wide range of spatial econometric application areas can be viewed as minor extensions of the basic approach we introduce here.

5.3.1 Sampling conditional distributions

An alternative to the analytical/numerical integration approach described in the previous section is to rely on a methodology known as Markov Chain Monte Carlo (MCMC) to estimate the parameters. MCMC is based on the idea that rather than work with the posterior density of our parameters, the same goal could be achieved by examining a large random sample from the posterior distribution. Let $p(\theta|\mathcal{D})$ represent the posterior, where θ denotes the parameters and $\mathcal{D}$ the sample data. If the sample from $p(\theta|\mathcal{D})$ were large enough, one could approximate the form of the probability density using kernel density estimators or histograms, eliminating the need to find the precise analytical form of the density.

The most widely used approach to MCMC is due to Hastings (1970) which generalizes the method of Metropolis et al. (1953), and is labeled *Metropolis-Hastings* sampling. Hastings (1970) suggests that given an initial value θ_0, we can construct a chain by recognizing that any Markov chain that has found its way to a state θ_t can be completely characterized by the probability distribution for time $t + 1$. His algorithm relies on a proposal or candidate distribution, $f(\theta|\theta_t)$ for time $t + 1$, given that we have θ_t. A candidate point θ^* is sampled from the proposal distribution and:

1. This point is accepted as $\theta_{t+1} = \theta^*$ with probability:

$$\psi_H(\theta_t, \theta^*) = \min\left[1, \frac{p(\theta^*|\mathcal{D})f(\theta_t|\theta^*)}{p(\theta_t|\mathcal{D})f(\theta^*|\theta_t)}\right] \qquad (5.23)$$

2. otherwise, $\theta_{t+1} = \theta_t$, that is, we stay with the current value of θ.

We can view the Hastings algorithm as indicating that we should toss a Bernoulli coin with probability ψ_H of "heads" and make a move to $\theta_{t+1} = \theta^*$ if we see a "head" coin toss, otherwise set $\theta_{t+1} = \theta_t$. Hastings demonstrated that this approach to sampling represents a Markov chain with the correct equilibrium distribution, capable of producing samples from the posterior $p(\theta|\mathcal{D})$.

An implication of this is that one can rely on Metropolis-Hastings (M-H) to sample from conditional distributions where the distributional form is unknown. This happens to be the circumstance with the conditional distribution for the spatial dependence parameters ρ, λ in our family of spatial econometric models from Chapter 2.

In other cases, the conditional distributions may take standard forms such as a multivariate normal, with a mean and variance that can be easily calculated using standard linear algebra required for ordinary linear regression. This is often true of the conditional distributions for the parameters β and σ in our family of spatial regression models. When the form of the conditional distributions are known, we can take an approach referred to as *Gibbs sampling* or *alternating conditional sampling*.

To illustrate MCMC sampling we consider the conditional distributions for the SAR model based on the NIG prior for the parameters β and σ^2 and a uniform prior for ρ. Beginning with the joint posterior for the model parameters $p(\beta, \sigma^2, \rho|\mathcal{D})$ from (5.16), we can find the conditional distributions for each of the parameters by considering expression (5.16) while treating the other parameters as known. For example, when considering the form taken by the conditional distribution for the parameters β, we treat the remaining parameters σ^2 and ρ as if they were known. We note that for the case where ρ is known, the conjugate NIG prior for β and σ^2 leads to a joint NIG (conditional on ρ) distribution for β and σ^2. Of course, the joint $NIG(c^*, T^*, a^*, b^*)$ leads to a conditional distribution for β that is a k-dimensional normal distribution, $N(c^*, T^*)$, and an $IG(a^*, b^*)$ conditional distribution for σ^2.

The remaining conditional distribution we require is that for the parameter ρ. For now, we will ignore the parameter ρ in this discussion, assuming it is fixed and known. We will discuss the conditional distribution for this parameter later.

This leaves us with only two sets of parameters β, σ to estimate. Let our parameter vector $\theta = (\beta_{(0)}, \sigma_{(0)})$, where the subscript zero indicates arbitrary initial values for the two sets of parameters. Given the initial value for $\sigma_{(0)}$ (and knowledge of ρ) we can calculate the mean, c^* and variance-covariance, T^*, for the multivariate normal conditional distribution of β using the expressions in (5.24). Note that we employ the value $\sigma_{(0)}^2$, for the parameter σ^2 in expression (5.24).

$$p(\beta|\rho, \sigma_{(0)}^2) \sim N(c^*, \sigma_{(0)}^2 T^*) \tag{5.24}$$
$$c^* = (X'X + T^{-1})^{-1}(X'Ay + T^{-1}c)$$
$$T^* = (X'X + T^{-1})^{-1}$$
$$A = I_n - \rho W$$

Given an algorithm that produces a vector of multivariate normal random deviates with the mean and variance-covariance shown in (5.24), we can replace the initial $\beta_{(0)}$ with the sampled values that we label $\beta_{(1)}$.

Alternating to the inverse gamma conditional distribution for σ^2 shown in (5.25), we use an algorithm to produce a random deviate from the $IG(a^*, b^*)$ distribution to update the parameter $\sigma_{(0)}^2$ and label it $\sigma_{(1)}^2$.

$$p(\sigma^2|\beta_{(1)}, \rho) \sim IG(a^*, b^*) \tag{5.25}$$
$$a^* = a + n/2$$
$$b^* = b + (Ay - X\beta_{(1)})'(Ay - X\beta_{(1)})/2$$
$$A = I_n - \rho W$$

Note that we used the "updated" value $\beta_{(1)}$ when producing the sample draw from the conditional distribution for σ^2 with which we update our parameter $\sigma_{(0)}^2$ to $\sigma_{(1)}^2$.

At this point, we return to the conditional distribution for β and produce an update $\beta_{(2)}$, based on using the updated $\sigma_{(1)}^2$ draw. This process of *alternating* sampling from the the two conditional distributions is continued until a large sample of "draws" for the parameters β and σ have been collected. This is not an ad hoc procedure as formal mathematical demonstrations have been provided by Geman and Geman (1984) and Gelfand and Smith (1990) that the stochastic process θ^t, representing our parameters is a Markov chain with the correct equilibrium distribution. Gibbs sampling is in fact a special case of the Hastings and Metropolis methods introduced earlier. An implication of this is that the drawn samples of parameters taken from the alternating sequential sampling of the complete sequence of *conditional* distributions for all parameters in the model represent samples from the *joint posterior* of the model parameters. Recall, this is the basis for all inference in Bayesian analysis.

Having a large sample of parameters from the posterior distribution allows us to proceed with inference regarding the model parameters β and σ, which would be based on statistics such as the mean and standard deviation computed from these sampled parameter draws. In fact, given a large enough sample of parameters, we could use *kernel density estimation* procedures to construct the entire posterior distribution of the parameters, not simply the mean and standard deviation point estimates.

5.3.2   Sampling for the parameter ρ

To this point, we have assumed unrealistically that the parameter ρ from our model is known. To complete our scheme for MCMC estimation of the SAR model we need to sample the parameter ρ from its conditional distribution. This takes the form:

$$p(\rho|\beta,\sigma) \propto \frac{p(\rho,\beta,\sigma|\mathcal{D})}{p(\beta,\sigma|\mathcal{D})}$$

$$\propto |I_n - \rho W| \exp\left(-\frac{1}{2\sigma^2}(Ay - X\beta)'(Ay - X\beta)\right) \quad (5.26)$$

This conditional distribution does not take a known form as in the case of the conditionals for the parameters β and σ where we had normal and inverse gamma distributions. Sampling for the parameter ρ must proceed using an alternative approach, such as Metropolis-Hastings. We will combine Metropolis-Hastings (M-H) sampling for the parameter ρ in our model and Gibbs sampling from the normal and inverse gamma distributions for the parameters β and σ to produce MCMC estimates for the SAR model (LeSage, 1997). This type of procedure is often labeled *Metropolis within Gibbs sampling*.

For (M-H) sampling we require a *proposal distribution* from which we generate a candidate value for the parameter ρ, which we label ρ^*. This candidate value as well as the current value that we label ρ^c are evaluated in expression (5.26) to calculate an *acceptance probability* using (5.27).

$$\psi_H(\rho^c, \rho^*) = \min\left[1, \frac{p(\rho^*|\beta,\sigma)}{p(\rho^c|\beta,\sigma)}\right] \quad (5.27)$$

We use a normal distribution as the proposal distribution along with a *tuned random-walk procedure* suggested by Holloway, Shankara, and Rahman (2002) to produce the candidate values for ρ. The procedure involves use of the current value ρ^c, a random deviate drawn from a *standard normal distribution*, and a *tuning parameter c* as shown in (5.28).

$$\rho^* = \rho^c + c \cdot N(0,1) \qquad (5.28)$$

Expression (5.28) should make it clear why this type of proposal generating procedure is labeled a random-walk procedure. The goal of *tuning* the proposals coming from the normal proposal distribution is to ensure that the M-H sampling procedure *moves* over the entire conditional distribution. We would like the proposal to produce draws from the dense part of this distribution and avoid a situation where the sampler is stuck in a very low density part of the conditional distribution where the density or support is low.

To achieve this goal, the tuning parameter c in (5.28) is adjusted based on monitoring the acceptance rates from the M-H procedure during the MCMC drawing procedure. Specifically, if the acceptance rate falls below 40%, we adjust $c' = c/1.1$, which decreases the variance of the normal random deviates produced by the proposal distribution, so that new proposals are more closely related to the current value ρ^c. This should lead to an increased acceptance rate. If the acceptance rate rises above 60%, we adjust $c' = (1.1)c$, which increases the variance of the normal random deviates so that new proposals range more widely over the domain of the parameter ρ. This should result in a lower acceptance rate. The goal is to achieve a situation where the tuning parameter settles to a fixed value resulting in an acceptance rate between 40 and 60 percent. At this point, no further adjustments to the tuning parameter take place and we continue to sample from the normal proposal distribution using the resulting tuned value of c.

There is a need to resort to tuning the proposal distribution because small sample sizes result in the parameter ρ having a conditional distribution that exhibits a wide dispersion, whereas large sample sizes usually produce a small dispersion since this parameter is estimated quite precisely in these circumstances. This means that a single setting for the tuning parameter will not work well in all circumstances, whereas this adaptive feedback tuning procedure will accommodate different samples arising from varying estimation problems. Figure 5.1 shows a plot of the acceptance rates along with the M-H draws for the parameter ρ associated with the first 250 draws to illustrate these issues. From the top half of the figure showing the monitored acceptance rates, we see that numerous adjustments to the tuning parameter take place during the first 50 passes through the MCMC sampling procedure. These adjustments stop after the first 100 draws and the M-H sampling procedure produces a relatively steady acceptance rate just under 50 percent.

The movement of the M-H sampler for the parameter ρ can be seen in the bottom half of the figure. We see sequences of draws for which the M-H procedure continues to reject candidate values in favor of keeping the current value. Both the acceptance rate as well as the sequence of draws can be examined as a way of detecting problems with the MCMC sampler. For example, a long sequence of rejections (a flat line in the plot of draws) would be indicative of the sampler getting stuck. In a wide range of applied situations

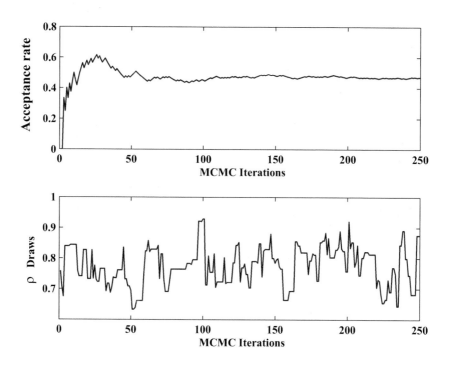

FIGURE 5.1:   Metropolis-Hastings acceptance rates and draws for ρ

involving samples ranging from 50 to over 100,000 observations this tuning procedure has never encountered problems for the standard family of spatial regression models from Chapter 2.

One consequence of the acceptance rate between 40 and 60 percent from M-H sampling with this tuning scheme is that we only collect draws for the parameter ρ half of the time. This may require that we carry out more passes through the MCMC sampler to collect a large enough sample of information from which to construct an accurate posterior distribution for the parameters in the model.

A more efficient alternative to the Metropolis-Hastings approach to obtaining samples for the parameter ρ is to rely on univariate numerical integration to obtain a normalizing constant and then construct a cumulative density function (CDF) for the conditional posterior distribution for the parameter ρ. Given this CDF, we can produce a draw from the conditional posterior distribution using *inversion*. This approach was introduced by Smith and LeSage (2004).

In our discussion of numerical integration over the parameter ρ, we noted the ability to express the posterior distribution $p(\rho|\mathcal{D})$ that resulted after analytical integration of the parameters β and σ as a vector using a grid of q values of ρ. We can exploit a similar expression that arises for the case of informative priors in a simple trapezoid rule integration scheme to rapidly calculate the normalizing constant and the associated CDF needed to generate draws using the inversion approach of Smith and LeSage (2004).

To illustrate this approach, we use Figure 5.2, that shows a cumulative conditional distribution function created using univariate numerical integration. For improved scaling of the figure, the domain of the spatial dependence parameter ρ was restricted to $(0, 1)$, and the univariate numerical integration procedure described in section 5.2.2 was based on a grid of $q = 2000$ values for ρ and the vectorization scheme from Pace and Barry (1997). The process of *drawing by inversion* involves a *uniform random deviate* drawn from the domain of support for ρ, which was restricted to $(0, 1)$ for this illustration. This random value is then evaluated using the numerically constructed conditional distribution function to produce a draw for ρ. The figure shows one such draw based on a single uniform deviate, which was collected on one pass through the MCMC sampling loop.

Despite the fact that this may seem complicated and time-consuming, it is not. For example, a sample of 2,500 draws for all three sets of parameters β, σ and ρ for the SAR model can be produced in 6.5 seconds on a laptop computer for a model containing 3,107 US county-level observations and five explanatory variables. This is faster than M-H sampling to produce the same number of draws. An advantage of this approach over Metropolis-Hastings sampling for the parameter ρ is that every pass through the MCMC sampling loop produces an *effective draw* for the parameter ρ. In the case of Metropolis-Hastings, given a rejection rate tuned to between 40 and 60 percent we would require around twice as many MCMC draws to produce the same effective sample of draws for the parameter ρ.

5.4   The MCMC algorithm

As noted in Chapter 2, we can use the approaches of either Pace and Barry (1997) or Barry and Pace (1999) to calculate a vector based on a grid of q values for ρ in the interval $(-1, 1)$ representing the log-determinant expression $(\ln |I_n - \rho W|)$ over this grid. Since this term arises in both the (log) conditional distribution for ρ needed for M-H sampling as well as the expression required for integration and draws via inversion, these computational innovations designed to assist in maximum likelihood estimation help here as well. We use a grid of 2,000 values and calculate this vector only once prior to beginning the MCMC sampling loop.

Another point to note is that we can impose the restriction that $-1 < \rho < 1$,

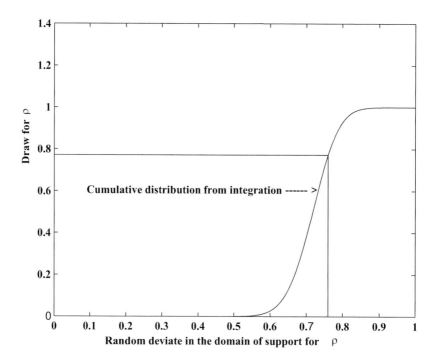

FIGURE 5.2: Draw by inversion using numerical integration to produce an empirical CDF

or any other desired interval using *rejection sampling* for the case of M-H sampling of the parameter ρ. This involves simply rejecting values of ρ outside the desired interval drawn from the proposal distribution and drawing another proposal. A formal measure of the posterior probability that the parameter ρ lies in the interval can be derived using a count of the proportion of candidate values that are rejected (Gelfand et al., 1990).

By way of summary, we formally present the MCMC sampler for the SAR model, with $NIG(c, T, a, b)$ priors for β and σ^2. Beginning with arbitrary values for the parameters $\beta_{(0)}, \sigma^2_{(0)}, \rho_{(0)}$, we sample sequentially from the following three conditional distributions.

1. Sample $p(\beta|\sigma^2_{(0)}, \rho_{(0)})$ using the $N(c^*, \sigma_{(0)}T^*)$ distribution. with a mean and variance calculated from:

$$c^* = (X'X + T^{-1})^{-1}(X'(I_n - \rho_{(0)}W)y + T^{-1}c)$$
$$T^* = (X'X + T^{-1})^{-1} \tag{5.29}$$

Label the sampled parameter vector $\beta_{(1)}$ and use this to replace the parameter vector $\beta_{(0)}$.

2. Sample $p(\sigma^2|\beta_{(1)}, \rho_{(0)})$, using an inverse gamma distribution $IG(a^*, b^*)$.

$$p(\sigma^2|\beta^{(1)}, \rho) \sim IG(a^*, b^*) \tag{5.30}$$
$$a^* = a + n/2$$
$$b^* = b + (Ay - X\beta_{(1)})'((Ay - X\beta_{(1)}))/2$$
$$A = I_n - \rho_{(0)}W$$

3. Sample $p(\rho|\beta_{(1)}, \sigma^2_{(1)})$, using either the M-H algorithm or integration and draw by inversion approach set forth in Section 5.3.2. Label this updated value $\rho_{(1)}$ and return to step 1.

One sequence of steps 1 to 3 constitute a single pass through the sampler. We carry out a large number of passes and after some initial *burn-in period* we collect the draws for the parameters from each pass. For example, we might carry out 7,500 draws excluding the first 2,500 and use the resulting sample to produce posterior estimates and inferences.

The first 2,500 are excluded to account for *start-up, or burn-in* of the sampler. We need to be confident that the MCMC sampling procedure has reached the *steady state* or equilibrium distribution motivated by Hastings (1970). In practice, one can produce samples from a short run of 2,500 draws with the first 500 excluded for burn-in and compare the means and standard deviations of the parameters from this run to those obtained from a longer run based on different starting values for the parameters. If the estimates and inferences are equivalent, then there are no likely problems with convergence of the MCMC sampler to a steady state. Once the sampler achieves a steady-state, we interpret the draws as coming from the posterior distribution. LeSage (1999) discusses a number of alternative statistical tests that can be applied to the sampled draws as a diagnostic check for convergence.

In the Bayesian MCMC literature, a great deal of attention is devoted to issues regarding convergence of samplers. However, the simple spatial regression models considered here do not encounter problems in this regard. This is not to say that attention should not be paid to issues of scaling transformations applied to variables and possible collinearity problems between explanatory variables in the model. However, these problems would likely exert an adverse impact on maximum likelihood estimates as well, especially on inferences based on variances calculated using a numerical estimate of the Hessian.

5.5   An applied illustration

We provide an illustration of Bayesian estimation and inference using a data sample from Pace and Barry (1997) containing voter turnout rates in 3,107 US counties during the 1980 presidential election. We use the (logged) proportion of voting age population that voted in the election as the dependent variable y, and measures of education, home ownership and income as explanatory variables, along with a constant term. The education and home ownership variables were expressed as (logged) population of voting age with high school degrees and (logged) population owning homes, and the median household income variable was also logged. Since all variables are logged, the coefficient estimates have an *elasticity* interpretation.

We wish to demonstrate that the Bayesian MCMC sampling procedures will produce nearly identical estimates and inferences as maximum likelihood methods when uninformative priors are assigned to the parameters β and σ. In this application we rely on a Beta prior distribution for ρ that we label $\mathcal{B}(d, d)$ introduced by LeSage and Parent (2007). This distribution is shown in (5.31) where $Beta(d, d), d > 0$ represents the Beta function, $Beta(d, d) = \int_0^1 t^{d-1}(1 - t)^{d-1}dt$. This prior distribution takes the form of a relatively uniform distribution centered on a mean value of zero for the parameter ρ. This represents an alternative to the uniform prior on the interval $(-1, 1)$.

$$\pi(\rho) \sim \frac{1}{Beta(d, d)} \frac{(1 + \rho)^{d-1}(1 - \rho)^{d-1}}{2^{2d-1}} \qquad (5.31)$$

Figure 5.3 depicts prior distributions associated with prior values $d = 1.01, 1.1$ and 2, for the $\mathcal{B}(d, d)$ prior. From the figure, we see that values of d near unity produce a relatively uninformative prior that places zero prior weight on end points of the interval for ρ, consistent with theoretical restrictions. In the figure, we use the interval $(-1, 1)$ for the parameter ρ which should incorporate the effective domain of support for the posterior distribution in most applied work, so this works well as a prior.

The prior mean of the multivariate normal distribution assigned to the parameters β was zero and a diagonal prior variance-covariance structure based on a scalar variance of $1e + 12$ was used. This creates an uninformative prior distribution that is centered on zero, but whose variance is extremely large resulting in a nearly uniform prior. Finally, the prior distribution assigned for the parameter σ^2 was based on an inverse gamma distribution, $IG(a, b)$, with the parameters $a = b = 0$, which results in a diffuse or non-informative prior for this parameter.

In addition to demonstrating equivalent estimates and inferences from maximum likelihood and Bayesian procedures when using relatively uninformative priors, we would also like to illustrate that equivalent posterior distributions

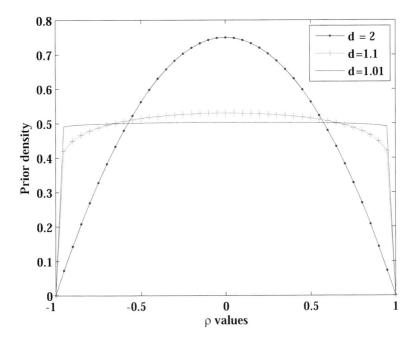

FIGURE 5.3:   The Beta(d, d) prior distribution for $d = 1.01, 1.1, 2.0$

for the parameter ρ will arise from the Metropolis-Hastings or *draw by inversion* schemes for sampling the parameter ρ.

Two posterior densities for the parameter ρ were constructed using a kernel density estimation routine applied to a sample of 5,000 draws that were retained from a run of 7,500 MCMC sampling draws, with the first 2,500 discarded for burn-in.

The kernel density estimates of the posterior distributions for the parameter ρ based on the two sampling procedures are presented in Figure 5.4. We see close agreement in the two resulting distributions.

Maximum likelihood estimates for this model and sample data are presented in Table 5.1 alongside Bayesian estimates based on both sampling schemes for the parameter ρ. To improve the accuracy of the t-statistics associated with the maximum likelihood estimates, these were based on variances calculated from the analytical information matrix rather than a numerical Hessian procedure. Contrary to Bayesian convention, we present calculated t-statistics using the posterior mean and standard deviation of the sampled MCMC draws for the parameters. This provides an easier comparison of the Bayesian estimation results with those from maximum likelihood estimation.

We note that conventional MCMC practice is to report 0.95 *credible inter-*

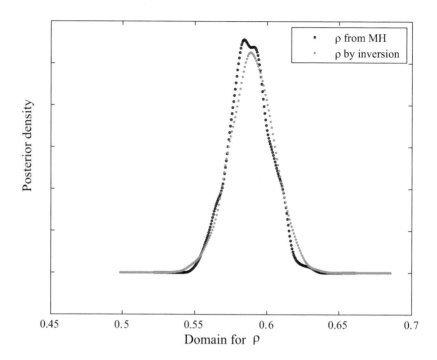

FIGURE 5.4: Kernel density estimates of the posterior distribution for ρ

vals constructed using the sample of draws from the MCMC sampler along with posterior means and standard deviations calculated from the sample of draws. This simply involves sorting the sampled draws from low to high and finding lower and upper 0.95 points. For example, given a vector of 10,000 sorted draws, we would use the 5,000-(9,500/2) and 5,000+(9,500/2) elements of this vector as the lower and upper 0.95 credible intervals. Inferences based on these should correspond to a 95% level of confidence from maximum likelihood.

From the table, we see that all estimates are nearly identical as are the ratios of the mean to standard deviation, suggesting they would produce similar inferences.

Summarizing our developments to this point, we have established that MCMC estimation can reproduce maximum likelihood estimates when we rely on uninformative priors. It was also noted that for large samples, it is unlikely that use of an informative prior will exert much impact on the posterior estimates and inferences, a conventional result concerning Bayesian versus maximum likelihood methods. The time required to produce the maximum likelihood and Bayesian estimates in our applied illustration were: 1/2 sec-

TABLE 5.1: Comparison of SAR model estimates

Variable	Max Likelihood $\hat{\beta}$	t-stat†	Bayesian Inversion mean $\hat{\beta}$	t-stat‡	Bayesian M-H mean $\hat{\beta}$	t-stat‡
constant	0.6265	15.59	0.6266	15.17	0.6243	14.91
education	0.2206	16.71	0.2202	14.18	0.2190	13.87
homeowners	0.4818	33.47	0.4816	33.28	0.4816	33.07
income	−0.0992	−6.35	−0.0993	−6.07	−0.0982	−5.94
$\hat{\rho}$	0.5869	41.92	0.5878	41.57	0.5892	40.87
$\hat{\sigma}$	0.0138		0.0138		0.0138	

† interpreted as an asymptotic t-statistic
‡ based on mean(draws) / std deviation(draws)

ond for maximum likelihood, 6.5 seconds for MCMC using draws by inversion and 12 seconds for M-H sampling. This suggests that the slower Bayesian estimation approach is not computationally competitive with maximum likelihood if our desire were to produce the same estimates and inferences. In the next section we provide illustrations of simple extensions of the Bayesian spatial regression models that hold advantages over conventional maximum likelihood methods that justify the increased computational time required to produce these estimates.

5.6 Uses for Bayesian spatial models

In this section we provide three uses for Bayesian MCMC estimation that can produce elegant and formal solutions to problems that arise in spatial regression modeling. One of these problems is that of heteroscedasticity and outliers that frequently arise in spatial data samples. In Section 5.6.1, we draw on work by Geweke (1993) to produce a heteroscedastic/robust variant of the spatial regression. This model subsumes the conventional spatial regression models that assume homoscedastic disturbances as a special case, and it is fast and simple to implement using MCMC methods. Section 5.6.2 shows how MCMC methods can be used to produce valid estimates and inferences regarding the *total, direct and indirect impacts* that are used to interpret the effect of changes in the explanatory variables on the dependent variable. A final example for use of MCMC methods is discussed in Section 5.6.3, where spatial regression models involving more than a single spatial weight matrix are discussed. These models require constrained multivariate optimization routines in a likelihood setting to produce estimates and inferences regarding the multiple spatial dependence parameters. In contrast, MCMC reliance on conditional distributions considerably simplifies estimation of these models.

5.6.1   Robust heteroscedastic spatial regression

One concrete illustration of the extensible nature of the MCMC estimation method is to extend our simple SAR model to include variance scalars that can accommodate heteroscedastic disturbances and/or outliers.

This type of prior information was introduced by Albert and Chib (1993) for the ordinary probit model and Geweke (1993) for a least-squares model. The prior pertains to assumed homoscedastic versus heteroscedastic disturbances. A set of variance scalars $(v_1, v_2, \ldots, v_n)$, is introduced that represent unknown parameters that need to be estimated. This allows us to assume $\varepsilon \sim N(0, \sigma^2 V)$, where V is a *diagonal* matrix containing parameters $(v_1, v_2, \ldots, v_n)$. The prior distribution for the v_i terms takes the form of a set of n *iid* $\chi^2(r)/r$ distributions, where r represents the single parameter of the χ^2 distribution. This allows us to estimate the additional n variance scaling parameters v_i by adding only a single parameter r, to our model. Use of a flexible family of distributions that is controlled by a single parameter such as r to specify a prior distribution is a common Bayesian approach. The parameter r that controls this family of prior distributions is labeled a *hyperparameter*. The notion here is that changes in this single parameter can potentially exert a great deal of influence on the nature of the prior distribution assigned to the model parameters it controls.

The specifics regarding the prior assigned to the v_i variance scaling parameters can be motivated by noting that a prior mean of unity will be assigned and a prior variance equal to $2/r$. This implies that as the *hyperparameter* r is assigned very large values, the prior variance becomes very small leading the variance scaling parameters v_i to approach their prior mean values of unity. This results in a prior specification that: $V = I_n$, the traditional assumption of constant variance across our observations or regions/points located in space. On the other hand, a small value assigned to the *hyperparameter* r will lead to a skewed prior distribution assigned to the variance scalar parameters v_i. The large prior variance leads to skew in the χ^2 prior distributions assigned to each variance scalar which will allow the estimates and posterior means for these parameters to deviate greatly from their prior mean values of unity.

Large values for the variance scalars v_i are associated with outliers or observations containing large variances. These observations will be down-weighted as in the case of generalized least-squares where large variances result in less weight assigned to an observation. In the context of spatial modeling, outliers or aberrant observations can arise due to *enclave effects*, where a particular area exhibits divergent behavior from nearby areas. As an example, we might see different crime rates in a "gated community" than in surrounding neighborhoods. Geweke (1993) shows that this approach to modeling the disturbances is equivalent to a model that assumes a Student-t prior distribution for the errors. We note that this type of distribution has frequently been used to deal with sample data containing outliers (Lange, Little and Taylor, 1989).

A formal statement of the Bayesian heteroscedastic SAR model is shown in (5.32), where we have added an *independent* normal and inverse-gamma prior for β and σ^2, and a uniform prior for ρ. This represents a departure from our previous use of the NIG prior for the parameters β and σ^2. This type of prior is generally considered more flexible, but does not have the advantage of being conjugate. Since we are relying on MCMC estimation, use of a conjugate prior is no longer important. These priors are in addition to the chi-squared prior for variance scalars, but as noted they are unlikely to exert much impact on the resulting estimates and inferences in large samples. As before, the prior distributions are indicated using $\pi()$.

$$y = \rho W y + X \beta + \varepsilon$$
$$\varepsilon \sim N(0, \sigma^2 V)$$
$$V_{ii} = v_i, i = 1, \ldots, n, \ V_{ij} = 0, \ i \neq j$$
$$\pi(\beta) \sim N(c, T)$$
$$\pi(r/v_i) \sim iid \ \chi^2(r), i = 1, \ldots, n$$
$$\pi(\sigma^2) \sim IG(a, b)$$
$$\pi(\rho) \sim U(1/\lambda_{\min}, 1/\lambda_{\max}) \tag{5.32}$$

We need the conditional posterior distributions for the parameters β, σ, and ρ as well as the variance scalars $v_i, i = 1, \ldots, n$ in this model to implement our MCMC sampling scheme. The conditional distribution for β takes the form of a multivariate normal shown in (5.33), which is a simple GLS variant of our previous expression, where the variance is known. This arises because we can condition on all other parameters in the model, including the diagonal matrix of variance scalars V.

$$p(\beta | \rho, \sigma, V) \propto N(c^*, T^*) \tag{5.33}$$
$$c^* = (X'V^{-1}X + \sigma^2 T^{-1})^{-1}(X'V^{-1}(I_n - \rho W)y + \sigma^2 T^{-1}c)$$
$$T^* = \sigma^2(X'V^{-1}X + \sigma^2 T^{-1})^{-1}$$

This illustrates an attractive feature of the MCMC method. Working with conditional posterior distributions greatly simplifies the calculations required to extend a basic model. If we have already developed computational code to implement a simpler homoscedastic model where $V = I_n$, the modifications required to implement this model are minor.

The expression needed to produce a draw from the conditional posterior distribution of σ^2 takes the form in (5.34). Using the relationship noted earlier involving the residuals, we see that we have a type of GLS expression, where the non-constant variance reflected by the diagonal matrix V can be assumed known.

$$p(\sigma^2|\beta,\rho,V) \propto IG(a^*,b^*) \qquad (5.34)$$
$$a^* = a + n/2$$
$$b^* = (2b + e'V^{-1}e)/2$$
$$e = Ay - X\beta$$
$$A = I_n - \rho W$$

The expression needed to produce draws for the parameter ρ takes the unknown distributional form shown in (5.35) (LeSage, 1997). This means we can still rely on either our numerical integration followed by a draw via inversion or the M-H approach.

$$p(\rho|\beta,\sigma^2,V) \propto |A| \exp\left(-\frac{1}{2\sigma^2}e'V^{-1}e\right) \qquad (5.35)$$

Geweke (1993) shows that the conditional distribution of V given the other parameters is proportional to a chi-square density with $r+1$ degrees of freedom. Specifically, we can express the conditional posterior of each v_i as in (5.36), where $v_{-i} = (v_1,\ldots,v_{i-1},v_{i+1},\ldots,v_n)$ for each i. That is, we sample each variance scalar conditional on all others. The term e_i represents the ith element of the vector $e = Ay - X\beta$.

$$p(\frac{e_i^2 + r}{v_i}|\beta,\rho,\sigma^2,v_{-i}) \propto \chi^2(r+1) \qquad (5.36)$$

We summarize by noting that estimation of this extended model requires adding a single conditional distribution for the new variance scalar parameters v_i introduced in the model to our MCMC sampling scheme. In addition, we made minor adjustments to the conditional posterior distributions for the other parameters in the model to reflect the presence of these new parameters. These adjustments would require only minor changes to any computational code already developed for the simpler model from Section 5.4.

A point to note is that introducing heteroscedastic disturbances in the context of a model estimated using maximum likelihood was proposed by Anselin (1988). However, that approach requires the modeler to specify a functional form as well as variables thought to model the non-constant variance over space. The approach introduced here does not require this additional model for the non-constant variance. In addition to automatically detecting and adjusting for non-constant variances, the MCMC method will also detect and automatically down-weight outliers or aberrant observations.

From a practitioner's viewpoint, it would seem prudent to use this method with a prior hyperparameter setting of $r=4$. This prior is consistent with a prior belief in heteroscedasticity, or non-constant variance as well as outliers. If the sample data does not contain these problems, the resulting posterior estimates for the variance scalar parameters v_i will take values near unity. A plot of the posterior mean of the variance scalar parameters can serve as a diagnostic for outliers or heteroscedasticity. A map of the posterior mean v_i values can be used to locate regions/observations with high and low variance as well as outliers.

It makes little sense to use this model with a large prior value assigned to the hyperparameter r. This reflects a prior belief in homoscedasticity. If a practitioner is confident regarding homoscedastic disturbances, then maximum likelihood estimates are much faster and easier to produce.

5.6.2 Spatial effects estimates

Gelfand et al. (1990) point out that the draws from MCMC sampling can be used to produce posterior distributions for functions of the parameters that are of interest. This makes testing complicated parameter relationships quite simple. For example, suppose we are interested in the hypothesis: $\gamma = \alpha \cdot \beta < 1$, where α, β are parameters in a model estimated with MCMC sampling. We can simply multiply the m draws for $\alpha_{(j)}$ and $\beta_{(j)}$, $j = 1, \ldots, m$, to produce $\gamma_{(j)}, j = 1, \ldots, m$. The posterior distribution of $\gamma = \alpha \cdot \beta$ can be used to find the posterior probability that $\gamma < 1$. This would be equal to the proportion of all draws in the vector γ that take values less than unity. If we find that 9,750 draws from a sample of 10,000 are less than unity, then the probability is 97.5%. Of course means, modes and standard deviations could also be constructed using the draws for γ.

If we are interested in conducting inference regarding the summary measures of the cumulative *total, direct* and *indirect impacts* associated with changes in the explanatory variables described in Chapter 2, we can construct these during the MCMC sampling process. On each pass through the sampler, we can use the current set of draws to produce a total and direct impact, as well as the indirect impact by subtracting the direct from the total effect. Saving these draws allows us to use these to construct the entire posterior distribution for the three types of impacts that arise from changing the explanatory variables X in the model.

In Chapter 2 we established the notion that each explanatory variable r has a multiplier impact on y that could be expressed as: $y = \sum_{r=1}^{p} S_r(W)x_r + \ldots$, where the multiplier term $S_r(W)$ takes different forms for the various members of the family of spatial models. For example, for the SDM model: $y = \rho W y + X\beta + W X\theta + \varepsilon$, we have $S_r(W) = (I_n - \rho W)^{-1}(I_n \beta_r + W\theta_r)$.

It should be easy to see that the sampled parameters β, θ, ρ could be directly entered into $S_r(W)$ on each pass through the MCMC sampling loop. This could be used to produce MCMC samples of the summary measures of the (average) cumulative direct impacts shown in (5.37), cumulative total impacts in (5.38) and cumulative indirect impacts in (5.39).

$$\text{direct: } \bar{M}_r(D) = n^{-1} tr(S_r(W)) \tag{5.37}$$

$$\text{total: } \bar{M}_r(T) = n^{-1} \iota'_n S_r(W) \iota_n \tag{5.38}$$

$$\text{indirect: } \bar{M}_r(I) = \bar{M}_r(T) - \bar{M}_r(D) \tag{5.39}$$

However, a more computationally astute approach would be to use the efficient trace computations set forth in Chapter 4 in conjunction with the MCMC draws. Use of the MCMC draws for problems involving small samples might produce a more accurate posterior parameter distribution than would arise from using maximum likelihood estimates to simulate from a multivariate normal distribution. This could occur because in small samples parameters may exhibit asymmetry or heavy tailed distributions that deviate slightly from normality.

5.6.3   Models with multiple weight matrices

We discuss how MCMC sampling can be used to estimate models such as the SAC from Chapter 2, that include more than a single spatial weight matrix. For models that contain numerous weight matrices and associated spatial dependence parameters, MCMC sampling from the conditional distributions leads to an important simplification.

The SAC model takes the form in (5.40) with the associated likelihood concentrated for the parameters β, σ^2 shown in (5.41).

$$
\begin{aligned}
y &= \rho W y + X\beta + u \\
u &= \lambda M u + \varepsilon \\
\varepsilon &\sim N(0, \sigma^2 I_n)
\end{aligned}
\tag{5.40}
$$

$$p(y|\beta,\sigma,\rho,\lambda) \propto |A||B| \exp\left(-\frac{1}{2\sigma^2}(BAy - BX\beta)'(BAy - BX\beta) \right) \tag{5.41}$$

$$A = I_n - \rho W$$

$$B = I_n - \lambda M$$

Maximizing the log of the likelihood function in (5.41) requires that we calculate two log-determinants $|A|$ and $|B|$, and the optimization problem involves solving a two-dimensional constrained optimization problem. The constraints are imposed to bound the spatial dependence parameters to their respective ranges based on the minimum and maximum eigenvalues of the spatial weight matrices W and M. This is not an extremely difficult optimization problem given current computational hardware and software, along with the efficient methods for computing log-determinants of sparse matrices described in Chapter 4.

The advantage of MCMC sampling the model from (5.40) is that the conditional distributions for ρ and λ take the forms shown in (5.42) and (5.43).

$$p(\rho|\beta,\sigma,\lambda) \propto \frac{p(\rho,\lambda,\beta,\sigma|y)}{p(\lambda,\beta,\sigma|y)} \tag{5.42}$$

$$\propto |A||B(\lambda^c)|\exp\left(-\frac{1}{2\sigma^2}(\tilde{B}Ay - \tilde{B}X\beta)'(\tilde{B}Ay - \tilde{B}X\beta)\right)$$

$$p(\lambda|\beta,\sigma,\rho) \propto \frac{p(\rho,\lambda,\beta,\sigma|y)}{p(\rho,\beta,\sigma|y)} \tag{5.43}$$

$$\propto |A(\rho^c)||B|\exp\left(-\frac{1}{2\sigma^2}(B\tilde{A}y - BX\beta)'(B\tilde{A}y - BX\beta)\right)$$

We note that when sampling for the parameter ρ, we rely on the current value/draw for λ in $|B|$, which we denote $|B(\lambda^c)|$, and $B(\lambda^c) = \tilde{B}$. Similarly, when sampling for the parameter λ we use the current value of ρ in $|A|$ and A, with similar notation. An implication of this is that we could still carry out our univariate numerical integration scheme to find a normalizing constant and produce a CDF from which to draw by inversion. Similarly, carrying out Metropolis-Hastings sampling for the parameter ρ is no more complicated than in the case of a model where the additional spatial dependence parameter λ does not exist. We can produce candidate values for the parameter ρ from a normal random-walk proposal density using the same procedure as described in section 5.3.2.

One can envision richer models that increase the number of spatial weight matrices. For example, a model similar to the model in (5.44) can be found in Lacombe (2004), where a sample of counties on borders of states was used to carry out an analysis of the state-level impact of public policy differences between states. The spatial weight matrix W was used to extract neighboring counties across the border in another state, whereas the weight matrix V was used to include dependence on neighboring counties within the same state.

$$y = \rho W y + \gamma V y + X\beta + u$$
$$u = \lambda M u + \varepsilon \tag{5.44}$$
$$\varepsilon \sim N(0, \sigma^2 I_n)$$

The likelihood for this model shown in (5.45) involves the log-determinant terms: $|A| = |I_n - \rho W - \gamma V|$ and $|I_n - \lambda M|$. Maximum likelihood estimation involves three-dimensional optimization of the concentrated likelihood with respect to the parameters β and σ^2 to produce estimates for the parameters ρ, γ and λ. This requires computing the log-determinant term: $|I_n - \rho W - \gamma V|$, over a two-dimensional grid of values for the parameters ρ and γ (see Chapter 4). There is also the log-determinant term $|B| = |I_n - \lambda M|$ defined in (5.45) that arises from the spatial autoregressive disturbance process, as well as the stability constraint $\rho + \gamma < 1$, which must be applied when solving the optimization problem.

$$p(y|\beta, \sigma, \rho) = (2\pi\sigma^2)^{-\frac{n}{2}} |A| |B| \exp\left(-\frac{1}{2\sigma^2}(BAy - BX\beta)'(BAy - BX\beta)\right)$$

$$A = |I_n - \rho W - \gamma V|$$
$$B = |I_n - \lambda M| \qquad\qquad (5.45)$$

MCMC estimation of this model allows us to fix the log-determinant A based on the current values of the parameters ρ and γ when sampling for the parameter λ from its conditional distribution. Similarly, we can fix the log-determinant B when sampling for the parameters ρ and γ. The stability restriction can be imposed using rejection sampling. This simply involves rejecting Metropolis-Hastings candidate values that violate the stability restriction. A count of the number of times these rejections occur during the sampling draws provides a posterior probability measure for consistency of the sample data with the stability restriction.

5.7   Chapter summary

Application of traditional Bayesian methods to estimation of spatial autoregressive models requires simple univariate numerical integration of the posterior distribution with respect to the parameter ρ over a closed interval. In contrast, recent advances in the area of Markov Chain Monte Carlo (MCMC) estimation allow Bayesian estimation of spatial autoregressive models as well as a host of useful variants on these models without the need to resort to numerical integration. We demonstrated that Bayesian methods in conjunction with MCMC estimation allow the basic family of spatial autoregressive models to be implemented in the usual case where disturbances are normally distributed with constant variances.

The greatest value of Bayesian MCMC methods lies in their ability to extend the basic spatial regression models to accommodate situations where the sample data exhibit outliers or heteroscedasticity. These methods are also useful for generating spatial impact estimates which take the form of functions of the model parameters. These functions can be used to determine the impact of changes in the explanatory variables of the model on the dependent variable. Use of MCMC draws in the functions allows a simple approach to inference regarding the dispersion of the impacts. Finally, MCMC methods allow estimation of models involving more than a single weight matrix, without resort to multivariate constrained optimization routines that are required for maximum likelihood estimation of these models. We will see other examples of places where MCMC methods can be applied to spatial regression models in other chapters.

A final point regarding the Bayesian methodology is that inference proceeds identically for all Bayesian models implemented with MCMC methods. The entire posterior distribution is available for all parameters in the model allowing means, medians, or modes to be used as point estimates, and measures of dispersion are easily constructed. This is in contrast to cases where alternatives to maximum likelihood estimation such as generalized method of moments are adopted to solve difficult spatial econometric problems. Here, inference may require adopting an alternative paradigm, or reliance on asymptotic approximations whose statistical operational characteristics are not well-understood.

Chapter 6

Model Comparison

This chapter describes model comparison procedures that allow practitioners to draw inferences regarding various aspects of spatial econometric model specifications. We focus on comparison of: 1) spatial versus non-spatial models, 2) models based on alternative spatial weight structures, and 3) models constructed using different sets of candidate explanatory variables.

A variety of strategies and statistical methods for comparing alternative model specifications are introduced in Section 6.1, with an applied illustration of these ideas provided in Section 6.2. Section 6.3 turns attention to Bayesian approaches to model comparison which provide a unified approach to the various types of model comparison issues that confront practitioners. A series of applied illustrations for these methods are provided in each section.

6.1 Comparison of spatial and non-spatial models

When maximum likelihood estimation is used for spatial regression models, inference on the spatial dependence parameter ρ can be based on a Wald test constructed using an asymptotic t-test from the estimated variance-covariance matrix, or a likelihood ratio test. These tests for spatial dependence versus the null hypothesis of no dependence require maximum likelihood estimation of the spatial model representing the alternative to the null hypothesis of no spatial dependence.

Since maximum likelihood estimation was cumbersome in the past, there is a great deal of literature on Lagrange Multiplier test statistics that require only estimation of the non-spatial model associated with the null hypothesis. For example, Burridge (1980) proposed an LM test for least-squares against the alternative SEM model taking the form shown in (6.1), where only least-squares residuals denoted by e, and a spatial weight matrix W are needed. The LM statistic in (6.1) follows an asymptotic $\chi^2(1)$ distribution.

$$LM = [e'We/(e'e/n)]^2/tr(W^2 + W'W) \qquad (6.1)$$

This statistic is related to an I-statistic proposed by Moran (1948), which has received a great deal of attention in the literature. This statistic also involves only use of the residuals from least-squares. Anselin (1988b) proposed LM tests for least-squares versus the SAR model, where again the appeal of these tests was that they did not require maximum likelihood estimation of the spatial model.

Given the current availability of software that makes estimation of the family of spatial regression models relatively simple and computationally fast, it is easy to test for spatial dependence using inference on the spatial dependence parameter ρ. This can be based on a t-test (constructed using the estimated variance-covariance matrix) for the null hypothesis that $\rho = 0$, or based on a likelihood ratio test that compares the spatial and non-spatial models.

An example of this would be comparison of a non-spatial regression model, which is nested within the spatial SAR model. Comparison of the non-spatial and spatial SAR model likelihoods, or use of a t-statistic on the spatial dependence parameter would allow one to draw an inference regarding the significance of spatial dependence in the dependent variable. A complication arises because this test ignores possible spatial dependence in the disturbances, since it conditions on a model specification involving spatial dependence in the dependent variable. It is also the case that a comparison of an ordinary non-spatial regression versus SEM models would ignore spatial dependence in the dependent variable, focusing only on dependence in the disturbances of the model. Joint tests that have power against the other alternative have been proposed by Anselin (1988b), as well as tests that are reported to be robust to misspecification involving the alternative form of dependence by Anselin, Bera, Florax and Yoon (1996).

This same issue arises when attempting to ascertain the appropriate model specification based on a comparison of likelihood function values, since SAR models ignore error dependence and SEM models do not account for spatial dependence in the dependent variable. The non-nested nature of these models greatly complicates formal testing for both spatial dependence, as well as an appropriate model specification.

However, as indicated in Chapter 2, the SDM model nests models involving dependence in both the disturbances as well as the dependent variable. There is too much emphasis in the spatial econometrics literature on use of statistical testing procedures to infer the appropriate model specification, and much of this literature ignores the SDM model. We make a number of observations regarding the benefits and costs associated with alternative spatial regression model specifications.

The cost of ignoring spatial dependence in the dependent variable is relatively high since biased estimates will result if this type of dependence is ignored. In addition, ignoring this type of dependence will also lead to an inappropriate interpretation of the explanatory variable coefficients as representing partial derivative impacts arising from changes in the explanatory variables. In contrast, ignoring spatial dependence in the disturbances will lead to a loss of efficiency in the estimates. As samples become large, efficiency becomes less of a problem relative to bias. Spatial data availability has increased dramatically for a number of reasons, including: increasing awareness of the need to consider variation in socio-economic relationships over space; improvements in geographical information system software and accompanying computational advances in geo-referencing sample data using postal addresses; improved software for working with Census data sets; and US government requirements that agencies make data available on the Internet. Because of this, efficiency of estimates may be of less concern as we begin analyzing larger spatial data samples. There is still the problem that ignoring spatial dependence in the disturbances will lead to bias in the inferences regarding dispersion of the estimates.

This line of reasoning suggests an asymmetric loss function for practitioners interested in unbiased estimates, since the costs of ignoring spatial dependence in the dependent variable is more likely to produce biased estimates than ignoring dependence in the disturbances. Alternatively, the benefits from accounting for dependence in the disturbances are increased efficiency of the estimates, whereas those arising from proper modeling of dependence in the dependent variable are reductions in bias of the explanatory variable coefficients, as well as improved efficiency.

In Chapter 2 we demonstrated that the presence of omitted variables in the SEM model will lead to the true data generating process being that associated with the SDM model. That is, use of the SDM model will help protect against omitted variables bias. It was also shown that the SDM model nests both spatial lag and spatial error models.

Finally, we note that inclusion of variables such as WX in the SDM model when the true DGP is the SAR model that does not include these variables will not lead to biased estimates for the explanatory variable parameters. In situations where omitted variables lead to the presence of WX in the model relationship, use of the SAR model that excludes these variables leads to omitted variables bias in the coefficient estimates.

Putting these ideas together allows us to consider reciprocal misspecification bias that can arise in the coefficient estimates from a costs versus benefits perspective. We enumerate the implications for biased coefficient estimates in the SEM, SAR, SDM and SAC models from a reciprocal misspecification viewpoint below.[1]

1. For cases where the true DGP is the SEM model, involving only spatial dependence in the disturbances, the SAR, SAC and SDM models will still produce unbiased but inefficient coefficient estimates. Inference regarding dispersion of the explanatory variables based on the asymptotic variance-covariance matrix for the SAR model will be mislead-

---

[1] We focus on the unbiasedness property of the coefficient estimates in our reciprocal misspecification considerations. In the presence of biased coefficient estimates, we are less concerned about correct inferences regarding the dispersion of the biased coefficient estimates.

ing, since error dependence is ignored when constructing the variance-covariance matrix. Error dependence is taken into account in the asymptotic variance-covariance matrix of the SDM and SAC models.

2. When the true DGP is the SAR model that includes spatial lag dependence, the SEM model would produce biased coefficient estimates, the SAR, SDM and SAC models would produce unbiased estimates, with measures of coefficient dispersion for the SDM and SAC models being correct. Recall that including variables such as WX in the model when their coefficients are zero does not produce bias in the explanatory variables estimates. Similarly, incorporating a model for spatial dependence in the disturbances where the dependence parameter is truly zero will not have an adverse impact on the SAC estimates for sufficiently large samples.

3. When the true DGP is the SDM model that includes both spatial lag dependence as well as spatial lags of the explanatory variables, the SEM, SAR and SAC coefficient estimates will suffer from omitted variables bias, since these models do not include the spatially lagged explanatory variables WX. The SEM model will suffer additional bias due to exclusion of the spatial lag of the dependent variable. The correctness of inference regarding the biased coefficient estimates for the SEM, SAR and SAC models becomes a moot issue here.

4. When the true DGP is the SAC model that includes both spatial lag and spatial error dependence, the SAR and SDM models will produce unbiased coefficient estimates, while the SEM model coefficients will be biased because this model ignores the spatial lag of the dependent variable. Incorrect inferences regarding dispersion of the estimates are likely to arise for the SAR model from ignoring spatial dependence in the disturbances. The SDM model does not ignore spatial dependence in the disturbances, but implies a different type of specification for error dependence from that in the true SAC DGP. The impact on inference regarding dispersion of the unbiased SDM coefficients in this type of situation is an issue that needs further exploration.

The conclusion we draw is that the SDM is the only model that will produce unbiased coefficient estimates under all four possible data generating processes. Inference about the dispersion of the unbiased SDM coefficient estimates in cases 1, 2 and 3 above will be correct, whereas case 4 is an issue that requires exploration.

We also reiterate our point from Chapter 2 that beginning with a DGP based on simple error dependence, the presence of omitted variables will lead to a model specification that conforms to the SDM. Since omitted variables are likely when dealing with regional data samples, this is another motivation for use of the SDM model.

6.2   An applied example of model comparison

There is a great deal of literature that examines regional production from the standpoint of the new economic geography (Duranton and Puga, 2001; Autant-Bernard, 2001; Autant-Bernard, Mairesse and Massard, 2007; Parent and LeSage, 2008). A question of interest is the role of regional differences in technology on production, Q. This is often explored using a constant returns regional production function shown in (6.2), where A represents technology, K capital and L labor inputs, and the shares parameter ψ is such that we have constant returns to scale.

$$Q = AK^{(1-\psi)}L^{\psi} \tag{6.2}$$

A total factor productivity relationship can be derived from the Cobb-Douglas production function as shown in (6.3).

$$
\begin{aligned}
\ln Q &= \psi \ln L + (1-\psi)\ln K + \ln A \\
\ln Q - \psi \ln L - (1-\psi)\ln K &= \ln A \\
\text{tfp} &= \ln A
\end{aligned}
\tag{6.3}
$$

Our interest centers on whether the regional stock of patents can play the role of the technology variable A in (6.3). Intuitively, the stock of corporate patents in each region should reflect a proxy for technology used in regional production.

Patent stocks which we label A act as an empirical proxy for technology, but these are unlikely to capture the true technology available to regions. We posit the existence of unmeasured technology, which we label A^* that is excluded from the (log) linear relationship in (6.3). It has become a stylized fact that empirical measures of regional technical knowledge A such as patent applications, educational attainment, expenditures or employment in research and development etc., exhibit spatial dependence (Autant-Bernard, Mairesse and Massard, 2007; Parent and LeSage, 2008). If both the measured variable A included in the empirical relationship (6.3) and the unmeasured excluded variable A^* exhibit spatial dependence, then our development in Chapter 2 indicates that a spatial regression relationship will result.

Specifically, let the spatial autoregressive processes in (6.4) and (6.5) govern spatial formation of technical knowledge stocks $a = \ln A$ and $a^* = \ln A^*$. The $n \times 1$ vector a reflects (logged) cross-sectional observations on regional technology in a sample of n regions, and we have introduced zero mean, constant variance disturbance terms u, v, ε, along with an $n \times n$ spatial weight matrix W reflecting the connectivity structure of the regions. The scalar parameters ϕ and ρ reflect the strength of spatial dependence in a and a^*.

$$a = \phi W a + u \tag{6.4}$$
$$a^* = \rho W a^* + v \tag{6.5}$$
$$v = u\gamma + \varepsilon \tag{6.6}$$
$$u \sim N(0, \sigma_u^2 I_n)$$
$$v \sim N(0, \sigma_v^2 I_n)$$
$$\varepsilon \sim N(0, \sigma_\varepsilon^2 I_n)$$

The relationship in (6.6) reflects simple (Pearson) correlation between shocks (u, v) to technology stocks a and a^* when the scalar parameter $\gamma \neq 0$. The particular type of spatial regression that results will depend on whether there is correlation between the shocks ($\gamma \neq 0$), or no correlation ($\gamma = 0$).

If we begin with the relationship in (6.7), where we use y to denote *tfp* which is logged, and use the definitions in (6.4) to (6.6), we arrive at (6.8).[2]

$$y = \beta a + a^* \tag{6.7}$$
$$y = \rho W y + a(\beta + \gamma) + W a(-\rho\beta - \phi\gamma) + \varepsilon \tag{6.8}$$
$$y = \rho W y + \eta_1 a + \eta_2 W a + \varepsilon$$

The expression in (6.8) represents a spatial Durbin model (SDM). This model subsumes the spatial error model SEM: $(I_n - \rho W)y = (I_n - \rho W)a\beta + \varepsilon$, as a special case when the parameter $\gamma = 0$ indicating no correlation in shocks to measured and unmeasured technical knowledge, and when the restriction $\eta_2 = -\rho\eta_1$ is true. A simple likelihood-ratio test of the SEM versus SDM model can be used to test the restriction $\eta_2 = -\rho\eta_1$.

Another way to view the condition $\gamma = 0$ is that the included variable a and excluded variable a^* are not correlated, since correlation between the shocks, u, v implies correlation between a and a^*. In the trivial case of no correlation, the omitted variable a^* does not exert bias on our model estimates, which can be seen from $\beta + \gamma = \beta$, when $\gamma = 0$. Conventional omitted variables treatment considers the non-trivial case where correlation exists between the included and excluded variables so that $\gamma \neq 0$. A related point is that $\gamma \neq 0$ will lead to a rejection of the common factor restriction since the coefficient $\eta_1 = (\beta + \gamma)$ will not be equal to $-\rho\eta_2$ when $\gamma \neq 0$, since $\eta_2 = (-\rho\beta - \phi\gamma)$. Intuitively, the only way an SEM model can be justified is if: 1) there are no omitted variables in the model or 2) if the included and excluded variables, a and a^* do not exhibit correlation. An omitted variable that is correlated with the stock of technical knowledge variable included in the model will lead to

---

[2]Without loss of generality we could include an intercept term in the model, but ignore this term in our discussion for simplicity.

a spatial regression model that must contain a spatial lag of the dependent variable.

In addition to testing the SDM versus SEM models, we can also test for spatial dependence in measured knowledge stocks. For dependence in measured knowledge stocks we can produce maximum likelihood estimates for the spatial autoregressive model: $a = \phi W a + u$, and rely on an asymptotic t-statistic to test whether the scalar parameter ϕ is different from zero.

This development provides a formal motivation for inclusion of what is known as a "spatial lag" of the explanatory variable Wa in regression relationships that seek to explore knowledge spillover effects in a spatial context. Our development arrives at a spatial regression specification as an econometric means of addressing omitted variables, unlike the theoretical models of Ertur and Koch (2007) which directly include spatial dependence structures in theoretical economic relationships to arrive at the same result. Our starting point is a non-spatial theoretical relationship where included and excluded explanatory variables exhibit spatial dependence, and the included and excluded variables are correlated by virtue of common (correlated) shocks to the spatial autoregressive processes governing these variables. In the event that the shocks are uncorrelated, we have an SEM model whereas correlation of the shocks leads to an SDM model as the spatial data generating process.

6.2.1 The data sample used

Following the same approach as in the application from Chapter 3, the dependent variable tfp in the relationship (6.3) was calculated using 2002 real gross value added data (deflated by 1995 Euro prices) as the measure of output Q, for a sample of 198 NUTS-2 European Union regions. The regions represent the 15 pre-2004 EU member states. See Fisher, Scherngell and Reisman (2008) for a complete description of the sample data.

Calculated regional shares of labor for each region during 2002 were used along with the assumption of constant returns to calculate regional tfp: $\ln Q - s \odot \ln L - (1 - s) \odot \ln K$, where $\odot$ represents the Haddamard (element-by-element) product of the $n \times 1$ vector of regional shares and the $n \times 1$ vectors of regional labor L and capital K.

6.2.2 Comparing models with different weight matrices

To implement our model we require a spatial weight matrix W, which can be constructed in a number of different ways. We could rely on a first-order contiguity weight matrix or a nearest-neighbor weight matrix based on m nearest-neighboring regions. Of course, we would row-normalize the weight matrices for reasons set forth in Chapter 4.

There is the question of which weight matrix is most appropriate for our model and sample data, and if we rely on a nearest-neighbor matrix W, the number m of neighbors to use must be specified. One approach to take would be to estimate models based on different spatial weight matrices and examine the log-likelihood function values. A Bayesian approach would rely on the log-marginal likelihood and associated model probabilities. Details regarding Bayesian model comparison are set forth in Section 6.3, but we note that given a set of models based on alternative spatial weight matrices posterior model probabilities for each model can be calculated. The model exhibiting the highest posterior model probability is that which best fits both the data and any prior distributions assigned for the parameters.

Table 6.1 shows the log-likelihood function values and posterior model probabilities associated with models based on nearest-neighbor weight matrices with $m = 3, 4, \ldots, 9$ and a spatial contiguity weight matrix. Both sets of results point to a seven nearest neighbor spatial weight matrix. In smaller spatial samples such as the 198 observation sample used here, there may be some uncertainty regarding the appropriate number of nearest neighbors to employ. The Bayesian posterior model probabilities point to this uncertainty regarding $m = 6, 7, 8$. We note that it is not in general possible to use formal tests for significant differences between the log-likelihood function values for models based on different weight matrices, since a model based on $m = 7$ does not generally nest one based on $m = 6$. This is one advantage of using Bayesian posterior model probabilities which do not require nested models to carry out these comparisons.

TABLE 6.1:   Spatial weights model comparison

Spatial Weights	Log-likelihood function values	Bayesian model probabilities
$m = 3$	−44.0759	0.0000
$m = 4$	−36.6702	0.0004
$m = 5$	−35.7321	0.0009
$m = 6$	−30.8378	0.1202
$m = 7$	−29.1900	0.6185
$m = 8$	−30.5063	0.1718
$m = 9$	−31.2139	0.0882
Contiguity	−51.0740	0.0000

One approach to dealing with the uncertainty regarding models based on alternative weight matrices is to test for similar estimates and inferences from models based on weight matrices reflecting $m = 6, 7, 8$, which we illustrate here. Table 6.2 presents maximum likelihood model estimates for the three different spatial weight structures, illustrating that the coefficients do not vary greatly for these three alternative spatial weight specifications. We will discuss a more formal approach based on Bayesian model averaging in Section 6.3.

TABLE 6.2: Estimates comparison for varying weights

Parameters	Coefficient	t-statistic	z-probability
		$m = 6$	
constant	0.6396	3.47	0.0005
a	0.1076	5.13	0.0000
$W \cdot a$	−0.0111	−0.34	0.7286
ρ	0.6219	8.86	0.0000
σ^2	0.1490	NA	NA
		$m = 7$	
constant	0.5684	3.10	0.0019
a	0.1109	5.33	0.0000
$W \cdot a$	−0.0160	−0.48	0.6267
ρ	0.6469	9.11	0.0000
σ^2	0.1470	NA	NA
		$m = 8$	
constant	0.5600	2.93	0.0033
a	0.1136	5.43	0.0000
$W \cdot a$	−0.0255	−0.76	0.4458
ρ	0.6629	9.07	0.0000
σ^2	0.1496	NA	NA

6.2.3 A test for dependence in technical knowledge

As indicated, we can test for dependence in measured knowledge stocks a using maximum likelihood estimates for the spatial autoregressive model: $a = \beta_0 \iota_n + \phi W a + u$, and rely on a t-statistic to test whether the scalar parameter ϕ is different from zero. Results from this spatial autoregression using a spatial weight matrix based on $m = 7$ nearest neighbors are shown in Table 6.3. From the table we see that $\phi = 0.7089$ with an associated t-statistic of 11.44 leading us to conclude that regional knowledge stocks exhibit spatial dependence, consistent with our assumption.

TABLE 6.3: Tests for spatial dependence in technical knowledge

Measurable knowledge: $a = \beta_0 \iota_n + \phi W a + u$			
Parameters	Coefficient	t-statistic	z-probability
β_0	1.7187	4.54	0.0000
ϕ	0.7089	11.44	0.0000

6.2.4   A test of the common factor restriction

The other test of interest is a comparison of the log-likelihood function values from SDM versus SEM models. A likelihood ratio test provides a test of the common factor restriction: $\eta_2 = -\rho\eta_1$. Given the results in Table 6.3 showing spatial dependence in observed knowledge stocks, we would expect a rejection of the SEM model in favor of the SDM model. The log-likelihood was -29.19 for the SDM model and -32.15 for the SEM model both based on a seven nearest neighbor spatial weight matrix. This leads to a difference of 2.96, and twice this difference in magnitude (5.92) represents a rejection of the SEM model in favor of the SDM model using the 95% critical value for $\chi^2(1)$ which equals 3.84, but not at the 99% level where the critical value is 6.635. The single degree of freedom reflects the single parameter restriction.

We can also test whether the more parsimonious SAR model that excludes the spatial lag of knowledge stocks is more consistent with the sample data than the SDM model using a likelihood ratio test. The log-likelihood for the SAR model was -34.42, leading to a difference with the SDM model of 5.23. Twice this difference (10.46) exceeds the 99% critical value of 6.635, allowing us to reject the SAR model in favor of the SDM model. Another way to test this restriction is to consider the distribution for the non-linear parameter combination $-\rho \cdot \eta_1$ versus that for η_2 based on the SDM model estimates. This can be accomplished using MCMC model estimation in conjunction with non-informative prior distributions for the model parameters. This will lead to posterior estimates for these parameters that should exhibit the same distribution as those from maximum likelihood estimation. Posterior means and 95% and 99% upper and lower credible intervals were constructed using 5,000 draws from MCMC estimation of the model, with the results shown in Table 6.4.

From the table we see that the posterior mean for the distribution of $-\rho\eta_1 = -0.0687$ is near the lower 0.95 credible interval for the parameter η_2, consistent with the likelihood ratio test results. As in the case of the likelihood ratio test, the lower 0.99 credible interval for the parameter η_2 (-0.0830) spans the posterior mean value for $-\rho\eta_1(-0.0687)$, precluding a 99% probability inference against the common factor restriction.

An important empirical implication is that if the included measures of knowledge stocks are not correlated with excluded knowledge available to regions, then no spatial lag of the dependent variable is implied in the resulting model. In the case found here, there is strong but not overwhelming evidence that included and excluded variables measuring regional knowledge stocks are correlated. This implies that a spatial lag of the dependent variable should be used in the model, as well as a spatial lag of the explanatory variable Wa. Since omitting either of these from the empirical model will lead to biased and inconsistent estimates for the parameters, it seems prudent to proceed using the SDM model to examine the impact of knowledge stocks on $y = \text{tfp}$, the focus of this example.

TABLE 6.4: Bayesian test for common factor restriction

Parameters	0.99 Lower	0.95 Lower	Mean	0.95 Upper	0.99 Upper
$-\rho\eta_1$	-0.1046	-0.0937	-0.0687	-0.0452	-0.0366
η_2	-0.0830	-0.0634	-0.0093	0.0431	0.0668

6.2.5 Spatial effects estimates

A second empirical implication is that calculation of the response of $y = \text{tfp}$ to changes in regional knowledge stocks, e.g., $\partial y/\partial a$ will differ depending on which model is appropriate. For the case of the SEM model, the coefficient estimates have the usual least-squares regression interpretation, so the log-log form of the relationship leads directly to elasticity estimates for the response of y to variation in the levels of knowledge stocks across the regional sample. For this case, there are no spatial spillover impacts and the response of y to changes in knowledge a is the same as that which would be inferred from a simple least-squares regression model.

In our case where the sample data was judged to be consistent with the SDM model, $\partial y/\partial a'$ takes a more complicated form and allows for spatial spillover impacts from changing a_i in one region i on y_j in other regions $j \neq i$. Specifically, (6.9) shows the partial derivative which takes the form of an $n \times n$ matrix (see Chapter 2 for a detailed derivation).

$$\partial y/\partial a' = (I_n - \rho W)^{-1}(I_n \eta_1 + W \eta_2) \tag{6.9}$$
$$\eta_1 = (\beta + \gamma)$$
$$\eta_2 = (-\rho\beta - \phi\gamma)$$

This important aspect of assessing the impact of spatial spillovers appears to have been overlooked in much of the spatial econometrics literature. Past empirical studies proxy knowledge available in other regions using either a spatial lag of innovation output from neighboring regions measured through their patents (Anselin, Varga and Acs, 1997), or by explanatory variables reflecting research effort in neighboring regions. They then proceed to assess the magnitude and significance of spatial spillovers using the parameters associated with these spatially lagged explanatory variables. It should be clear from the partial derivative in (6.9) that the coefficient η_2 used in past studies is an incorrect representation of the impact of changes in the variable a on y. In fact, the parameter η_2 in our model is negative and statistically insignificant, but we will see that positive and statistically significant spatial spillovers exist based on a correct measure.

As motivated in Chapter 2, we can calculate scalar summary measures for the $n \times n$ matrix of partial derivatives that represent direct and indirect (spatial spillover) impacts on the dependent variable (total factor productivity) that arise from changing the explanatory variable a. The main diagonal of the matrix: $(I_n - \rho W)^{-1}(I_n \eta_1 + W \eta_2)$ represents own partial derivatives (direct impacts), while the off-diagonal elements correspond to cross-partial derivatives (indirect impacts). These are averaged to produce scalar summary measures using the average of the main diagonal elements from the matrix and the row- or column-sums of the matrix elements excluding the diagonal. In addition to these scalar measures of the mean direct and indirect impacts, we also construct measures of dispersion that can be used to draw inferences regarding the statistical significance of the direct and indirect effects. These are based on simulating parameters from the normally distributed parameters η_1, η_2 and ρ, using the maximum likelihood estimates and associated variance-covariance matrix. The simulated draws are then used in the computationally efficient formulas from Chapter 4 to calculate the implied distribution of the scalar summary measures.

An alternative to simulating draws based on maximum likelihood estimates of the parameters and variance-covariance matrix is to rely on MCMC draws from Bayesian estimation of the model using a diffuse prior for all model parameters. This will produce results that are centered on the maximum likelihood estimates which were reported in the applied illustration of Chapter 3. Table 6.5 presents these estimates based on 5,000 MCMC draws from Bayesian estimation of the model. The 5,000 draws were used to construct empirical estimates of the lower and upper 0.95 and 0.99 credible intervals, reported in the table. The scalar summaries shown in the table reflect *cumulative impacts* aggregated over space, since: $(I_n - \rho W)^{-1} = I_n + \rho W + \rho^2 W^2 + \ldots$, we are examining effects that fall on first-order neighbors (W), second-order neighbors (W^2), and so on, cumulatively. The table reports impact estimates based on alternative spatial weight matrices constructed using $m = 6, 7, 8$ nearest neighbors.

From the table, we see that the cumulative impact estimates are not very sensitive to the particular spatial weight matrix used, producing similar estimates and identical inferences regarding the significance of the impacts. The direct effects from changing knowledge stocks on regional total factor productivity are positive and significantly different from zero using the 0.99 lower and upper bounds. The indirect effects estimates are positive and different from zero using the 0.95 bounds, but the 0.99 lower bound spans zero, suggesting we cannot be 99% confident that positive spatial spillovers exist. The mean indirect estimates for the case of $m = 7$ are around 1.5 times the size of the direct effects, suggesting a possible role for spatial spillovers arising from regional patent stocks. As noted in Chapter 2, we interpret the effects parameters in relation to movements from one steady-state equilibrium to another. Given the log-transformations applied to both the dependent variable total factor productivity and the explanatory variable patent stocks, we can interpret the effects magnitudes as elasticities. This implies that a 10% increase in the average stock of regional (corporate) patents would lead to a

TABLE 6.5: Cumulative knowledge stocks effects estimates

Effects	0.99 Lower	0.95 Lower	Mean	0.95 Upper	0.99 Upper
	$m = 6$ Neighbors weight matrix				
direct effect	0.0655	0.0804	0.1156	0.1508	0.1667
indirect effect	−0.0158	0.0244	0.1479	0.3003	0.3853
total effect	0.0903	0.1358	0.2635	0.4242	0.5215
	$m = 7$ Neighbors weight matrix				
direct effect	0.0682	0.0837	0.1184	0.1537	0.1681
indirect effect	−0.0176	0.0290	0.1613	0.3314	0.4361
total effect	0.0998	0.1396	0.2798	0.4553	0.5677
	$m = 8$ Neighbors weight matrix				
direct effect	0.0717	0.0848	0.1207	0.1565	0.1727
indirect effect	−0.0356	0.0090	0.1496	0.3244	0.4519
total effect	0.0783	0.1247	0.2703	0.4561	0.5855

2.8% increase in total factor productivity (based on the $m = 7$ model). This is the cumulative effect after enough time has elapsed to move the relationship to a new steady state equilibrium. Of the 2.8% increase in factor productivity, around 1.2 percent would result from direct impacts and 1.6 percent from indirect impacts (spatial spillovers). These results indicate that factor productivity is inelastically related to regional knowledge stocks, with a 0.28 implied elasticity coefficient.

This example shows that likelihood-based model comparison tests can be useful in providing guidance to practitioners, and there is a large spatial econometrics literature devoted to these. However, we wish to caution against some testing practices. Most tests are developed against specific alternatives and performing multiple tests that have not been designed to work together coherently may not yield desirable outcomes. This is a particular problem with tests applied to non-nested models and when decisions are made on the basis of sequential tests since results may vary with the sequence used. In addition, exclusive reliance on statistical tests to determine (1) which model specification is appropriate, (2) which explanatory variables exert a significant impact and (3) what type of spatial weight matrix to use is likely to constitute overuse of the sample data.

Specifically, we would like to caution against the common practice of testing to choose between the SAR and SEM. In the early days of the spatial literature where sample sizes were very small, possible efficiency gains may have justified this practice. However, current spatial data sets contain sufficient observations to examine more general alternatives such as the SDM. For larger data sets issues such bias and interpretation may be more important than the variance of the estimates. We have argued that there are strong econometric motivations for the SDM model, which subsumes the SAR and SEM models as a special case. Further, this model arises quite naturally in the presence of omitted variables that are correlated with included variables.

In the next section we discuss Bayesian approaches to model comparison, which hold some advantages over likelihood-based methods. One advantage is the ability to compare non-nested models that create difficulties for likelihood-ratio tests. A second advantage is that likelihood-based tests depend on the quality of the point estimates used to evaluate the likelihood function. Bayesian model comparison methods integrate over the model parameters to produce inferences regarding alternative models that are unconditional on specific values taken by the model parameters.

6.3   Bayesian model comparison

Zellner (1971) sets forth the basic Bayesian theory behind model comparison for non-spatial regression models where a discrete set of m alternative models are under consideration. The approach involves specifying prior probabilities for each model[3] as well as prior distributions for the regression parameters. Posterior model probabilities are then calculated and used for inferences regarding the consistency of alternative regression models with the sample data and prior.

When we compare models based on alternative spatial weight structures, we typically have a small number m of alternative models, and the models differ only in terms of the type of spatial weight matrix used. A Bayesian approach for comparing spatial regression models based on differing spatial weights is set forth in Section 6.3.1.

An alternative scenario arises when comparing models based on different sets of explanatory variables. In these situations a small set of 15 candidate explanatory variables will lead to $2^{15} = 32,768$ possible models, and a larger set of 50 candidate variables results in $2^{50} = 1.1259e + 015$ models. This makes it infeasible to calculate posterior model probabilities for the large number of possible models. A Markov Chain Monte Carlo model composition methodology known as MC^3 proposed by Madigan and York (1995) has gained popularity in the non-spatial regression literature (e.g. Dennison, Holmes, Mallick and Smith (2002); Fernández, Ley, and Steel (2001)). Since the question of which explanatory variables are most important often arises in applied regression modeling, the MC^3 methods have gained popularity in

---

[3]The alternative models are often taken as equally likely, so each model is assigned the same prior probability equal to $1/m$, where m is the number of models under consideration.

the regression literature. In Section 6.3.2 we describe an extension to the case of spatial regression models proposed in LeSage and Parent (2007).

6.3.1 Comparing models based on different weights

We assume that a small set of m alternative spatial regression models $M = M_1, M_2, \ldots, M_m$ are under consideration, each based on a different spatial weight matrix. Other model specification aspects such as the explanatory variables and type of model, (e.g., SAR, SDM) are held constant. Prior probabilities are specified for each model, which we label $\pi(M_i), i = 1, \ldots, m$, as well as prior distributions for the parameters $\pi(\eta)$, $\eta = (\rho, \alpha, \beta, \sigma^2)$, where α represents the intercept term, β the k parameters associated with the explanatory variables, ρ the spatial dependence parameter and σ^2 the constant, scalar noise variance parameter.

If the sample data are to determine the posterior model probabilities, the prior probabilities should be set equal to $1/m$, making each model equally likely a priori. These are combined with the likelihood for y conditional on η as well as the set of models M, which we denote $p(\mathcal{D}|\eta, M)$. The joint probability for the set of models, parameters and data takes the form in (6.10), where $\mathcal{D}$ represents the sample data.

$$p(M, \eta, \mathcal{D}) = \pi(M)\pi(\eta|M)p(\mathcal{D}|\eta, M) \tag{6.10}$$

Application of Bayes' rule produces the joint posterior for both models and parameters as shown in (6.11).

$$p(M, \eta|\mathcal{D}) = \frac{\pi(M)\pi(\eta|M)p(\mathcal{D}|\eta, M)}{p(\mathcal{D})} \tag{6.11}$$

The posterior model probabilities take the form in (6.12), which requires integration over the parameter vector η. Numerical integration over the $(k + 3) \times 1$ parameter vector could be difficult in cases where k is even moderately large. We use k to represent the number of explanatory variables, and we have three additional parameters for the intercept, spatial dependence and noise variance.

$$p(M|\mathcal{D}) = \int p(M, \eta|\mathcal{D})d\eta, \tag{6.12}$$

As a specific example, consider the SAR model, where the likelihood function for the parameters $\eta = (\alpha, \beta, \sigma^2, \rho)$, based on the data $\mathcal{D} = \{y, x, W\}$ takes the form shown in (6.13), where we include the spatial weight matrix W to indicate that the likelihood is conditional on the particular weight matrix employed in the model. That is, the weight matrix is taken as given and treated in the same manner as the sample data information in y, X.

$$L(\eta|\mathcal{D}) \propto (\sigma^2)^{-n/2}|I_n - \rho W|\exp\{-\frac{1}{2\sigma^2}e'e\} \qquad (6.13)$$
$$e = (I_n - \rho W)y - \alpha \iota_n - X\beta$$

An essential part of any Bayesian analysis is assigning prior distributions for the parameters in η. This can be accomplished using different approaches. We use the NIG prior for β and σ^2, but rely on Zellner's g-prior for the normal distribution parameters assigned for β in the model. An uninformative prior is assigned to the intercept parameter α, and the $\mathcal{B}(d,d)$ prior introduced in Chapter 5 is assigned to the parameter ρ.

LeSage and Parent (2007) point out that a great deal of computational simplicity arises if we employ Zellner's g-prior (Zellner, 1986) for the parameters β in the NIG prior for the SAR model. This normal prior distribution takes the form shown in (6.14), and we assign the same prior for the parameters β in all models. We can simply assign zero values for the prior mean vector β_0, but must pay attention to scale model variables so this prior is relatively consistent with zero values for the model parameters. For example, we would not want to use this type of prior if coefficient estimates took on very large magnitudes that were far from zero.

The parameter g controls the dispersion of the prior which reflects our uncertainty regarding the prior mean setting for β_0. A smaller value of g leads to greater dispersion in the prior, so an automatic setting of $1/n$, where n is the sample size works to create a relatively uninformative prior, and $1/n^2$ is even more uninformative.[4] For the case we consider here involving comparison of alternative spatial weight matrices, one can rely on a completely uninformative prior, but we will reuse the g-prior in our discussion of the MC^3 method in the next section.

$$\pi_b(\beta|\sigma^2) \sim N[\beta_0, \sigma^2(gX'X)^{-1}] \qquad (6.14)$$

Using the NIG prior for β and σ^2 with a normal prior, $N(\beta_0, \sigma^2(gX'X)^{-1})$, for the parameters β, and inverse gamma prior, $IG(a,b)$, for σ^2 shown in (6.15) allows us to draw on the conjugate nature of these two prior distributions. We note that the parameterization for the inverse gamma prior takes a slightly different form than that used in Chapter 5, which proves helpful in this development. The case of a non-informative prior on σ^2 arises when $a = b = 0$.

$$\pi_s(\sigma^2) \sim \frac{(ab/2)^{a/2}}{\Gamma(a/2)}(\sigma^2)^{-(\frac{a+2}{2})}\exp(-\frac{ab}{2\sigma^2}) \qquad (6.15)$$

---

[4]Setting $g = 1/(1e+15)$ would in effect produce a totally uninformative prior.

As noted, we rely on the $\mathcal{B}(1.01, 1.01)$ prior from Chapter 5 with these prior parameter settings which produce a relatively uninformative prior that places very little prior weight on end points of the $(-1, 1)$ interval for ρ.

Using Bayes' theorem, the log marginal likelihood $\int p(M, \eta | \mathcal{D}) d\eta$ where $\eta = (\alpha, \beta, \sigma^2, \rho)$ for the SAR model can be written as the integral in (6.16), and associated definitions in (6.17) (LeSage and Parent, 2007).

$$\int \pi_b(\beta | \sigma^2) \pi_s(\sigma^2) \pi_r(\rho) p(\mathcal{D} | \alpha, \beta, \rho, \sigma^2) \, d\beta \, d\sigma^2 \, d\rho \qquad (6.16)$$

$$= \kappa_1 (2\pi)^{-(n+k)/2} |C|^{1/2} \int |I_n - \rho W| \frac{1}{\sigma^{n+a+k+2}}$$

$$\times \exp\{-\frac{1}{2\sigma^2}[ab + S(\rho) + \beta'C\beta$$

$$+ (\beta - \hat{\beta}(\rho))'(X'X)(\beta - \hat{\beta}(\rho))]\} \pi_r(\rho) \, d\beta \, d\sigma^2 \, d\rho,$$

with,

$$\kappa_1 = \Gamma\left(\frac{a}{2}\right)^{-1} \left(\frac{ab}{2}\right)^{a/2} \qquad (6.17)$$

$$S(\rho) = e(\rho)'e(\rho)$$

$$e(\rho) = (I_n - \rho W)y - X\hat{\beta}(\rho) - \hat{\alpha}\iota_n$$

$$C = gX'X$$

$$\hat{\beta}(\rho) = (X'X)^{-1}X'(I_n - \rho W)y$$

$$\hat{\alpha} = \bar{y} - \rho \overline{Wy}$$

$$\overline{Wy} = (1/n) \sum_i (Wy)_i$$

$$\bar{X} = 0$$

Using the properties of the multivariate normal pdf and the inverted gamma pdf to analytically integrate with respect to β and σ^2, we can arrive at an expression for the log marginal that will be required for model comparison purposes. We note that since the intercept term is common to all models, this leads to $n - 1$ as the degrees of freedom in the posterior (6.18).

$$p(\rho | \mathcal{D}) = \kappa_2 (\frac{g}{1+g})^{k/2} \qquad (6.18)$$

$$\times |A|[ab + S(\rho) + Q(\rho)]^{-\frac{n+a-1}{2}} \pi_r(\rho)$$

Where the terms used in the posterior expression are defined below.

$$A = I_n - \rho W$$

$$\kappa_2 = \frac{\Gamma\left(\frac{n+a-1}{2}\right)}{\Gamma\left(\frac{a}{2}\right)}(ab)^{\frac{a}{2}}\pi^{-\frac{n-1}{2}}$$

$$S(\rho) + Q(\rho) = \frac{1}{g+1}[Ay - X\hat{\beta}(\rho) - \hat{\alpha}\iota_n]'[Ay - X\hat{\beta}(\rho) - \hat{\alpha}\iota_n]$$
$$+ \frac{g}{g+1}[Ay - \hat{\alpha}\iota_n]'[Ay - \hat{\alpha}\iota_n].$$

An important point regarding expression (6.18) is that we must rely on univariate numerical integration over the parameter ρ to convert this to the scalar expression necessary to calculate $p(M|y)$ needed for model comparison purposes, where we use y to represent the data. This is a contrast with conventional regression models where analytical integration over the parameters β and σ leads to a scalar expression that can be used to compare models (Fernández, Ley, and Steel, 2001). However, we provide details regarding a computationally simple approach to carrying out the univariate numerical integration in the chapter appendix following LeSage and Parent (2007).[5] The case of the SDM model is identical to the SAR model presented here with the explanatory variables matrix X replaced by the matrix $\tilde{X} = \begin{pmatrix} X & WX \end{pmatrix}$.

We draw upon the Bayesian theory of model comparison to consider comparison of a set of models based on m alternative weight matrices $W_{(i)}, i = 1, \ldots, m$. Each of these is considered a different model denoted by a likelihood function and prior for the parameters $\theta = (\rho, \beta, \sigma^2)$.

$$p(\theta^{(i)}|y, W_{(i)}) = \frac{p(y|\theta^{(i)}, W_{(i)})p(\theta^{(i)}|W_{(i)})}{p(y|W_{(i)})} \qquad (6.19)$$

Use of Bayes' rule set forth in Chapter 5 to explode terms like $p(y|W_{(i)})$ produces posterior model probabilities, the basis for inference about different models/spatial weight matrices, given the sample data.

$$p(W_{(i)}|y) = \frac{p(y|W_{(i)})p(W_{(i)})}{p(y)} \qquad (6.20)$$

$p(y|W_{(i)})$ is the marginal likelihood for this model comparison situation. As we have seen, the key quantity needed for model comparison is the marginal likelihood:

$$p(y|W_{(i)}) = \int p(y|\theta^{(i)}, W_{(i)})p(\theta^{(i)}|W_{(i)})d\theta^{(i)} \qquad (6.21)$$

Bayesians often avoid dealing with $p(y)$, by relying on the *posterior odds ratio* for model i versus model j:

---

[5]Analogous expressions for the SEM model are presented in LeSage and Parent (2007).

$$PO_{ij} = \frac{p(W_{(i)}|y)}{p(W_{(j)}|y)} = \frac{p(y|W_{(i)})p(W_{(i)})}{p(y|W_{(j)})p(W_{(j)})} \tag{6.22}$$

A virtue of this approach is that posterior model probabilities or Bayes' factors can be used to compare non-nested models. This allows the method to be used for comparing models based on: 1) different spatial weight matrices; 2) different model specifications (including those that may not be members of the family of models set forth in Chapter 2; and 3) models based on different sets of explanatory variables contained in the matrix X.

An issue that arises with this approach is the need to avoid a paradox pointed out by Lindley (1957). He noted that when comparing models with different numbers of parameters that rely on diffuse priors, the simpler model is always favored over a more complex one, irrespective of the sample data information. An implication is that two models with an equal number of parameters can be compared using diffuse priors, but for model comparisons that involve changes in the number of parameters, strategic priors must be developed and used.

A strategic prior would recognize that flat priors can in fact be highly informative because assigning a diffuse prior over a parameter value assigns a large amount of prior weight to values of the parameter that are very large in absolute value terms. An implication is that there is no natural way to encode complete prior ignorance about parameters. A strategic prior in the context of model comparison involving alternative spatial weight matrices would be one that explicitly recognizes the role that parameters and priors play in controlling model complexity. Given this, we could explore how prior settings impact posterior model selection regarding the alternative spatial weight matrices.

For the case of homoscedastic disturbances in the SAR and SEM spatial regression models from Chapter 2, we have seen that we can analytically integrate out the parameters β and σ to arrive at an expression for the marginal likelihood that depends only on the parameter ρ. This is the only prior distribution we need be concerned with, since our comparison of models based on different weight matrices does not depend on the parameters β and σ, because these have been integrated out. Hepple (1995b) provides expressions for the log-marginal likelihood when non-informative priors are used for a number of spatial regression models including the SAR and SEM.

6.3.2 Comparing models based on different variables

A large literature on Bayesian model averaging over alternative linear regression models containing differing explanatory variables exists (Fernández, Ley, and Steel, 2001). The Markov Chain Monte Carlo model composition (MC^3) approach introduced in Madigan and York (1995) is set forth here for the case of spatial regression models. For a regression model with k possible explanatory variables, there are 2^k possible ways to select regressors to be included or excluded from the model. For $k = 15$, we have 32,768 possible models, ruling out computation of the log-marginal for all possible models as impractical.

The motivation for this literature is the classic trade-off between attempting to include a sufficient number of explanatory variables in our models to overcome potential omitted variables bias and inclusion of redundant variables that decrease precision of the estimates. It is precisely this trade-off that model averaging seeks to address.

The MC^3 method of Madigan and York (1995) devises a strategic stochastic Markov chain process that can move through the potentially large model space and sample regions of high posterior support. This eliminates the need to consider all models by constructing a sampler that explores relevant parts of the very large model space. If we let M denote the current model state of the chain, models are proposed using a neighborhood, nbd(M) which consists of the model M itself along with models containing either one more variable (labeled a 'birth step'), or one less variable (a 'death step') than M. A transition matrix, q, is defined by setting $q(M \rightarrow M') = 0$ for all $M' \notin$ nbd(M) and $q(M \rightarrow M')$ constant for all $M' \in$ nbd(M). The proposed model M' is compared to the current model state M using the acceptance probability shown in (6.23).

$$\min \left[1, \frac{p(M'|y)}{p(M|y)} \right] \tag{6.23}$$

Use of univariate numerical integration methods described in the chapter appendix allows us to construct a Metropolis-Hastings sampling scheme that implements the MC^3 method. A vector of the log-marginal values for the current model M is stored during sampling along with a vector for the proposed model M'. These are then scaled and integrated to produce the ratio $p(M'|y)/p(M|y)$ in (6.23) that determines acceptance or rejection of the proposed model. In contrast to conventional regression models, there is a need to store log-marginal density vectors for each unique model found during the MCMC sampling to calculate posterior model probabilities over the set of all unique models visited by the sampler.

Although the use of birth and death processes in the context of Metropolis-Hastings sampling will theoretically produce samples from the correct posterior, Richardson and Green (1997) among others advocate incorporating a "move step" in addition to the birth and death steps into the algorithm. We rely on this approach as there is evidence that combining these move steps improves convergence of the sampling process (Dennison, Holmes, Mallick and Smith, 2002; Richardson and Green, 1997). The move step takes the form of replacing a randomly chosen single variable in the current explanatory variables matrix with a randomly chosen variable not currently in the model. Specifically, we might propose a model with one less explanatory variable (death step) and then add an explanatory variable to this new model proposal (birth step). This leaves the resulting model proposal with the same dimension as the original one with a single component altered. This type of sampling process is often labeled *reversible jump* MCMC. The model proposals that result from birth, death and move steps are all subjected to the Metropolis-Hastings accept/reject decision shown in (6.23), which is valid so long as the probabilities of birth, death and move steps have equal probability of $1/3$.

The Bayesian solution to incorporating uncertainty regarding specification of the appropriate explanatory variables into the estimates and inferences is to *average* over alternative model specifications. This is in contrast with much applied work that relies on a single model specification identified using various model comparison criterion that lead to a "most preferred model." The averaging involves weighting alternative model specifications by their posterior model probabilities. We note that the MC^3 procedure identifies models associated with particular explanatory variables and assigns a posterior model probability to each of these models. Like all probabilities, the posterior model probabilities sum to unity, so they can be used as weights to form a linear combination of estimates from models based on differing explanatory variables. This weighted combination of sampling draws from the posterior are used as the basis for posterior inference regarding the mean and dispersion of the individual parameter estimates.

Typically tests are performed with the aim of selecting a single best model that excludes irrelevant variables. This approach ignores *model uncertainty* which arises in our spatial regression model from two sources. One aspect of model uncertainty is the appropriate spatial weight matrix describing connectivity between regions used to specify the structure of spatial dependence. The second aspect of model uncertainty arises from variable selection, which sequential testing procedures ignore (Koop, 2003). As is typical in all regression models, we are also faced with *parameter uncertainty*. Fernández, Ley, and Steel (2001) point to MC^3 in conjunction with Bayesian model averaging as a way to accommodate both model and parameter uncertainty in a straightforward and formal way.

6.3.3 An applied illustration of model comparison

An important point to note about all spatial model comparison methods is that their performance will depend on the strength of spatial dependence in the sample data. We illustrate this point using the latitude-longitude coordinates from a sample of 258 European Union regions to produce a data-generated spatial sample. The location coordinates were used to construct a spatial weight matrix based on the five nearest neighboring regions. This weight matrix was used to generate a y-vector based on the spatial autoregressive model: $y = \rho W y + X\beta + \varepsilon$. The explanatory variables matrix X was generated as a three column matrix of standard normal random deviates, and the three β parameters were all set to unity. The scalar noise variance

TABLE 6.6: Posterior probabilities for models with differing ρ and weight matrices

m/ρ	-0.5	-0.2	-0.1	0.0	0.1	0.2	0.5
1	0.00	0.0000	0.0001	0.1091	0.0000	0.0000	0.00
2	0.00	0.0000	0.0007	0.0936	0.0003	0.0000	0.00
3	0.00	0.0000	0.0011	0.0956	0.0364	0.0000	0.00
4	0.00	0.0001	0.0559	0.1168	0.0907	0.0008	0.00
5*	1.00	0.9998	0.8694	0.1369	0.4273	0.9872	1.00
6	0.00	0.0001	0.0612	0.1402	0.2374	0.0116	0.00
7	0.00	0.0000	0.0086	0.1484	0.1479	0.0003	0.00
8	0.00	0.0000	0.0029	0.1594	0.0599	0.0000	0.00

parameter σ^2 was also set to one. The operational characteristics of any specification test to detect the true model structure will usually depend on the signal/noise ratio in the data generating process, determined by the variance of the matrix X relative to the noise variance, which we hold constant in this data-generated illustration.

A proper uniform prior was placed on the parameter ρ. For this example, all models contain the same matrix X, differing only with respect to the spatial weight matrix.

A series of seven models were generated based on varying ρ values ranging from -0.5 to 0.5, with the parameters described above held constant. It took around 4 seconds to produce posterior probabilities for a set of 8 models based on spatial weight matrices constructed using 1 to 8 nearest neighbors for this sample of 258 observations. Most of the time (3.3 seconds) was spent computing the log-determinant term for the 8 different weight matrices, using the method of Pace and Barry (1997).

The posterior model probabilities are presented in Table 6.6, for models associated with $m = 1, \ldots, 8$ neighbors and values of ρ ranging from -0.5 to 0.5. From the table, we see that positive or negative values of 0.2 or above for ρ lead to high posterior probabilities associated with the correct model, that based on $m = 5$ nearest neighbors. Absolute values of 0.1 or less for ρ lead to less accurate estimates of the true data generating model, with the posterior probabilities taking on a fairly uniform character for the case of $\rho = 0$. Intuitively, when ρ is small or zero, it will be difficult to assess the proper spatial weight matrix specification, since the spatial lag term, Wy, in the model is associated with a zero coefficient.

Another example is taken from the public choice literature (Turnbull and Geon, 2006), where the dependent variable representing county government services provision takes a form involving a Box-Cox type transformation. Specifically, let $g = GP^\phi$ denote the median voter's public good consumption, where G is government expenditures and P represents county population. The scalar $0 < \phi < 1$ is a consumption congestion parameter. This parameter reflects the degree of publicness with 0 representing a purely public good and 1 a private good. If we model $g = \rho W g + X\beta + \varepsilon$, we would be interested in comparing models based on varying numbers of neighbors as well as the parameter ϕ. A sample of government expenditures for 950 US counties located in metropolitan areas, and 1,741 counties located outside of metropolitan areas was used to form a SAR model with g as the dependent variable and various explanatory variables (such as taxes, intergovernmental aid and population in- and out-migration over the previous five years).

Table 6.7 shows posterior model probabilities from a limited range of values for ϕ, the congestion parameter and m, the number of neighbors that were used in the model comparison. The range shown in the table is where posterior probability mass was non-zero. These calculations required bivariate evaluation of models over both parameters, but are still reasonably simple to carry out.

TABLE 6.7: Posterior probabilities for varying values of m and ϕ

M/ϕ	Metropolitan county sample				
	$\phi = 0.4$	$\phi = 0.5$	$\phi = 0.6$	$\phi = 0.7$	$\phi = 0.8$
$m = 5$	0.0000	0.0000	0.0000	0.0000	0.0000
$m = 6$	0.0000	0.0038	0.0039	0.0000	0.0000
$m = 7$	0.0000	0.1424	0.0980	0.0000	0.0000
$m = 8$	0.0000	0.0763	0.0936	0.0001	0.0000
$m = 9$	0.0000	0.1295	0.4380	0.0001	0.0000
$m = 10$	0.0000	0.0031	0.0055	0.0000	0.0000
$m = 11$	0.0000	0.0007	0.0027	0.0000	0.0000
$m = 12$	0.0000	0.0002	0.0012	0.0000	0.0000
$m = 13$	0.0000	0.0001	0.0005	0.0000	0.0000
$m = 14$	0.0000	0.0000	0.0000	0.0000	0.0000
M/ϕ	Non-metropolitan county sample				
	$\phi = 0.3$	$\phi = 0.4$	$\phi = 0.5$	$\phi = 0.6$	$\phi = 0.7$
$m = 6$	0.0000	0.0000	0.0000	0.0000	0.0000
$m = 7$	0.0000	0.0000	0.0004	0.0000	0.0000
$m = 8$	0.0000	0.0024	0.8078	0.0008	0.0000
$m = 9$	0.0000	0.0006	0.1064	0.0007	0.0000
$m = 10$	0.0000	0.0001	0.0803	0.0006	0.0000
$m = 11$	0.0000	0.0000	0.0001	0.0000	0.0000
$m = 12$	0.0000	0.0000	0.0000	0.0000	0.0000

The results point to a model based on $m = 9$ and $\phi = 0.6$ for the metropolitan sample and $m = 8$, $\phi = 0.5$ for the non-metropolitan sample. The high posterior model probabilites for the parameter ϕ near the midpoint of the 0 to 1 range indicate that county government services are viewed as midway between the extremes of pure public and private goods. For the US counties, the average number of first-order contiguous neighbors (those with borders that touch each county) is around 6, so the number of neighbors chosen from the model comparison exercise represents slightly more than just the contiguous counties.

6.3.4   An illustration of MC^3 and model averaging

We use a 49 neighborhood data set for Columbus, Ohio from Anselin (1988) that contains observations on the median housing values (*hvalue*) for each neighborhood and household income (*income*) as explanatory variables. Neighborhood crime is the dependent variable in the model that we use to illustrate MC^3 and model averaging.

As already noted, if the data generating process is the SAR model, then $\hat{\beta}_{SAR} = (X'X)^{-1}X'(I_n - \rho W)y$, and least-squares estimates for β are biased and inconsistent. In these cases, we would expect that regression-based MC^3 procedures would not produce accurate estimates and inferences regarding which variables are important.

To illustrate differences between non-spatial and spatial MC^3 and model averaging results, we estimate an SDM model as well as the SLX model using standard methods. The SDM model is shown in (6.24). Intuitively, housing values and household income levels in nearby neighborhoods might contribute to explaining variation in neighborhood crime rates, y.

$$
\begin{aligned}
y &= \alpha \iota_n + \rho W y + \beta_1 hvalue + \beta_2 income \\
&\quad + \beta_3 W \cdot hvalue + \beta_4 W \cdot income + \varepsilon \\
y &= \rho W y + X\beta + W X \theta + \varepsilon
\end{aligned}
\tag{6.24}
$$

The contiguity-based weight matrix from Anselin (1988) was used to produce standard estimates in Table 6.8. The greatest disagreement in the two sets of estimates is with respect to the two spatially lagged explanatory variables, which could be a focus of model comparison and inference. For example, it might be of interest whether housing values and household income levels in nearby neighborhoods contribute to explaining variation in neighborhood crime rates. For later reference, we note that the sign of the spatially lagged house value variable is different in the SLX (least-squares) and SDM regressions, and the significance of the spatial lag of household income is different.

For our MC^3 procedure, the intercept term and spatial lag of the dependent variable are included in all models. This leads to four candidate variables and $2^4 = 16$ possible models. This makes it simple to validate our MC^3 algorithms by comparison with exact results based on posterior model probabilities for the set of 16 models. The explanatory variables were put in deviation from the means form and scaled by their standard deviations, and the Zellner *g*-prior was used in the MC^3 procedure.

TABLE 6.8: SLX and SDM model estimates

Variable	SLX estimates			SDM estimates		
	Estimate	$t-stat$	Prob	Estimate	t−stat	Prob
Constant	75.028	11.3	0.00	43.52	3.4	0.00
income	−1.109	−2.9	0.00	−0.91	−2.7	0.00
hvalue	−0.289	−2.8	0.00	−0.29	−3.2	0.00
$W \cdot$ income	−1.370	−2.4	0.01	−0.53	−0.9	0.34
$W \cdot$ hvalue	0.191	0.9	0.34	−0.24	−1.3	0.17
$W \cdot y$				0.41	2.6	0.00

TABLE 6.9: SLX BMA model selection information

Variables/models	M1	M2	M3	M4	M5
income	1	1	1	1	1
hvalue	0	0	1	1	1
$W \cdot$ income	0	1	1	0	1
$W \cdot$ hvalue	0	0	1	0	0
Model Probs	0.056	0.088	0.091	0.209	0.402

TABLE 6.10: SDM BMA model selection information

Variables/models	M1	M2	M3	M4	M5
income	1	1	1	1	1
hvalue	1	0	1	1	1
$W \cdot$ income	1	0	1	0	0
$W \cdot$ hvalue	1	0	0	1	0
Model Probs	0.033	0.071	0.098	0.128	0.486

Since the number of possible models here is $2^4 = 16$, it would have been possible to simply calculate the log-marginal posterior for these 16 models to find posterior model probabilities. Instead, we applied our MC^3 algorithm to the SLX and SDM models. A run of 10,000 draws was sufficient to uncover all 16 unique models, requiring 15 and 27 seconds respectively for the SLX (least-squares) and SAR MC^3 procedures.[6] Information regarding the top five models is provided in Table 6.9 for the SLX MC^3 procedure and Table 6.10 for the SDM MC^3 procedure, with the posterior model probabilities shown in the last row of the two tables. These tables use '1' and '0' indicators for the presence or absence of variables in each of the models presented.

From the tables we see that the disagreement regarding the spatial lag of

---

[6]MATLAB version 7 software was used in conjunction with a Pentium III M laptop computer.

household income between the two models appears in the MC^3 results as it did in the basic model estimates. The model with the highest posterior model probability from the SLX MC^3 procedure includes household income from neighboring regions, whereas this variable appears in a model having the third highest posterior probability in the SDM results.

It is instructive to see how Bayesian model averaging can help to resolve the issue regarding the significance of the spatial lag of household income. Model averaged estimates based on a posterior model probability weighted combination of MCMC draws from estimation of all 16 models are shown in Table 6.11. The results from the SLX model averaging procedure were constructed using Bayesian MCMC sampling for the SLX model with a diffuse prior. A similar MCMC procedure with a diffuse prior was used to construct draws for the SDM model. Having draws from MCMC is convenient because these can be weighted by the posterior model probabilities to form a posterior distribution that reflects the model uncertainty that model averaging procedures attempt to capture. The table reports means as well as 0.95 *credible intervals* constructed using the simulation draws.

The SLX model resolved the question of importance for the spatial lag of *income* in favor of this variable being included in the model, whereas the spatial lag of neighboring house values (*hvalue*) does not appear important in this model. It is interesting to note that the averaged coefficients for both *income* and $W\cdot$ *income* are smaller in value than those from standard least-squares estimation. This reduction in magnitude arises from taking into account our uncertainty regarding the appropriate model specification. The reduction in coefficient magnitude due to model uncertainty also makes it clearer that the spatial lag of *hvalue* is not an important variable when one takes into account alternative model specifications. However, we will have more to say about this later.

We see a similar reduction in magnitude for the averaged coefficients versus those from standard estimation of the SDM model that ignores model uncertainty. The reduction in magnitudes here points to a lack of importance for the spatial lags of *income* and *hvalue*. However, we need to calculate direct and indirect effects estimates to draw conclusions about the magnitude of impact on neighborhood crime associated with changes in these variables. In fact, in the SLX model we have direct and indirect effects that result in a total effect as well. This can be seen by considering the partial derivative of y for this model with respect to the explanatory variables X. This would take a form involving both the coefficients on *income* and *hvalue* as well as the spatial lags of these variables. Specifically, the matrix $S_r(W) = I_n\beta_r + W\theta_r$ arises from the partial derivative calculation. In a way, these effects do not require calculation, since the mean direct effects are the coefficients on the non-spatial variables and the mean indirect effects are those associated with the spatial lags of the explanatory variables. The mean total effects are simply the sum of these two coefficients. However, if we wish to construct confidence intervals on these impacts we can use the MCMC draws to do this.

TABLE 6.11: SLX and SDM model averaged estimates

Variable	SLX estimates			SDM estimates		
	Mean	0.95 Lower	0.95 Upper	Mean	0.95 Lower	0.95 Upper
income	−0.9879	−1.0832	−0.8829	−1.0697	−1.4305	−0.7192
hvalue	−0.1499	−0.1821	−0.1162	−0.2454	−0.3399	−0.1543
$W\cdot$ income	−0.9583	−1.1197	−0.7984	−0.0645	−0.1688	0.0359
$W\cdot$ hvalue	−0.0547	−0.1341	0.0184	0.0338	−0.0078	0.0742
$W\cdot y$				0.4046	0.2648	0.5294

Since the SDM model effects estimates are a non-linear function of the co-efficients, there is some question about how to construct our effects estimates in a model averaging setting. We follow Dennison, Holmes, Mallick and Smith (2002, p. 234-235) who relate models that have different nonlinear basis sets that describe the relationship between the response and covariates. In our model the response is y, the covariates are X, WX, and the matrix inverse $(I_n - \rho W)^{-1}$ that relates the response to changes in X could be viewed as a non-linear basis set.

Dennison, Holmes, Mallick and Smith (2002) discuss analysis of situations involving a non-linear smooth of the data, e.g., $E(y|\text{data}, \text{parameters}) = Sy$ where S is an $n \times n$ smoothing or hat matrix that transforms responses to fitted values. Specifically, their model takes the form in (6.25), where $\mathcal{D}$ represents the sample data and Σ, ϕ are variance-covariance and parameters from a seemingly unrelated VAR model that contains time-lag interactions between the elements in the matrix Y.

$$E(Y|\mathcal{D}, \Sigma, \phi) = B\tilde{\beta} \qquad (6.25)$$
$$= B(B'\phi B + \Sigma^{-1})^{-1}B'\phi Y$$
$$= SY \qquad (6.26)$$

In an illustration they calculate the smoothing matrix S for each sampled basis set B from the MCMC simulation and argue that averaging over these draws produces $E(S|\mathcal{D})$, an expected smoothing matrix. Predictions are then made using an average over two data sets Y_1, Y_2 contained in the matrix Y.

This suggests we should proceed by calculating a posterior probability weighted average of our smoothing matrix $(I_n - \rho W)^{-1}(I_n\beta_r + W\theta_r)$, us-ing MCMC draws for ρ, β, θ arising from each set of 16 models. The main diagonal elements of this matrix would reflect direct impacts, and off-diagonal elements would reflect indirect impacts that could be transformed to the scalar summary measures described in Chapter 2.

We note that for linear model relationships, model averaging relies on a lin-ear combination of the parameter draws from MCMC simulation constructed using the posterior model probabilities. In the linear model case, this constitutes the posterior distribution for the parameters ρ, β, θ, which should provide the basis for all Bayesian inference. Taking this approach with our non-linear spatial model relationship would produce different estimates and inferences regarding the impact estimates. This is because an average of non-linear terms is not the same as the non-linear terms averaged. To illustrate the difference in outcomes, we compare these two approaches. Specifically, this second approach applied the posterior model probabilities to the 16 sets of parameter draws, then used the single linear combination of draws to construct the effects estimates matrix, $(I_n - \rho W)^{-1}(I_n \beta_r + W \theta_r)$. The main diagonal elements of this single matrix were treated as direct impacts and off-diagonal elements as indirect impacts that were transformed to the scalar summary measures.

TABLE 6.12: SLX and SDM model averaged impact estimates

Variable	0.99 Lower	Mean	0.99 Upper	Std.
	SLX impacts			
direct income	−1.1237	−0.9879	−0.8435	0.0601
direct hvalue	−0.1981	−0.1499	−0.1013	0.0205
indirect income	−1.1885	−0.9583	−0.7286	0.0957
indirect hvalue	−0.1691	−0.0547	0.0585	0.0489
total income	−2.1755	−1.9463	−1.6959	0.0983
total hvalue	−0.3292	−0.2046	−0.0924	0.0505
	SDM Linear matrix impacts			
direct income	−1.6586	−1.1299	−0.5839	0.2268
direct hvalue	−0.3932	−0.2533	−0.1218	0.0589
indirect income	−1.6744	−0.8081	−0.2756	0.2973
indirect hvalue	−0.2976	−0.1063	0.0285	0.0686
total income	−3.2186	−1.9380	−0.9749	0.4630
total hvalue	−0.6651	−0.3597	−0.1158	0.1126
	SDM Non-linear matrix impacts			
direct income	−1.1816	−0.6525	−0.1714	0.2157
direct hvalue	−0.2961	−0.1641	−0.0316	0.0566
indirect income	−1.5945	−0.4435	−0.0318	0.3040
indirect hvalue	−0.4045	−0.1111	−0.0061	0.0779
total income	−2.5626	−1.0960	−0.2472	0.4653
total hvalue	−0.6504	−0.2752	−0.0447	0.1199

For the case of the SLX model, the non-linearity issue does not arise since the effects matrix takes the form: $(I_n \beta_r + W \theta_r)$. Table 6.12 shows the impact estimates calculated both ways, as well as impact estimates calculated for the

SLX model. The SLX effects were re-calculated using 0.99 credible intervals for consistency with the SDM impacts reported in the table. As noted, the total impacts for the SLX model are simply the sum of the draws for the parameters β associated with the matrix X and θ associated with WX.

In the table, we label the impacts calculated based on the non-linear expected smoothing matrix as *Non-linear matrix impacts* and those based on the (linear) posterior probability combination of draws for the parameters as *Linear matrix impacts*. From the table, we see much greater dispersion in the SDM model effects estimates calculated using both the *non-linear matrix* and *linear matrix* than for the SLX model effects, as can be seen from the standard deviations reported.

There are also differences between the SLX and SDM impact estimates that are likely to be statistically different, as well as differences between the impacts reported using the two different model averaging calculation approaches. For example, the direct effect for *hvalue* from the SLX model is around two standard deviations away from that from the SDM *linear matrix* effects, using the larger standard deviation from the SDM effect estimate. The same is true of the indirect effect for this variable. The direct effects of *income* from the *linear matrix* versus *non-linear matrix* approach are around three standard deviations apart. It is interesting that the standard deviations from the two approaches to calculating model averaged impact estimates are very similar, but the means diverge. This is what we might expect since an average of non-linear terms is not the same as the non-linear terms averaged.

TABLE 6.13: SLX and SDM single model effects estimates

Variable	0.99 Lower	Mean	0.99 Upper	Std.
SLX single model effects				
direct income	−2.0406	−1.1338	−0.2320	0.3893
direct hvalue	−0.5415	−0.2837	−0.0226	0.1070
indirect income	−2.7249	−1.3744	−0.0166	0.5721
indirect hvalue	−0.2966	0.1881	0.6654	0.2061
total income	−3.6954	−2.5082	−1.2688	0.5066
total hvalue	−0.5935	−0.0955	0.4015	0.2078
SDM single model effects				
direct income	−2.0129	−1.0624	−0.1784	0.3879
direct hvalue	−0.5230	−0.2831	−0.0527	0.1007
indirect income	−6.0508	−1.6823	0.7551	1.4323
indirect hvalue	−0.7393	0.1846	0.9578	0.3332
total income	−7.5905	−2.7447	−0.0894	1.5900
total hvalue	−1.0907	−0.0986	0.8142	0.3685

Finally, it is of interest to compare the impact estimates from a single model versus those based on the model averaging procedure. These are shown in Table 6.13, for a *saturated* version of both the SLX and SDM models, where *all variables* were included during estimation. Focusing on the SLX effects estimates, we see the same pattern of shrinkage towards zero for the model averaged estimates relative to those from the saturated model, reflecting the role of model uncertainty. We also see that the saturated model produced much larger standard deviations, or dispersion in the effects estimates, presumably due to the inclusion of all variables in the saturated model. This reflects the classic trade-off between attempting to include a sufficient number of variables to overcome potential omitted variables bias and inclusion of redundant variables that decrease precision of the estimates. It is precisely this trade-off that model averaging seeks to address. A comparison of the two sets of SLX effects estimates suggests that model averaging was successful in this regard. It is also of interest that the model averaged effects lead us to infer that the total impacts from both variables are negative and significant. In contrast, the single saturated model estimates suggest that the total effect of *hvalue* is not significant.

We compare the *non-linear matrix* model averaged SDM impact estimates to those from the single saturated model in Table 6.13. Again, the mean impact estimates are relatively smaller for the model averaged estimates than the saturated single model. Again, this suggests that model averaging is capturing our uncertainty about the model specification. With regard to the dispersion of the single model estimates and that of the model averaged estimates we also see the same pattern as with the SLX effects estimates. The model averaged estimates exhibit less dispersion, which can be seen by comparing the standard deviations reported in both tables. This implies that the saturated model is suffering from the classic over-inclusion of redundant variables, and model averaged effects estimates improve on this situation. As in the case of the single model SLX effects estimates, the SDM model effects from the single saturated model would lead to the conclusion that the total effect of *hvalue* is not significant, whereas both sets of model averaged estimates show a negative and significant total effect for this variable.

6.4 Chapter summary

A number of issues arise in applied modeling regarding model specification. In the case of spatial regression models these include questions regarding the type of spatial weight matrix to use as well as the usual uncertainty about explanatory variables.

A desirable extension of the MC^3 methodology and model averaging would be to determine *both* the spatial weight matrix and explanatory variables. We note that explanatory variables determined using the MC^3 methodology presented here were conditional on the specific spatial weight matrix employed. LeSage and Fischer (2008) extend the MC^3 approach of LeSage and Parent (2007) presented here to accomplish this. The extension introduces a "birth step" and "death step" that can be used to increase or decrease the number of nearest neighbors in the spatial weight matrix.

There are other approaches to Bayesian model comparison that represent approximations to the log-marginal likelihood needed to calculate posterior model probabilities. These methods are useful where it is difficult or impossible to carry out integration of the parameters that arise in the log-marginal likelihood expression. For example, Parent and LeSage (2008) use a method proposed by Chib (1995), and Chib and Jeliazkov (2001) to compare a host of alternative non-nested spatial model specifications based on varying types of spatial, technological and transport connectivity of European regions. This method approximates the log-marginal likelihood using MCMC draws of the parameters to "integrate" these out of the expression for the log-marginal likelihood.

A simple procedure proposed by Newton and Raftery (1994) is to evaluate the log-likelihood function on each pass through the MCMC sampler and calculate a harmonic mean of these values as an approximation to the log-marginal likelihood. This approach is illustrated in LeSage and Polasek (2008) to compare models of the type discussed in Chapter 8 based on two different spatial weight matrices.

6.5 Chapter appendix

In this appendix, we describe a computationally efficient approach to evaluating four separate terms involved in the univariate integration problem over the range of support for the parameter ρ in the SAR model. In the context of the MC^3 described in Section 6.3 there is a need for a computationally fast scheme for the univariate integration. This must be carried out on every pass through the MCMC sampler which occurs thousands of times.

The four terms in (6.18) for the SAR model that vary with ρ are shown in (6.27) as T_1, T_2, T_3 and T_4.

$$T_1(\rho) = |I_n - \rho W| \tag{6.27}$$
$$T_2(\rho) = [(I_n - \rho W)y - X\hat{\beta}(\rho) - \hat{\alpha}\iota_n]'[(I_n - \rho W)y - X\hat{\beta}(\rho) - \hat{\alpha}\iota_n]$$
$$T_3(\rho) = (\frac{g}{1+g})\hat{\beta}(\rho)'X'X\hat{\beta}(\rho)$$
$$T_4(\rho) = \frac{1}{Beta(d,d)}\frac{(1+\rho)^{d-1}(1-\rho)^{d-1}}{2^{2d-1}}$$

A log transformation can be applied to all terms $T_1, \ldots, T_4$, allowing us to rely on computationally fast methods presented in Pace and Barry (1997) and Barry and Pace (1999) to compute the log-determinant in T_1 (see Chapter 4).

Pace and Barry (1997) also suggest a vectorization of the terms in T_1 and T_2 that we used in Chapter 3 for maximum likelihood estimation of the SAR model. This involves constructing log-determinant values over a grid of q values of ρ, which is central to our task of integration for the terms $T_1(\rho)$ and $T_2(\rho)$. In applied work involving the SAR model, we typically rely on a restriction of ρ to the $(-1, 1)$ or $[0, 1)$ interval to avoid the need to compute eigenvalues.

Turning attention to the term $T_2(\rho)$, we follow Chapter 3 and write the term, $[(I_n - \rho W)y - X\hat{\beta}(\rho) - \hat{\alpha}\iota_n]'[(I_n - \rho W)y - X\hat{\beta}(\rho) - \hat{\alpha}\iota_n]$ as a vector in q values of ρ. For our problem we have the expression shown in (6.28).

$$T_2(\rho_i) = e(\rho_i)'e(\rho_i), \quad i = 1, \ldots, q \tag{6.28}$$

With:

$$e(\rho_i) = e_o - \rho_i e_d$$
$$e_o = y - X\beta_o - \alpha_o\iota_n$$
$$e_d = Wy - X\beta_d - \alpha_d\iota_n$$
$$\beta_o = (X'X)^{-1}X'y$$
$$\beta_d = (X'X)^{-1}X'Wy$$
$$\alpha_o = \bar{y}$$
$$\alpha_d = \overline{Wy} \tag{6.29}$$

The term T_3 can be vectorized using a loop over ρ_i values along with the expression $\hat{\beta}(\rho) = \beta_o - \rho_i\beta_d$. Finally, the term $T_4(\rho)$ representing the prior on the parameter ρ is simple to compute over a grid of q values for ρ, and transform to logs.

One important point to note is that we do not need to estimate the model parameters $\eta = (\alpha, \beta, \sigma, \rho)$ to carry out numerical integration leading to posterior model probabilities. Intuitively, we have analytically integrated the parameters α, β and σ out of the problem, leaving only a univariate integral in ρ. Given any sample data y, X along with a spatial weight matrix W, we can rely on the Pace and Barry (1997) vectorization scheme applied to our task. This involves evaluating the log-marginal density terms $T_1, \ldots, T_4$ over a fine grid of q values for ρ ranging over the interval $(-1, 1)$. Given a matrix of vectorized log-marginal posteriors, integration can be accomplished using Simpson's rule.

Further computational savings can be achieved by noting that the grid can be rough, say based on 0.01 increments in ρ, which speeds the direct sparse matrix approach of Pace and Barry (1997) or Barry and Pace (1999) computations. Spline interpolation can then be used to produce a much finer grid very quickly, as the log-determinant is typically quite well-behaved for reasonably large spatial samples in excess of 250 observations.

Another important point concerns scaling which is necessary to carry out numerical integration for the anti-log of the log-marginal posterior density. Our approach allows one to evaluate log-marginal posteriors for each model under consideration and store these as vectors ranging over the grid of ρ values. Scaling then involves finding the maximum of these vectors placed as columns in a matrix, (e.g. the maximum from all columns in the matrix). This maximum is then subtracted from all elements in the matrix of log-marginals, producing a value of zero as the largest element, so the anti-log is unity. This approach to scaling provides an elegant solution that requires no user-intervention and works for all problems.

Chapter 7

Spatiotemporal and Spatial Models

Unlike the previous chapters, this chapter is more theoretical and concentrates on the spatiotemporal foundations of spatial models. To achieve this goal, we assume that regions are only influenced by their own and other regions' past variables (no simultaneous influence). We show that this strict spatiotemporal framework results in a long-run equilibrium characterized by simultaneous spatial dependence. Note, we specifically avoid assuming spatial simultaneity in the spatiotemporal process as this would be assuming what we are trying to show. To keep the exposition as simple as possible and to expose relations among some of the common models we employ a number of assumptions such as symmetric W, constant or deterministically growing X, and no structural change over time.

Strictly temporal models provide our starting point, and econometrics provides a rich set of non-spatial temporal models grounded in economic theory. Partial adjustment models provide a classic example of this type of model. Partial adjustment models as well as other motivations give rise to specifications that employ temporal lags of both the dependent and explanatory variables.

In the context of regional data, conventional temporal models allow the dependent variable y_t for each region to be temporally dependent on past period values $y_{t-j}, j = 1, \ldots, j - 1$ of the own region. These conventional temporal models can be reasonably modified to allow for spatial dependence on other regions through time using spatial lags of the time lags (space-time lags) $W y_{t-1}$ and $W X_{t-1}$. These can be incorporated into the model in addition to conventional temporal lags, y_{t-1} and X_{t-1}, leading to a form of spatiotemporal model.

We have already noted that cross-sectional spatial lag models such as the SAR exhibit *simultaneous dependence* which may seem counterintuitive in some applied settings. However, cross-sectional spatial dependence can arise from a diffusion process working over time rather than occurring simultaneously. In this chapter we explore how spatiotemporal processes working over time can lead to equilibrium outcomes that exhibit spatial dependence. Our focus is on the spatiotemporal underpinnings of the cross-sectional spatial dependence that we often observe in regional data samples. We show how spatiotemporal data generating processes are related to many of the cross-sectional models popular in spatial econometrics and statistics. In addition,

189

this analysis suggests more complicated spatial models for further exploration.

7.1 Spatiotemporal partial adjustment model

We provide a simple generalization of the well-known partial adjustment model to illustrate how temporal models can be adapted to a spatiotemporal setting. The temporal partial adjustment development in Greene (1997, p. 698-799) serves as a starting point, but we extend this to a spatiotemporal setting. The basic equations for our spatial partial adjustment model are in (7.1)–(7.3).

$$y_t^* = U_t\psi + WU_t\gamma + \alpha\iota_n \tag{7.1}$$
$$y_t = (1 - \phi)y_t^* + \phi G_1 y_{t-1} + \varepsilon_t \tag{7.2}$$
$$G_1 = \theta I_n + \pi W \tag{7.3}$$

Let y_t^* denote the equilibrium value of the dependent variable, y_t. The $n \times p$ matrix U_t contains non-constant exogenous explanatory variables, and the $n \times 1$ vector of disturbances ε_t are distributed $N(0, \sigma^2 I_n)$. The parameter ϕ governs the degree of partial adjustment between previous values of the dependent variable, y_{t-1} and the equilibrium values y_t^*. The parameters ψ capture the effect of own-region explanatory variables, γ captures the effects of explanatory variables at nearby locations, and α is an intercept parameter. The scalar parameters θ, π measure the extent of temporal and spatial dependence captured by the $n \times n$ matrix G_1.

The equilibrium level of the dependent variable, y_t^*, depends upon the explanatory variables of the own observations (U_t), nearby observations reflected in the spatial lag (WU_t), and an intercept (ι_n). The parameters associated with these explanatory variables are ψ, γ, and α. This type of model specifies observed y_t as a linear combination (governed by ϕ) of the equilibrium levels y_t^* and past values of the dependent variable (y_{t-1}) as well as nearby dependent variables reflected by the spatiotemporal lag vector (Wy_{t-1}).

Manipulating (7.1)–(7.3) yields (7.4) indicating that y_t depends on: temporal and space-time lags of the dependent variable (Gy_{t-1}), spatial lags of the explanatory variables (WU_t), in addition to the conventional relationship involving the explanatory variables U_t.

$$y_t = U_t(1 - \phi)\psi + WU_t(1 - \phi)\gamma + \iota_n(1 - \phi)\alpha + \phi G_1 y_{t-1} + \varepsilon_t \tag{7.4}$$

As a concrete example, consider the situation faced by retailers. The sales performance of retail stores is often modeled as a function of the store's size and the sizes of competitor stores. For retail activities involving commodities such as groceries, hardware, and clothing, stores that are located nearby represent the competition.

For this retailing example, suppose the variable U_t represents store size and the spatial lag variable WU_t the average size of nearby stores (competitors). If y_t measures store sales, these should be positively related to own store size and inversely to competitor store sizes. The desired level y_t^* is the expected store sales given the size of the store and that of competitor stores. Previous store sales and previous sales of competitors also influence current store sales. Lee and Pace (2005) fitted a spatiotemporal model (with simultaneous spatial components) to store sales in Houston and found strong spatial dependence. Store size was found to be important and store sales exhibited strong temporal as well as spatial dependence.

We can simplify (7.4) using the symbols defined in (7.6) to represent the underlying structural parameter combinations and combining the explanatory variables into a single matrix X_t. This results in a classic spatiotemporal model in (7.5).

$$y_t = X_t\beta + Gy_{t-1} + \varepsilon_t \tag{7.5}$$
$$G = \tau I_n + \rho W, \quad \tau = \phi\theta, \quad \rho = \phi\pi, \quad \beta = (1-\phi)\begin{bmatrix} \psi & \gamma & \alpha \end{bmatrix}' \tag{7.6}$$
$$X_t = \begin{bmatrix} U_t & WU_t & \iota_n \end{bmatrix}$$

To summarize, a minor change in some of the models used to motivate temporal lags of the explanatory and dependent variables in non-spatial econometrics can lead to a spatiotemporal specification such as (7.5). Section 7.2 discusses the relation between spatiotemporal models and cross-sectional spatial models.

7.2 Relation between spatiotemporal and SAR models

As briefly discussed in Chapter 1, we can relate cross-sectional spatial models to long-run equilibria associated with spatiotemporal models. In this section, we generalize the simple motivational example from Chapter 1 to explore the relation between spatiotemporal and spatial models. We will use the relationship derived here to show how a spatiotemporal mechanism (perhaps arising from the partial adjustment mechanism discussed in the previous section) can yield many of the cross-sectional models discussed in the spatial statistics and spatial econometrics literature. We note that the partial adjustment motivation from the previous section is not the only way to motivate a spatiotemporal generating process. Our developments apply more generally to the relation between spatiotemporal processes and cross-sectional spatial models, and we use these relationships to motivate new spatial specifications.

We begin with the model in (7.7), where y_t is an $n \times 1$ dependent variable vector at time t $(t \geq 0)$, and the $n \times k$ matrix X_t represents explanatory variables. As illustrated in the previous section, X_t could contain spatial lags of the explanatory variables. This is a generalization of the spatial temporal autoregressive model (STAR) (Pfeifer and Deutsch, 1980; Cressie, 1993; Pace et al., 2000) that relies on past period dependent variables and contains no simultaneous spatial interaction. We will show that this dynamic relationship implies a cross-sectional steady state that can be viewed as a simultaneous spatial interaction.

Note, we specifically avoid assuming any form of simultaneous spatial dependence in the spatiotemporal process itself as this would be assuming what we are trying to demonstrate (how simultaneous spatial dependence arises).

$$y_t = Gy_{t-1} + X_t\beta + v_t \tag{7.7}$$
$$X_t = \varphi^t X_0 \tag{7.8}$$
$$G = \tau I_n + \rho W \tag{7.9}$$
$$d_t = X_t\gamma \tag{7.10}$$
$$v_t = r + d_t + \varepsilon_t \tag{7.11}$$

The scalar parameter τ governs dependence between each region at time t and $t-1$, while the scalar parameter ρ reflects spatial dependence between each region at time t and neighboring regions at time $t-1$. The scalar parameter φ allows the explanatory variables to grow at a constant rate (φ) per period (as opposed to holding explanatory variables constant over time as in Chapter 1). A value of $\varphi = 1$ represents no growth in X_0 over time and values $\varphi > 1$ allow for growth in the explanatory variables. We assume $\varphi > \tau$.

As before, the spatial weight matrix W is an $n \times n$ exogenous non-negative matrix. We assume W is symmetric and scaled to have a maximum eigenvalue of 1 with a minimum eigenvalue that is greater than or equal to -1. Scaling any symmetric weight matrix by its maximum eigenvalue provides one way of obtaining a symmetric W with a maximum eigenvalue of 1. Alternatively, a symmetric doubly stochastic W has a maximum eigenvalue of 1.

Given that G is composed of the identity matrix and a symmetric weight matrix W, it is symmetric as well. Since G is real and symmetric, it has n real eigenvalues and a full rank set of n orthogonal real eigenvectors. The largest magnitude eigenvalue of G equals $\tau + \rho$. We assume the following stability restrictions in (7.12)

$$(\tau + \rho)^t < \kappa, \quad \rho \in [0, 1), \quad \tau \in [0, 1) \tag{7.12}$$

where κ is a small positive constant. This will ensure that for sufficiently large values of t, we can assume that G^t takes on the small values required for our analysis. Although we could examine negative τ and ρ, we choose to look only at positive ρ and τ to simplify the exposition and because negative ρ and τ are of minor interest.

In Chapter 2 we considered the role of omitted variables which we generalize here. The generalization involves assuming the overall $n \times 1$ disturbance vector v_t can be partitioned into three components. These components represent omitted variables independent of the explanatory variables (r), omitted variables correlated with the explanatory variables (d_t), and a random noise term (ε_t). The first component is an $n \times 1$ vector r that captures omitted variables uncorrelated with X_t that remain constant over time. These might be amenities, region specific attributes such as land or water area, border lengths of the regions or difficult-to-specify locational discounts and premia. For simplicity, we assume that r is distributed $N(0, \sigma_r^2 I_n)$. The second component is an $n \times 1$ vector $d_t = X_t \gamma$ representing the effect of omitted variables that are correlated with X_t, where $\gamma \neq 0$ reflects the strength of correlation. This component can grow or decrease over time since $X_t = \varphi^t X_0$. The third component is a random $n \times 1$ vector ε_t that we assume is distributed $N(0, \sigma_\varepsilon^2 I_n)$ and independent of ε_{t-i} for $i \in (0, t]$. We further assume that ε_{t-i} for $i \in [0, t]$ is independent of r and X_t.

The STAR model uses only past dependent variables and current independent variables to explain variation in the current dependent variable vector. Following Elhorst (2001) we use the recursive relation: $y_{t-1} = G y_{t-2} + X_{t-1}\beta + r + d_{t-1} + \varepsilon_{t-1}$ implied by the model in (7.7) to consider the state of this dynamic system after passage of t time periods, which is shown in (7.13)–(7.17).

$$y_t = (I_n \varphi^t + G\varphi^{t-1} +, \ldots, + G^{t-1}\varphi)X_0\beta + G^t y_0 + z \qquad (7.13)$$

$$z = z_1 + z_2 + z_3 \qquad (7.14)$$

$$z_1 = (I_n + G +, \ldots, + G^{t-1})r \qquad (7.15)$$

$$z_2 = (I_n \varphi^t + G\varphi^{t-1} +, \ldots, + G^{t-1}\varphi)X_0\gamma \qquad (7.16)$$

$$z_3 = \varepsilon_t + G\varepsilon_{t-1} + G^2\varepsilon_{t-2} +, \ldots, + G^{t-1}\varepsilon_1 \qquad (7.17)$$

Taking the expectation of the dependent variable in (7.13) for sufficiently large t yields the long-run equilibrium as shown in (7.18)–(7.21). Note, the terms involving r and ε vanish from the expectation of y_t since these both have expectations of zero and multiplication of a matrix function by these zero vectors yields zero vectors. We assume t is large enough for convergence as this is the result of a long-run process. This ensures the vector $G^t y_0$ from (7.13) will approximately vanish, and therefore the long-run equilibrium will not depend upon the initial values of y_0.

In addition, we require that $G^t \varphi^{-t}$ also vanishes in order to proceed from the finite series in (7.19) to the simpler expression in (7.20) (using the geometric series definition $(1 - a)^{-1} = 1 + a + a^2 +, \ldots$ for $\mathrm{abs}(a) < 1$). If $\varphi = 1$, this is the same convergence criteria as needed to ensure that the vector $G^t y_0$ from (7.13) will approximately vanish. In the case of a growing explanatory variable ($\varphi > 1$), this aids convergence of $G^t \varphi^{-t}$.

$$E(y_t) \approx \left(I_n \varphi^t + G\varphi^{t-1} +, \ldots, +G^{t-1}\varphi\right) X_0(\beta + \gamma) \tag{7.18}$$

$$\approx \left(I_n + G\varphi^{-1} +, \ldots, +G^{t-1}\varphi^{-(t-1)}\right) \varphi^t X_0(\beta + \gamma) \tag{7.19}$$

$$\approx (I_n - \varphi^{-1}G)^{-1} X_t(\beta + \gamma) \tag{7.20}$$

$$\approx \left(I_n - \frac{\rho}{\varphi - \tau} W\right)^{-1} \left(\frac{\varphi}{\varphi - \tau}\right) X_t(\beta + \gamma) \tag{7.21}$$

There is a relation between the expression in (7.21) and a cross-sectional spatial regression based on a set of time t cross-sectional observations shown in (7.22) (where ξ_t are the disturbances), with the associated expectation shown in (7.23).

$$y_t = \rho^* W y_t + X_t \beta^* + \xi_t \tag{7.22}$$

$$E(y_t) = (I_n - \rho^* W)^{-1} X_t \beta^* \tag{7.23}$$

The relation between (7.21) and (7.23) is such that, for a sufficiently large sample n, a consistent estimator applied to the spatiotemporal model in (7.7) and the cross-sectional model in (7.22) would produce estimates that exhibit the relations (7.24) and (7.25).

$$\rho^* = \frac{\rho}{\varphi - \tau} \tag{7.24}$$

$$\beta^* = \frac{\varphi(\beta + \gamma)}{\varphi - \tau} \tag{7.25}$$

Standard non-spatial models with omitted variables that are correlated with the included explanatory variables yield parameters $\beta + \gamma$, part of the numerator of β^*. However, the long-run temporal multiplier, $m_t = \varphi(\varphi - \tau)^{-1}$ amplifies $\beta + \gamma$ resulting in β^*. Furthermore, m_t involves the temporal autoregressive parameter τ as well as the parameter φ that governs the growth trend in X. If $\varphi = 1$ (no growth in the explanatory variables), this yields the classic temporal multiplier $(1 - \tau)^{-1}$.

The long-run spatial multiplier in (7.23) resembles the traditional spatial multiplier of $(I_n - \rho W)^{-1}$, but the spatial dependence parameter ρ is amplified by $(\varphi - \tau)^{-1}$. Consequently, values of $\varphi > 1$ associated with growth in X reduce the spatial dependence of the system as measured by ρ^*, all else equal. This arises because this process gives more weight to the present, whereas spatial influences in this model require time to develop. Conversely, values of $\varphi < 1$ give more weight to past values allowing more time for spatial influences to develop. The temporal dependence parameter also affects the overall spatial dependence, since greater levels of temporal dependence increase the role of the past, and the role of space via diffusion.

An interesting implication of this development is that cross-sectional spatial regressions and spatiotemporal regressions could produce very different estimates of dependence even when both types of models are correctly specified. For example, a cross-sectional spatial regression could result in estimates pointing to high spatial dependence while a spatiotemporal regression would produce estimates indicating relatively high temporal dependence and low spatial dependence. Despite the fact that the estimates from these two types of models are seemingly quite different, both regressions could be correct since they are based on different information sets. Use of a cross-sectional sample taken at a point in time reflects a different information set that will focus the estimates and inferences on a long-run equilibrium result arising from evolution of the spatiotemporal process. In contrast, use of a space-time panel data set will lead to estimates and inferences that place more emphasis on the time dynamics embodied in time dependence parameters.

In applied practice, use of a space-time panel data set might produce parameter estimates indicating low spatial dependence and high temporal dependence. This could lead to an erroneous inference that a pure temporal regression without any spatial component is appropriate. Care must be taken because these two regression model specifications have very different implications. A process with low spatial dependence and high positive temporal dependence implies a long-run equilibrium with high levels of spatial dependence. In contrast, use of a purely temporal regression specification implies a long-run equilibrium that is non-spatial. An implication is that spatial dependence estimates that are small in magnitude could dramatically change inferences about the underlying spatiotemporal process at work and interpretation of model estimation results.

This same cautionary note applies to parameters for the explanatory variables β^* arising from a single cross-section. These parameter values are inflated by the long-run multiplier when $\varphi(\varphi - \tau)^{-1} > 1$ relative to β from the spatiotemporal model. This is a well-known result from time-series analysis for the case of long-run multiplier impacts in autoregressive models. As previously mentioned, β^* also picks up omitted variable effects represented by the parameter γ that reflects the strength of correlation between included and omitted variables.

The relation in (7.24) underlies the interpretation of cross-sectional spatial autoregressive models. Since these models provide no explicit role for passage of time, they need to be interpreted as reflecting an equilibrium or steady state outcome. This also has implications for the impact from changes in the explanatory variables of these models. The model in (7.22) and (7.23) literally states that y_i and y_j $(i \neq j)$ simultaneously affect each other. However, viewing changes in X as setting in motion a series of changes that will lead to a new steady-state equilibrium at some unknown future time seems more intuitive in many situations.

7.3   Relation between spatiotemporal and SEM models

In Section 7.2 we discussed the relation between spatiotemporal and spatial autoregressive models. In this section, we extend that analysis to error models. We begin with the set of equations (7.26)–(7.30).

$$y_t = G(y_{t-1} - X_{t-1}\beta) + X_t\beta + v_t \tag{7.26}$$
$$X_t = \varphi^t X_0 \tag{7.27}$$
$$G = \tau I_n + \rho W \tag{7.28}$$
$$d_t = X_t\gamma \tag{7.29}$$
$$v_t = r + d_t + \varepsilon_t \tag{7.30}$$

We now use the recursive relation: $y_{t-1} = G(y_{t-2} - X_{t-2}\beta) + X_{t-1}\beta + r + d_{t-1} + \varepsilon_{t-1}$ implied by the model in (7.26) to consider the state of this dynamic system after passage of t time periods from some initial time period 0, as shown in (7.31)–(7.35).

$$y_t = \varphi^t X_0\beta + G^t(y_0 - X_0\beta) + z \tag{7.31}$$
$$z = z_1 + z_2 + z_3 \tag{7.32}$$
$$z_1 = (I_n + G +, \ldots, +G^{t-1})r \tag{7.33}$$
$$z_2 = (I_n\varphi^t + G\varphi^{t-1} +, \ldots, +G^{t-1}\varphi)X_0\gamma \tag{7.34}$$
$$z_3 = \varepsilon_t + G\varepsilon_{t-1} + G^2\varepsilon_{t-2} +, \ldots, +G^{t-1}\varepsilon_1 \tag{7.35}$$

Considering the expectation of the dependent variable in (7.31) when t becomes large yields the long-run equilibrium shown in (7.36).

$$E(y_t) \approx X_t\beta + (I_n - \frac{\rho}{\varphi - \tau}W)^{-1}X_t(\frac{\gamma\varphi}{\varphi - \tau}) \tag{7.36}$$

The long-run equilibrium of the spatiotemporal system shown in (7.36) is non-spatial when $\gamma = 0$ so that no omitted variables that are correlated with the explanatory variables are present. In the presence of an omitted variable correlated with included variables ($\gamma \neq 0$), the form becomes more complicated. We note that terms involving r and ε_{t-i} ($i \in [0,t]$) vanish from the expectation of y_t, since these both have zero expectations and multiplication of a matrix function by these zero vectors yields zero vectors.

Summarizing these developments, there are important relationships between spatiotemporal models and cross-sectional spatial models. These relationships should further our understanding and interpretation of both cross-sectional and spatiotemporal models. These relationships have been ignored by much of the literature on spatial panel data models. This literature has largely focused on augmenting error covariance structures from conventional panel data models to account for spatial dependence.

7.4 Covariance matrices

In Section 7.3 we presented a spatiotemporal model with various disturbance components that were governed by a set of parameters related to assumptions about the model and disturbances. We derived an expression for $E(y_t)$ as a function of the explanatory variables, but did not examine the covariance matrix implied by the various assumptions regarding the model and disturbances. We now address this, and show that various values of the model parameters result in common spatial covariance models proposed in the literature. In particular, we show that the simultaneously specified Gaussian (SSG) and conditionally specified Gaussian (CSG) models arise from different circumstances that can be related to the model parameters.

The general spatiotemporal model in (7.7)–(7.11) results in a steady state in (7.13)–(7.17) with three error components: a location-specific time-persistent component which we view as a locational omitted variable (z_1), a component associated with an omitted variable that is correlated with explanatory variables (z_2), and a time-independent component (z_3).

We use z to represent the overall covariance. The assumed independence of ε_{t-i} and r, the zero expectation of these terms $(E(\varepsilon_{t-i}) = E(r) = 0)$, and the deterministic nature of X_t leads to a covariance for z that is the sum of the two expressions in (7.38).

$$\Omega_z = E(zz') - E(z)E(z') \tag{7.37}$$
$$= E(z_1 z_1') + E(z_3 z_3') \tag{7.38}$$

If $\sigma_\varepsilon^2 = 0$, $\Omega_z = E(z_1 z_1')$, and if $\sigma_r^2 = 0$, we have $\Omega_z = E(z_3 z_3')$. For simplicity, we examine individual error components first, and then look at them in combination.

We begin by finding the variance-covariance structure that arises from a model containing only the locational omitted variable component z_1. Assuming approximate convergence, (7.39) can be expressed in terms of G as shown in (7.40). We use the assumed symmetry of G to square the term in brackets in (7.40) rather than use the more cumbersome outer product. The assumption of symmetry also results in simpler forms for τ and ρ in (7.41), and ρ_1 in (7.42).

$$E(z_1 z_1') = [I_n + G + G^2 + \ldots]^2 \sigma_r^2 \tag{7.39}$$
$$\approx (I_n - G)^{-2} \sigma_r^2 \tag{7.40}$$
$$\approx (1 - \tau)^{-2} (I_n - \frac{\rho}{1 - \tau} W)^{-2} \sigma_r^2 \tag{7.41}$$
$$\approx (1 - \tau)^{-2} (I_n - \rho_1 W)^{-2} \sigma_r^2 \tag{7.42}$$
$$\rho_1 = \frac{\rho}{1 - \tau} \tag{7.43}$$

From this development we conclude that an error component associated with the locational omitted variable component z_1 leads to a covariance structure that matches that of the SSG model. The SSG model covariance takes the form: $\sigma^2 (I_n - A_s)^{-2}$, where $\sigma^2 = \sigma_r^2 (1 - \tau)^{-2}$ and $A_s = \rho_1 W$.

We now examine the variance-covariance structure that arises from a model containing an error component that represents disturbances that are independent over time. Using equation (7.17) representing the z_3 component we can form the outer product which in conjunction with symmetry of G leads to (7.44).

$$z_3 z_3' = \sum_{i=0}^{t-1} \sum_{j=0}^{t-1} G^i \varepsilon_{t-i} \varepsilon_{t-j}' G^j \tag{7.44}$$

Since $E(\varepsilon_{t-i} \varepsilon_{t-j}') = 0_n$ when $i \neq j$ and $\sigma_\varepsilon^2 I_n$ when $i = j$, the expectation of $z_3 z_3'$ has the form in (7.45) where symmetry allows use of terms such as G^2 rather than GG'. Using the approximate convergence of $G^{2(t-1)}$ allows simplifying (7.45), and this leads to expressions involving G as shown in (7.46) and (7.47). In terms of τ and ρ this has the form (7.48), and finally in terms of ρ_2 and ρ_3 this has the form (7.49).

$$E(z_3 z_3') = [I_n + G^2 + G^4 + \ldots]\sigma_\varepsilon^2 \tag{7.45}$$
$$\approx (I_n - G^2)^{-1} \sigma_\varepsilon^2 \tag{7.46}$$
$$\approx ((I_n - G)(I_n + G))^{-1} \sigma_\varepsilon^2 \tag{7.47}$$
$$\approx (1 - \tau)^{-1} (I_n - \frac{\rho}{1 - \tau} W)^{-1} \cdot$$
$$(1 + \tau)^{-1} (I_n + \frac{\rho}{1 + \tau} W)^{-1} \sigma_\varepsilon^2 \tag{7.48}$$
$$\approx (1 - \tau^2)^{-1} (I_n - \rho_2 W)^{-1} (I_n - \rho_3 W)^{-1} \sigma_\varepsilon^2 \tag{7.49}$$
$$\rho_2 = \frac{\rho}{1 - \tau} \tag{7.50}$$
$$\rho_3 = -\frac{\rho}{1 + \tau} = -\rho_2 \left[\frac{1 - \tau}{1 + \tau} \right] \tag{7.51}$$

These expressions can be simplified by considering special cases. For example, consider the case where the time dependence parameter $\tau = \tau_o$ where τ_o equals a large value such as 0.95. To meet the stability restrictions $\rho < 1 - \tau_o$ or less than 0.05. For concreteness, we let $\rho = \rho_o = 0.04$. Using (7.50) shows that $\rho_2 = 0.04/(1 - 0.95) = 0.8$ and using (7.51) shows that $\rho_3 = -0.04/(1 + 0.95) = -0.0205$. The term involving ρ_3 is small relative to the term involving ρ_2, and therefore it does not materially affect the approximation in (7.52).

$$E(z_3 z_3') \approx ((1 - \tau_o^2)^{-1} \sigma_\varepsilon^2)(I_n - \rho_2 W)^{-1} \tag{7.52}$$

The CSG model of covariance equals $\sigma^2 (I_n - A_c)^{-1}$. Therefore for large values such as τ_o, the covariance associated with the time independent disturbances approaches the CSG model, where $\sigma^2 = (1 - \tau_o^2)^{-1} \sigma_\varepsilon^2$ and $A_c = \rho_2 W$.

We now examine the overall variance-covariance matrix, Ω_z that arises from a model containing all disturbance components.

$$\Omega_z \approx (I_n - G)^{-2} \sigma_r^2 + ((I_n - G)(I_n + G))^{-1} \sigma_\varepsilon^2 \tag{7.53}$$

To give this more structure, we examine the scenario when τ is large (τ_o). Combining the individual covariance expressions for this case results in (7.54).

$$\Omega_z \approx \frac{\sigma_r^2}{(1 - \tau_o)^2}(I_n - A_s)^{-2} + \frac{\sigma_\varepsilon^2}{1 - \tau_o^2}(I_n - A_c)^{-1} \tag{7.54}$$

For fixed positive levels of σ_r^2, σ_ε^2 and fixed non-singular matrices A_s, A_c, increasing τ_o will inflate the SSG component faster than the CSG component. Consequently, SSG will dominate for large τ_o. For concreteness, using $\tau_o = 0.95$ results in (7.55), where the SSG component has a far greater weight than the CSG component (assuming the disturbance variances do not materially offset the relative contributions).

$$\Omega_z \approx 400 \sigma_r^2 (I_n - A_s)^{-2} + 10.2564 \sigma_\varepsilon^2 (I_n - A_c)^{-1} \tag{7.55}$$

We conclude that the general covariance expression in (7.53) leads to a potentially complicated covariance specification. However, simpler models emerge when we consider special cases. For example, if $\sigma_r^2 = 0$ and τ is large, the CSG covariance specification emerges. When $\sigma_r^2 > 0$, $\sigma_\varepsilon^2 > 0$, and τ is large, the SSG covariance specification emerges. If $\sigma_\varepsilon^2 = 0$, the SSG specification emerges.

A practical implication of these results is that models with excellent explanatory variables that address spatial effects might be able to greatly reduce the magnitude of impact that can potentially arise from omitted locational premia and discounts $(\sigma_r^2 = 0)$. This type of applied modeling situation would tend to favor a CSG specification.

Conversely, parsimonious models or those lacking important spatial explanatory variables so that $\sigma_r^2 > 0$ might lead to the SSG specification. This would be especially true when the dependent variable exhibits high temporal dependence.

7.4.1 Monte Carlo experiment

An interesting feature of the development of the covariance matrix for the general spatiotemporal model is the relation between SSG and CSG specifications and how these arise. In addition, the possibility of recovering the temporal dependence parameter, τ, from a single cross-sectional sample has many ramifications. To examine these issues, we carry out an experiment using the spatiotemporal model without the presence of latent effects. Our experiment will focus on dependence captured by $E(z_3 z_3')$.

A simple Monte Carlo experiment generated data using the spatiotemporal model in (7.56) setting $t = 1, \ldots, 250$, and $n = 50,000$ observations in each time period.

$$y_t = G y_{t-1} + X\beta + \varepsilon_t \tag{7.56}$$

The parameters $\beta = \begin{bmatrix} 0 & 1 \end{bmatrix}'$, and X included a constant vector ι_n as well as a *standard normal* vector (mean zero and variance of unity). The matrix X was constant for all time periods, and thus $\varphi = 1$. A standard normal vector was used for ε_t, and the last period's observations were used as the cross-sectional sample. That is, the n cross-sectional observations from the last time period ($t = 250$) from the spatiotemporal process became the dependent variable y for the spatial regression.

We considered three cases when composing the matrix G used to generate the experimental data. The first case had larger temporal and smaller spatial dependence ($\tau = 0.75$, $\rho = 0.15$), while the second case reversed this situation using ($\tau = 0.15$, $\rho = 0.75$). The third case used settings ($\tau = 0.4$, $\rho = 0.4$) reflecting moderate levels of spatial and temporal dependence. The values of τ and ρ in the spatiotemporal model imply various parameters in the cross-sectional spatial model. From the relations in the preceding section, $\rho^* = \rho/(\varphi - \tau)$, $\rho_2 = \rho/(1-\tau)$, $\rho_3 = -\rho/(1+\tau)$, and $\beta^* = \varphi\beta/(\varphi - \tau)$. Since X was held constant over time, $\varphi = 1$ for this experiment, and there were no omitted variables so $\gamma = 0$. For each of the three cases, we simulated and estimated 25 trials. Each trial took under 2.4 minutes to compute. The mean and standard deviation of the model parameters calculated on the basis of the 25 trials are shown in Table 7.1. The table also shows the theoretical parameter values based on values used in the spatiotemporal generating process in the rows labeled *true*.

Table 7.1 shows estimates for the parameters that were on average correct with varying levels of dispersion across the set of 25 outcomes for the three cases considered. In particular, the underlying parameters ρ^*, ρ_2, and β_2^* were estimated very accurately. However, estimates of ρ_3 displayed higher

TABLE 7.1: Experimental estimates

Cases	τ	ρ	ρ^*	ρ_2	ρ_3	β_2^*
1 true	0.7500	0.1500	0.6000	0.6000	−0.0857	4.0000
1 mean	0.7632	0.1447	0.6000	0.5992	−0.0846	3.9989
1 s.d.	0.1060	0.0694	0.0000	0.0283	0.0442	0.0071
2 true	0.1500	0.7500	0.8824	0.8824	−0.6522	1.1765
2 mean	0.1518	0.7465	0.8800	0.8800	−0.6484	1.1853
2 s.d.	0.0153	0.0171	0.0000	0.0071	0.0232	0.0045
3 true	0.4000	0.4000	0.6667	0.6667	−0.2857	1.6667
3 mean	0.4131	0.3889	0.6668	0.6608	−0.2768	1.6670
3 s.d.	0.0528	0.0476	0.0048	0.0222	0.0443	0.0075

variability and this degraded the estimation accuracy of τ and ρ, especially for the higher values of τ. Although ρ^* and ρ_2 had very similar values in the table, this results from assuming that X is constant over time. These parameters would differ when X changes over time.

The results from this experiment should be viewed as a demonstration that it is reasonable to rely on cross-sectional spatial regression models to analyze sample data generated by spatiotemporal processes. In particular, estimates of the regression parameters, β^*, were very accurate and differed only by a factor of proportionality relative to estimates of β from a spatiotemporal model.

The experiment validates our Chapter 2 spatiotemporal motivation for observed spatial dependence in cross-sections involving regional data samples. It is important that we understand how spatial dependence arises in cross-sectional regional data samples, and the next section pursues this topic further.

7.5 Spatial econometric and statistical models

The most common models in spatial econometrics are the autoregressive model (SAR), the spatial Durbin model (SDM), and the SSG error model (SEM). In spatial statistics, the CSG error model or CAR is also common. This section discusses how particular values for parameters in the general spatiotemporal model lead to many of these popular cross-sectional models.

Taking the DGP implied by (7.21), (7.42), and (7.43) yields a SAR DGP. This model specification arises when: 1) X remains constant over time and 2) the disturbances are time-persistent location-specific disturbances taking the form we have labeled r, and 3) there are no omitted variables correlated with the explanatory variables. These restrictions yield (7.57) and the simpler form shown in (7.58).

$$y = (I_n - \frac{\rho}{1-\tau}W)^{-1}(1-\tau)^{-1}X\beta +$$

$$(I_n - \frac{\rho}{1-\tau}W)^{-1}(1-\tau)^{-1}r \tag{7.57}$$

$$y = (I_n - \rho^*W)^{-1}X\beta^* + (I_n - \rho^*W)^{-1}r^* \tag{7.58}$$

In the context of a SAR model where y represents variation in home prices, r could capture amenities. These might include water views, tree shade, bike paths, landscaping, and sidewalks. We might also have externalities reflecting proximity to sewerage treatment plants, hazardous waste sites, nearby houses with garish, discordant colors, or noise from roads. Other environmental factors could include microclimates such as frost pockets, locations on the south or north side of a hill as well as the direction and strength of wind. Also, Catholic school districts, Catholic parishes, mosquito abatement districts, special assessment areas, and parade routes are ignored in most applied modeling situations. Any or all of these serve as examples of potentially omitted variables that have a spatial character. In fact, almost every location is influenced by variables that change slowly over time and the relevant variables differ across locations.

The SEM or error model DGP in (7.59) arises when the only error component is r and no omitted variables exist.

$$y = X\beta + (I_n - \frac{\rho}{1-\tau}W)^{-1}(1-\tau)^{-1}r \tag{7.59}$$

$$y = X\beta + (I_n - \rho^*W)^{-1}r^* \tag{7.60}$$

Therefore, the standard SSG error model emerges from a spatiotemporal error process with time-persistent, location-specific disturbances r and no omitted variables that are correlated with X (i.e., $\gamma = 0$). Unlike the autoregressive case, X does not need to be constant over time to arrive at this model specification.

However, if omitted variables that are correlated with X are present (i.e., $\gamma \neq 0$) and no growth in X occurs, this results in a more complicated expression in (7.61).

$$y = X\beta + (I_n - \frac{\rho}{1-\tau}W)^{-1}X(\gamma(1-\tau)^{-1}) +$$

$$(I_n - \frac{\rho}{1-\tau}W)^{-1}(1-\tau)^{-1}r \tag{7.61}$$

To estimate (7.61) we can transform y by $(I_n - \frac{\rho}{1-\tau}W)$ to yield (7.63) that has *iid* disturbances.

$$(I_n - \frac{\rho}{1-\tau}W)y = (I_n - \frac{\rho}{1-\tau}W)X\beta +$$

$$X\gamma(1-\tau)^{-1} + (1-\tau)^{-1}r \qquad (7.62)$$

$$(I_n - \rho^*W)y = X\beta_1 + WX\beta_2 + r^* \qquad (7.63)$$

The new equation (7.63) is the SDM. This specification arises when: 1) omitted variables that are correlated with X are present (i.e., $\gamma \neq 0$) and 2) no growth in X occurs over time.

Suppose interest centers on an error model with *iid* disturbances over time with no omitted variables ($\gamma = 0$). Using an earlier expression (7.49) for the covariance Ω_{z3} shows that the error model DGP in (7.64) and (7.65) arises for large τ (τ_o).

$$y = X\beta + \epsilon \qquad (7.64)$$

$$\epsilon \sim N(0, \Omega_{z3}) \approx N(0, (1-\tau_o^2)^{-1}\sigma_\epsilon^2(I_n - \frac{\rho}{1-\tau_o}W)^{-1}) \qquad (7.65)$$

Therefore, the standard CSG error model emerges from a spatiotemporal error process with *iid* disturbances ε, when no omitted variables are present that are correlated with X (i.e., $\gamma = 0$), and τ is large. Unlike the autoregressive case, X does not need to be constant over time to arrive at this form.

In conclusion, the general spatiotemporal autoregressive and error models presented in Sections 7.2 and 7.3 subsume many of the well-known models from spatial econometrics and spatial statistics. Assumptions regarding specific parameters of the more general model lead to spatial autoregressive models (SAR), the spatial Durbin model (SDM), the SSG error model (SEM), and the CSG error model (CAR).

7.6 Patterns of temporal and spatial dependence

Although autoregressive dependence is the most commonly used form of temporal dependence, other forms exist such as moving average and exponential. We provide a brief development that shows how alternative types of temporal dependence in the context of spatiotemporal processes relate to alternative spatial models. In particular, this development suggests that spatial models can often inherit the form of the underlying spatiotemporal process.

We begin with a specification in (7.66) that assumes the model variables are in long-run equilibrium. In the last time period t, the dependent variable y_t depends on space-time lags as well as on $X\beta$ and disturbances u_t.

$$\begin{bmatrix} \omega_0 I_n & \omega_1 G & \omega_2 G^2 & \cdots & \omega_t G^t \\ 0_n & \omega_0 I_n & \omega_1 G & \cdots & \omega_{t-1} G^{t-1} \\ 0_n & 0_n & \omega_0 I_n & \omega_1 G \\ \vdots & \vdots & & 0_n & \ddots & \vdots \\ 0_n & & & 0_n & & \omega_0 I_n \end{bmatrix} \begin{bmatrix} y_t \\ y_{t-1} \\ y_{t-2} \\ \vdots \\ y_0 \end{bmatrix} = \begin{bmatrix} X\beta \\ X\beta \\ X\beta \\ \vdots \\ X\beta \end{bmatrix} + \begin{bmatrix} u_t \\ u_{t-1} \\ u_{t-2} \\ \vdots \\ u_0 \end{bmatrix} \quad (7.66)$$

We rewrite this in more compact form in (7.67).

$$Hy = (\iota_{t+1} \otimes X\beta) + u \quad (7.67)$$

We assume that the weights ω_i are associated with some analytic function $F(\cdot)$ and that an inverse function $F^{-1}(\cdot)$ exists with weights π_i. The models discussed in earlier sections of this chapter reflect restricted versions of this more general specification. For example, the autoregressive model uses $\omega_0 = 1$, $\omega_1 = -1$, and $\omega_2, \ldots, \omega_t = 0$ while $\pi_0, \ldots, \pi_t = 1$. However, we can represent other forms of dependence using different values for the parameters ω_i and π_i.

Our interest focuses on the long-run equilibrium, and so we examine the last period. We assume the system has converged so $\omega_t G^t$ is very small in magnitude. Solving for y_t and taking its expectation yields (7.68), where $(H^{-1})_1$ represents the first row of H^{-1}. Given the constant mean structure $X\beta$ which does not change over time sets up the simpler form in (7.69), with an associated matrix function version in (7.70).

$$E(y_t) \approx (H^{-1})_1(\iota_{t+1} \otimes X\beta) \quad (7.68)$$

$$\approx \sum_{i=0}^{t-1} \pi_i G^i X\beta \quad (7.69)$$

$$\approx F^{-1}(G)X\beta \quad (7.70)$$

Some examples may clarify the relation between different types of time-series dependence and the resulting spatial equilibria.

Autoregressive Case If $F(G) = (I_n - G)$,

$$E(y_t) \approx ((1-\tau)I_n - \rho W)^{-1} X\beta \quad (7.71)$$

$$\approx (I_n - \frac{\rho}{1-\tau} W)^{-1} X \frac{\beta}{1-\tau} \quad (7.72)$$

$$\approx (I_n - \rho^* W)^{-1} X\beta^* \quad (7.73)$$

Matrix Exponential Case If $F(G) = e^{-\alpha G}$,

$$E(y_t) \approx e^{\alpha(\tau I_n + \rho W)} X\beta \tag{7.74}$$
$$\approx e^{\alpha \rho W} X(e^{\alpha \tau}\beta) \tag{7.75}$$
$$\approx e^{\alpha^{**}W} X\beta^{**} \tag{7.76}$$

To make this last case more concrete, for the matrix exponential spatial specification described in Chapter 9, $w_i = (i!)^{-1}\alpha^i$ and $\pi_i = (i!)^{-1}(-\alpha)^i$. We chose $\alpha = 0.2$ to ensure quick convergence (to fit the tables on the page). The matrix H in (7.66) was populated using the definition of w_i which produced (7.77).

$$H = \begin{bmatrix} 1.0000 & -0.2000 & 0.0200 & -0.0013 & 0.0001 & -0.0000 & 0.0000 \\ 0.0000 & 1.0000 & -0.2000 & 0.0200 & -0.0013 & 0.0001 & -0.0000 \\ 0.0000 & 0.0000 & 1.0000 & -0.2000 & 0.0200 & -0.0013 & 0.0001 \\ 0.0000 & 0.0000 & 0.0000 & 1.0000 & -0.2000 & 0.0200 & -0.0013 \\ 0.0000 & 0.0000 & 0.0000 & 0.0000 & 1.0000 & -0.2000 & 0.0200 \\ 0.0000 & 0.0000 & 0.0000 & 0.0000 & 0.0000 & 1.0000 & -0.2000 \\ 0.0000 & 0.0000 & 0.0000 & 0.0000 & 0.0000 & 0.0000 & 1.0000 \end{bmatrix} \tag{7.77}$$

The numerical inverse of the matrix H is shown in (7.78) as H^{-1}, where we see table entries that correspond very closely to the formula for π_i.

$$H^{-1} = \begin{bmatrix} 1.0000 & 0.2000 & 0.0200 & 0.0013 & 0.0001 & 0.0000 & 0.0000 \\ 0.0000 & 1.0000 & 0.2000 & 0.0200 & 0.0013 & 0.0001 & 0.0000 \\ 0.0000 & 0.0000 & 1.0000 & 0.2000 & 0.0200 & 0.0013 & 0.0001 \\ 0.0000 & 0.0000 & 0.0000 & 1.0000 & 0.2000 & 0.0200 & 0.0013 \\ 0.0000 & 0.0000 & 0.0000 & 0.0000 & 1.0000 & 0.2000 & 0.0200 \\ 0.0000 & 0.0000 & 0.0000 & 0.0000 & 0.0000 & 1.0000 & 0.2000 \\ 0.0000 & 0.0000 & 0.0000 & 0.0000 & 0.0000 & 0.0000 & 1.0000 \end{bmatrix} \tag{7.78}$$

Therefore, taking the first row of (H^{-1}) from (7.78) and using (7.68) shows that $E(y_t) = (I_n + 0.2G + 0.02G^2 + 0.0013G^3 + , \ldots)X\beta$ or $E(y_t) = e^{0.2G}X\beta$.

To further explore the relation between the matrix exponential spatiotemporal model and the matrix exponential spatial model we conducted a simple Monte Carlo experiment. We generated the data using the spatiotemporal equation (7.66) where w_i represent terms from the Taylor series expansion of the matrix exponential. See both Chapter 4 and Chapter 9 for more details regarding the matrix exponential spatial specification.

In the Monte Carlo experiment we used 200 periods ($t = 1, \ldots, 200$), with $5,000$ observations for each period. The parameters $\beta = \begin{bmatrix} 0 & 1 \end{bmatrix}'$, and the matrix X consisted of a constant term and a standard normal vector (mean zero, variance of unity) that was held constant over all time periods. The vector

u represented a standard normal deviate that was held constant over time to reflect locational omitted variables. We used the last period's observations as the cross-sectional sample. That is, the cross-sectional $n \times 1$ dependent variable vector y was taken from the last period of a spatiotemporal process.

We used three settings for the strength of temporal and spatial dependence to determine parameters used to form αG. The first case had moderate temporal dependence and no spatial dependence ($\alpha\tau = 1.00$, $\alpha\rho = 0.0$), while the second case reversed this and had no temporal dependence with moderate spatial dependence ($\alpha\tau = 0.0$, $\alpha\rho = 1.0$). The third case represents moderate spatial and temporal dependence ($\alpha\tau = 1.0$, $\alpha\rho = 1.0$). For each case we simulated and estimated the model parameters using 100 trials via the matrix exponential methods described in Chapter 9.

TABLE 7.2: Matrix exponential Monte Carlo results

Cases	α^{**}	β_0^{**}	β_1^{**}
1 true	0.0000	0.0000	2.7183
1 mean	0.0049	0.0042	2.7111
1 s.d.	0.0249	0.0599	0.0371
2 true	1.0000	0.0000	1.0000
2 mean	0.9976	0.0087	0.9954
2 s.d.	0.0227	0.0172	0.0125
3 true	1.0000	0.0000	2.7183
3 mean	0.9981	0.0145	2.7097
3 s.d.	0.0186	0.0320	0.0204

The experimental results appear in Table 7.2 where α^{**} is the overall dependence parameter, β_0^{**} is the intercept parameter, and β_1^{**} is the parameter associated with the non-constant explanatory variable. These results show close agreement between the spatiotemporal theory and experimental estimates.

As the matrix exponential example illustrates, the relation between spatial and spatiotemporal models is not specific to the autoregressive model specification. Many alternative forms of temporal dependence could lead to spatial dependence model specifications. This has the potential to broaden and expand spatial modeling. These developments also generalize our spatiotemporal motivation for observed cross-sectional spatial dependence.

7.7   Chapter summary

This chapter dealt with spatiotemporal models as well as the relation between spatial and spatiotemporal models. We began with Section 7.1 and showed that the well-known partial adjustment model can be easily augmented to include previous values of the dependent and explanatory variables from nearby observations or regions. This results in a spatiotemporal model that contains both time as well as space-time lags of the model variables, but no contemporaneous spatial lags.

In Section 7.2 we introduced a general spatiotemporal autoregressive process involving explanatory variables with deterministic growth, temporal lags of variables, temporal lags of spatial lags of variables, and a disturbance term comprised of three components. One component was a locational omitted variable (not correlated with the included variables), the second component was an omitted variable correlated with the included variables, and the third component was an *iid* disturbance.

Although this was a more general process, it did not contain any form of simultaneous spatial dependence in the spatiotemporal process itself as this would be assuming what we are trying to demonstrate (how simultaneous spatial dependence arises).

We showed that the resulting long-run equilibrium arising from this strict spatiotemporal process took the form: $E(y_t) \approx (I_n - \rho^* W)^{-1} X_t \beta^*$, where $\rho^* = \rho(\varphi - \tau)^{-1}$, and $\beta^* = (\beta + \gamma)\varphi(\varphi - \tau)^{-1}$. The scalar ρ is the spatial dependence parameter, τ is the temporal dependence parameter, γ reflects dependence between the included and omitted variable, and φ equals 1 plus the growth rate of X over time.

The relation between the long-run equilibrium parameters and the spatiotemporal model parameters has important implications concerning the interpretation and use of these models. In particular, cross-sectional spatial regressions and spatiotemporal regressions could produce very different estimates of dependence even when both types of models are correctly specified. For example, a cross-sectional spatial regression could result in estimates indicating high spatial dependence while a spatiotemporal regression could produce estimates indicating relatively high temporal dependence and low spatial dependence. Despite the fact that the estimates from these two types of models are seemingly quite different, both regressions could be correct since they are based on different information sets. In some applications, a large and significant temporal parameter estimate coupled with a small and less significant spatial parameter estimate could entice a practitioner into deleting the spatial variable. Although this strategy may seem fine from a goodness-of-fit perspective, interpretation of the two models would differ greatly. A small spatial parameter estimate and large temporal parameter estimate in a spatiotemporal model implies a long-run equilibrium with material spatial dependence, whereas a purely temporal regression implies a long-run equilibrium with no spatial dependence. This would make a large difference in how we interpret the impacts arising from changes in the explanatory variables of the model.

Section 7.3 showed a similar analysis for a spatiotemporal error model where, in the absence of omitted variables correlated with the included variables, the long-run equilibrium resulted in $E(y_t) \approx X_t\beta$. Although the disturbances exhibit spatial dependence, the actual long-run equilibrium is nonspatial. However, in the presence of omitted variables correlated with included variables, the error model yields a form of SDM, and this is fundamentally a spatial model.

Section 7.4 addressed the form of covariance structure that arises from using a spatiotemporal process to motivate a cross-sectional model specification. This development considered the case where the disturbances consisted of components related to a locational omitted variable as well as *iid* disturbances. Various parameter combinations resulted in CSG and SSG covariance structures. In particular, the presence of time-persistent, spatially dependent disturbances and high levels of time dependence τ yield the SSG covariance structure.

Section 7.4.1 provided a Monte Carlo demonstration that the CSG specification can arise in the presence of strong temporal dependence as discussed in Section 7.4. Under the correct model specification, the regression parameters were well estimated by a purely spatial model and comparable (proportionally) to those estimated from a spatiotemporal model using panel data. The Monte Carlo experiment corroborated the relation between the spatial and spatiotemporal approaches. However, for actual data the relation between these may be difficult to confirm as these relations take the form of ratios and are very sensitive to minor estimation errors, especially those pertaining to the temporal parameter τ.

Section 7.6 showed how specific assumptions within the general spatiotemporal framework could lead to standard cross-sectional spatial models. We showed that other temporal DGPs such as the matrix exponential process also implied various spatial long-run equilibria. This provides a means of extending various results from time series analysis to spatial econometrics.

The relation between spatiotemporal and spatial models motivates some modeling strategies. First, economic theory underlies many time series models and these may have spatial analogs. Therefore, theory may suggest more specific functional forms.

Second, allowing for growth in the explanatory variables X over time implies that the dependence parameter governing the mean could differ from the dependence parameter governing the covariance structure of the model. For example, in the spatial autoregressive model that allows X to grow over time the DGP is: $y = (I_n - \rho_a W)^{-1} X\beta + (I_n - \rho_b W)^{-1} r$. Only the special case in which X is constant over time leads to $\rho_a = \rho_b$. In large samples, efficiency becomes less of an issue relative to bias, and so getting the mean part of the model correct (i.e., $E(y_t) = F^{-1}(G)X\beta$) becomes more important and greatly affects model interpretation.

More generally, the processes governing X over time enter into the form of the spatial model. This suggests more elaborate lag structures involving the explanatory variables. For example, variants of the SDM that involve some form of distributed lags might be considered (Byron, 1992).

A third implication for modeling strategy would be that the covariance structure may be more complicated than in the standard models. Although modeling covariance may not be as important in large samples, it obviously affects the validity of inference for marginal variables and could aid in prediction as well as imputation.

Finally, using different parameters for modeling the mean of the process versus the disturbance process should provide some insurance against contamination of the mean model parameters that can arise from misspecification of the disturbance process. As discussed in Chapter 3 in the context of the Hausman test, misspecification of the model for the disturbance process will not affect parameter estimates associated with the (correctly specified) mean model in large samples. Therefore, using different parameters to specify the model for the mean versus the model for the disturbances may lead to more robust spatial modeling. The SDEM introduced in Section 2.7 represents one of the simplest models with this property.

The development here should also benefit the literature on space-time panel data models. This literature relies on traditional spatial regression models augmented with random effects parameters and space-time covariance structures. Using spatiotemporal processes of the type explored here could make these models more intuitive. This would ensure that space-time panel model specifications could be justified as arising from underlying space-time interactions that evolve over time to a steady state equilibrium. It would also promote understanding of the properties associated with the observed panel of cross-sectional sample data used to estimate the parameters of these models.

Chapter 8

Spatial Econometric Interaction Models

Gravity models have often been used to explain origin-destination (OD) flows that arise in fields such as international and regional trade, transportation economics, population migration research, modeling of commodity flows, communication and other types of information flows along a network and journey-to-work studies.

A large literature on theoretical foundations for these models in the specific context of international trade models exists (Anderson, 1979; Anderson and van Wincoop, 2004). In the regional science literature the gravity model has been labeled a *spatial interaction model* (Sen and Smith, 1995), because the regional interaction is directly proportional to the product of regional size measures. In the case of interregional commodity flows, the measure of regional size is typically gross regional product or regional income. The model predicts more interaction in the form of commodity flows between regions of similar (economic) size than regions dissimilar in size. In other contexts such as knowledge flows between regions (LeSage, Fischer and Scherngell, 2007) the size measure of regions might be the stock of patents, so that regions with similar knowledge stocks would exhibit more spatial interaction taking the form of knowledge flows.

These models rely on a function of the distance between an origin and destination as well as explanatory variables pertaining to characteristics of both origin and destination regions. Spatial interaction models assume that using distance as an explanatory variable will eradicate the spatial dependence among the sample of OD flows between pairs of regions. The notion that use of distance functions in conventional spatial interaction models effectively captures spatial dependence in interregional flows has long been challenged. Griffith (2007) provides an historical review of regional science literature on this topic in which he credits Curry (1972) as the first to conceptualize the problem of spatial dependence in flows. Griffith and Jones (1980) in a study of Canadian journey-to-work flows noted that flows from an origin are "enhanced or diminished in accordance with the propensity of emissiveness of its neighboring origin locations." They also stated that flows associated with a destination are "enhanced or diminished in accordance with the propensity of attractiveness of its neighboring destination locations."

LeSage and Pace (2008) make the point that assuming independence between flows is heroic since OD flows are fundamentally spatial in nature. They extend the traditional gravity model to allow for spatial lags of the dependent variable, which represent flows from neighboring regions in these models. In contrast to typical spatial econometric models where the sample involves n regions, with each region being an observation, these models involve $n^2 = N$ origin-destination pairs with each origin-destination pair being an observation. Spatial interaction modeling seeks to explain variation in the level of flows across the sample of N OD pairs.

This chapter introduces maximum likelihood estimation procedures for spatial interaction models set forth in LeSage and Pace (2008) along with Bayesian MCMC estimation procedures that have not appeared elsewhere. Section 8.1 introduces the notation, and develops a general family of spatial econometric interaction models that accommodate spatial dependence. Section 8.2 sets forth the maximum likelihood estimation approach from LeSage and Pace (2008) as well as Bayesian MCMC estimation procedures. An illustration is provided in Section 8.3 using population migration flows between metropolitan areas. Section 8.4 discusses extensions to the spatial econometric interaction model as well as alternative spatial modeling approaches to dealing with OD flows.

8.1   Interregional flows in a spatial regression context

Let Y denote an $n \times n$ square matrix of interregional flows from n origin regions to n destination regions where the n columns represent different origins and the n rows represent different destinations as shown in Table 8.1. The flows considered here reflect a closed system that consists of an equal number of origin and destination regions.

TABLE 8.1:   Origin-destination flow matrix

Destination /Origin	Origin 1	Origin 2	...	Origin n
Destination 1	$o_1 \to d_1$	$o_2 \to d_1$	...	$o_n \to d_1$
Destination 2	$o_1 \to d_2$	$o_2 \to d_2$	...	$o_n \to d_2$
$\vdots$		$\vdots$		
Destination n	$o_1 \to d_n$	$o_2 \to d_n$	...	$o_n \to d_n$

Given the organization of the OD flow matrix in Table 8.1, we can use $n^{-1}Y\iota_n$ to form an $n \times 1$ vector representing an average of the flows from all of the n origins to each of the n destinations, where ι_n is an $n \times 1$ vector of ones. Similarly, $n^{-1}Y'\iota_n$ would produce an $n \times 1$ vector that is an average of flows from all of the n destinations to each of the n origins.

We can produce an $N(= n^2) \times 1$ vector of these flows from the flow matrix in Table 8.1 in two ways, one reflecting an *origin-centric* ordering as shown in Part A of Table 8.2, and the other reflecting a *destination-centric* ordering as in Part B of the table.

TABLE 8.2: Origin- and destination-centric OD flow arrangements

Part A: Origin-centric scheme for OD flows

Origin-centric index $l^{(o)}$	Origin-index $o^{(o)}$	Destination-index $d^{(o)}$
1	1	1
$\vdots$	$\vdots$	$\vdots$
n	1	n
$\vdots$	$\vdots$	$\vdots$
$N - n + 1$	n	1
$\vdots$	$\vdots$	$\vdots$
N	n	n

Part B: Destination-centric scheme for OD flows

Destination-centric index $l^{(d)}$	Origin-index $o^{(d)}$	Destination-index $d^{(d)}$
1	1	1
$\vdots$	$\vdots$	$\vdots$
n	n	1
$\vdots$	$\vdots$	$\vdots$
$N - n + 1$	1	n
$\vdots$	$\vdots$	$\vdots$
N	n	n

In the tables, the indices $l^{(o)}$, $l^{(d)}$ denote the overall index from $1, \ldots, N$ for the origin-centric and destination-centric orderings respectively. The origin and destination indices o, d range from $1, \ldots, n$, indicating the region of origin and destination respectively.

Beginning with a matrix Y whose columns reflect origins and rows destinations, we obtain the origin-centric ordering using the *vec* operator that transforms a matrix into a column vector by stacking columns sequentially, $y^{(o)} = \text{vec}(Y)$. The destination-centric ordering is produced using $y^{(d)} =$

vec(Y'). These two orderings are related by the vec-permutation matrix P so that $Py^{(o)} = y^{(d)}$. Based on the properties of permutation matrices, it is also true that $y^{(o)} = P^{-1}y^{(d)} = P'y^{(d)}$. For our discussion in this chapter we will focus on the origin-centric ordering where the first n elements in the stacked vector $y^{(o)}$ reflect flows from origin 1 to all n destinations. The last n elements of this vector represent flows from origin n to destinations 1 to n. We will refer to this OD flow vector as simply y, which represents the dependent variable vector in our spatial econometric interaction model.

A conventional gravity or spatial interaction model relies on a single vector or an $n \times k$ matrix of explanatory variables that we label X, containing k characteristics for each of the n regions. The matrix X is repeated n times to produce an $N \times k$ matrix representing destination characteristics that we label X_d. LeSage and Pace (2008) note that X_d equals $\iota_n \otimes X$, where ι_n is an $n \times 1$ vector of ones. A second matrix can be formed to represent origin characteristics that we label X_o. This would repeat the characteristics of the first region n times to form the first n rows of X_o, the characteristics of the second region n times for the next n rows of X_o and so on, resulting in an $N \times k$ matrix that we label $X_o = X \otimes \iota_n$. International trade models typically rely on a single explanatory variable vector X such as income to reflect the size of regions. This would result in $N \times 1$ vectors X_d, X_o rather than matrices of explanatory variables.

We note that the vec-permutation matrix P can be used to translate between origin-centric and destination-centric ordering of the sample data. For example, if we adopted the destination-centric ordering (as opposed to the origin-centric ordering), specification of the destination explanatory variables matrix would be $X_d^{(d)} = X \otimes \iota_n$. This can be seen using the relation: $P'X_dP = P'(\iota_n \otimes X)P = X_d^{(d)}$, to translate the origin-centric destination covariates X_d to the destination-centric ordering scheme $X_d^{(d)}$. Rules for multiplication using Kronecker products allow us to simplify the expression $P'(\iota_n \otimes X)P$, (Horn and Johnson, 1994, Corollary 4.3.10, p. 260), so that $P'(\iota_n \otimes X)P = X \otimes \iota_n$, and thus $X_d^{(d)} = X \otimes \iota_n$, under the destination-centric ordering of the sample data.

The distance from each origin to destination is also included as an explanatory variable vector in the gravity model. If we let G represent the $n \times n$ matrix of distances between origins and destinations, $g = $ vec(G) is an $N \times 1$ vector of distances from each origin to each destination formed by stacking the columns of the origin-destination distance matrix into a variable vector.

This results in a regression model of the type shown in (8.1). This is identical to the model that arises when applying a log transformation to the standard gravity model (Sen and Smith, 1995, c.f., equation (6.4)).

$$y = \alpha \iota_N + X_d\beta_d + X_o\beta_o + \gamma g + \varepsilon \qquad (8.1)$$

The vectors β_d and β_o are $k \times 1$ parameter vectors associated with the destination and origin region characteristics. If a log transformation is applied to the dependent variable y and explanatory variables matrix X, the coefficient estimates would reflect elasticity responses of OD flows to the various origin and destination characteristics. The scalar parameter γ reflects the effect of distance g, and α denotes the constant term parameter. The $N \times 1$ vector ε has a zero mean, constant variance and zero covariance between disturbances.

LeSage and Pace (2008) propose a spatial autoregressive extension of the non-spatial model in (8.1) shown in (8.2). This model can be viewed as *filtering* for spatial dependence related to the destination and origin regions.

$$(I_N - \rho_d W_d)(I_N - \rho_o W_o)y = \alpha \iota_N + X_d \beta_d + X_o \beta_o + \gamma g + \varepsilon \qquad (8.2)$$

The $N \times N$ matrix W_d is constructed from the typical row-stochastic $n \times n$ matrix W that describes spatial connectivity between the n regions. We assume that W is similar to a symmetric matrix so that it has real eigenvalues and n orthogonal eigenvectors. The matrix W_d can be written using the Kronecker product shown in (8.3).

$$W_d = I_n \otimes W = \begin{pmatrix} W & 0_n & \cdots\cdots & 0_n \\ 0_n & W & & \vdots \\ \vdots & & W & \\ & & & \ddots \\ 0_n & \cdots & & W \end{pmatrix} \qquad (8.3)$$

The motivation for this construction is given by the origin-centric notation from Table 8.2. Let Y_1 be origin-destination flows from the first origin to all destinations. The spatial lag WY_1 would then contain a spatial average of flows from this origin to neighbors of each destination $i = 1, \ldots, n$. Similarly, the spatial lag WY_2 would produce a spatial average of flows from the second origin to neighbors of each destination, and so on. This motivates use of the Kronecker product to repeat the spatial lags n times, resulting in an $N \times N$ spatial weight matrix that captures *destination-based* dependence.

This type of dependence reflects the intuition that forces leading to flows from an origin to a destination may create similar flows to nearby or neighboring destinations, which is captured by the spatial lag created using the matrix-vector product $W_d y$. This spatial lag formally captures the notion set forth in Griffith and Jones (1980) that flows associated with a destination are "enhanced or diminished in accordance with the propensity of attractiveness of its neighboring destination locations."

Taking a similar approach to that used in developing the matrix W_d, we can also create an $N \times N$ row-standardized spatial weight matrix that we label $W_o = W \otimes I_n$. This follows by noting that $W(Y_1')$ provides spatial averages around each origin of flows to the first destination. Doing this for all destinations yields WY', and $\text{vec}(WY') = (W \otimes I_n)\text{vec}(Y) = (W \otimes I_n)y$. The spatial lag formed by the matrix product $W_o y = (W \otimes I_n)y$ captures *origin-based spatial dependence* using an average of flows from neighbors to the origin regions to each of the destinations. This type of dependence reflects the notion that forces leading to flows from any origin to a particular destination region may create similar flows from nearby neighboring origins. The spatial lag $W_o y$ formally captures the point of Griffith and Jones (1980) that flows from an origin are "enhanced or diminished in accordance with the propensity of emissiveness of its neighboring origin locations."

As already noted for the case of the explanatory variables matrices, the vec-permutation matrix P can be used to translate between origin-centric and destination-centric ordering of the sample data. For example, if we adopt the destination-centric ordering (as opposed to the origin-centric ordering used here), specification of the destination weight matrix would be $W_d^{(d)} = W \otimes I_n$. As in the case of the matrix X_d, we can use the relation: $P' W_d P = P'(I_n \otimes W)P = W \otimes I_n = W_d^{(d)}$, to produce the destination weight matrix for the destination-centric ordering scheme.

The model in (8.2) is motivated by the fact that both types of dependence are likely to exist in our spatial specification for origin-destination flows. This model can be viewed as a successive spatial filter that filters the OD flows in y successively by $(I_N - \rho_d W_d)$ and $(I_N - \rho_o W_o)$. Remarkably one can change the order in which the filter is applied and arrive at the same model. That is, we could remove origin dependence first and destination dependence second, using the filter $(I_N - \rho_o W_o)(I_N - \rho_d W_d)$. This is true because the cross-product term $(W \otimes I_n)(I_n \otimes W) = W \otimes W$ is the same as the cross-product $(I_n \otimes W)(W \otimes I_n)$ via the mixed-product rule for Kronecker products.

Expanding the product $(I_N - \rho_d W_d)(I_N - \rho_o W_o) = I_N - \rho_d W_d - \rho_o W_o + \rho_d \rho_o W_d \cdot W_o = I_N - \rho_d W_d - \rho_o W_o - \rho_w W_w$, leads us to consider a third type of dependence reflected in the product $W_w = W_o \cdot W_d = (I_n \otimes W) \cdot (W \otimes I_n) = W \otimes W$.[1] This spatial weight matrix reflects an average of flows from neighbors to the origin to neighbors of the destination, which LeSage and Pace (2008) label *origin-to-destination dependence* to distinguish it from *origin-based dependence* and *destination-based dependence*.

This leads LeSage and Pace (2008) to propose the general spatial autoregressive interaction model in (8.4) that takes into account origin, destination, and origin-to-destination dependence.

$$y = \rho_d W_d y + \rho_o W_o y + \rho_w W_w y + \alpha \iota_N + X_d \beta_d + X_o \beta_o + \gamma g + \varepsilon \qquad (8.4)$$

The omitted variables and space-time dynamic motivations used to produce spatial regression models that contain spatial lags of the dependent variables

---

[1] We note that this specification implies a restriction that $\rho_w = -\rho_o \rho_d$, but this restriction need not be enforced in applied work. Of course, restrictions on the values of the scalar dependence parameters ρ_d, ρ_o, ρ_w must be imposed to ensure stationarity in the case where ρ_w is free of the restriction.

set forth in Chapter 2 and Chapter 7 also apply to this model. One can begin with a non-spatial theoretical relationship such as the utility theory used to motivate non-spatial interaction models for migration, or the monopolistic competition model in conjunction with a CES utility function used to derive a non-spatial gravity equation for trade flows. If we posit the existence of omitted variables that exhibit spatial dependence and are correlated with included variables, a model such as (8.4) that contains spatial lags of the dependent variable (as well as the explanatory variables) will arise. That is, an SDM variant of the model in (8.4) could be appropriate for spatial econometric modeling of OD flows. Similarly, a space-time dynamic can be used to motivate the model in (8.4) as the long-run steady state equilibrium of a process where OD flows exhibit time dependence as well as space-time dependence.

In the trade literature, Anderson and van Wincoop (2004) argue that it is important to include interaction terms that capture the fact that bilateral trade flows depend not only on bilateral barriers to trade, but also on trade barriers across all trading partners. Trade barriers or multilateral trade resistance are usually modeled as arising from price differentials between regions taking the form of 'cost-in-freight' (c.i.f.) and 'free-on-board' (f.o.b.) prices at the destination and origin regions. The argument is essentially that bilateral predictions do not readily extend to a multilateral world because these ignore indirect interactions that link all trading partners. Also of note is the work of Behrens, Ertur and Koch (2007) who extend the monopolistic competition model in conjunction with a CES utility function to derive a gravity equation for trade flows that contains spatial lags of the dependent variable. They accomplish this using a quantity-based version of the CES model and exploiting the fact that price indices (that represent multilateral resistance to trade) implicitly depend on trade flows. This in conjunction with the fact that bilateral trade flows in their model depend on flows from all other trading partners, leads to a model that displays a spatial autoregressive structure in trade flows. Intuitively, they argue that when goods are gross substitutes, trade flows from any origin to a particular destination will depend on the entire distribution of bilateral trade barriers (prices of substitute goods).

A problem that plagues the empirical trade literature is the lack of reliable regional price information, (Anderson and van Wincoop, 2004). Because of this, Anderson and van Wincoop (2004) suggest a non-statistical computational approach to replace the unobservable prices. As we have already motivated, the presence of latent unobservable variables that exhibit spatial dependence would lead to a model of the type in (8.4) which accounts for the unobserved variables using spatial lags of the dependent variable.

LeSage and Pace (2008) point out that this general model leads to nine more specific models that may be of interest in empirical work. We enumerate four of these models that result from various restrictions on the parameters $\rho_i, i = d, o, w$. The model comparison methods from Chapter 6 could be used to test these parameter restrictions. However, the nested nature of the family of nine models associated with the parameter restrictions allows the use of conventional likelihood ratio tests. The four models are enumerated below.

Non-spatial model. The restriction: $\rho_d = \rho_o = \rho_w = 0$ produces the non-spatial model where no spatial autoregressive dependence exists.

Model 1. The restriction: $\rho_w = 0$ leads to a model with separable origin and destination autoregressive dependence embodied in the two weight matrices W_d and W_o, while ruling out origin-to-destination based dependence between neighbors of the origin and destination locations that would be captured by W_w.

Model 2. The restriction: $\rho_w = -\rho_d\rho_o$ results in a successive filtering model involving both origin W_d, and destination W_o dependence as well as product separable interaction W_w, constrained to reflect the filter $(I_N - \rho_d W_d)(I_N - \rho_o W_o) = (I_N - \rho_o W_o)(I_N - \rho_d W_d) = (I_N - \rho_d W_d - \rho_o W_o + \rho_d\rho_o W_w)$.

Model 3. The unrestricted model shown in (8.4) involves three matrices W_d, W_o, and W_w, which represents the most general member of the family of models. Appropriate restrictions on ρ_d, ρ_o, and ρ_w can thus produce the other more specialized models.

8.2 Maximum likelihood and Bayesian estimation

The likelihood provides the starting point for both maximum likelihood and Bayesian estimation. We note that the concentrated log-likelihood function for the model specifications will take the form in (8.5).

$$\ln L = \kappa + \ln |I_N - \rho_d W_d - \rho_o W_o - \rho_w W_w| - \frac{N}{2}\ln(S(\rho_d, \rho_o, \rho_w)) \quad (8.5)$$

where $S(\rho_d, \rho_o, \rho_w)$ represents the sum of squared errors expressed as a function of the scalar dependence parameters alone after concentrating out the parameters $\alpha, \beta_o, \beta_d, \gamma$ and σ^2, and the constant κ that does not depend on ρ_d, ρ_o, ρ_w (LeSage and Pace, 2008).

LeSage and Pace (2008) show that the log-determinant of the $N \times N$ matrix that appears in (8.5) can be calculated using only traces of the $n \times n$ matrix W, which greatly simplifies estimation of these models (see Chapter 4). Further computational savings can be achieved by noting that we need not reproduce the $n \times k$ data matrix X using the Kronecker products $\iota_n \otimes X, X \otimes \iota_n$, if we exploit the special structure of this model. The algebra of Kronecker products can be used to form moment matrices without dealing directly with

N by N matrices. Given arbitrary, conformable matrices A, B, C, then $(C' \otimes A)\text{vec}(B) = \text{vec}(ABC)$ (Horn and Johnson, 1994, Lemma 4.3.1, p. 254-255). Using $Z = \begin{pmatrix} \iota_N & X_d & X_o & g \end{pmatrix}$ yields the moment matrix $Z'Z$ shown in (8.6), where the symbol 0_k denotes a $1 \times k$ vector of zeros.[2]

$$
Z'Z = \begin{pmatrix}
N & 0_k & 0_k & \iota'_n G \iota_n \\
0'_k & n X'X & 0'_k 0_k & X'G \iota_n \\
0'_k & 0'_k 0_k & n X'X & X'G \iota_n \\
\iota'_n G \iota_n & \iota'_n G'X & \iota'_n G'X & tr(G^2)
\end{pmatrix}
\tag{8.6}
$$

Using the algebra of Kronecker products also allows us to avoid forming the N by N matrices W_d, W_o, or W_w. Since $W_d y = (I_n \otimes W)\text{vec}(Y)$, it follows that $W_d y = \text{vec}(WY)$, using the relation, $(C' \otimes A)\text{vec}(B) = \text{vec}(ABC)$. Similarly, $W_o y = \text{vec}(YW')$, and $W_w y = \text{vec}(WYW')$. We use these forms to rewrite the model from (8.4) as shown in (8.7), where E is an $n \times n$ matrix of theoretical disturbances.

$$
\text{vec}(Y) - \rho_d \, \text{vec}(WY) - \rho_o \, \text{vec}(YW') - \rho_w \, \text{vec}(WYW') = Z\delta + \text{vec}(E) \tag{8.7}
$$

The expression on the left-hand-side of (8.7) is a linear combination of four components, one involving the dependent variable vector $\text{vec}(Y)$, and the other three representing spatial lags of this vector that reflect destination-based dependence $\text{vec}(WY)$, origin-based dependence $\text{vec}(YW')$ as well as origin-to-destination based dependence, $\text{vec}(WYW')$. This allows us to express the parameter estimates as a linear combination of four separate components which we label $\hat{\delta}^{(t)} = (Z'Z)^{-1}Z' \, \text{vec}(F^{(t)}(Y))$, where $F^{(t)}(Y)$ equals Y, WY, YW', or WYW' when $t = 1, \ldots, 4$. These components can be used to determine the parameter estimate $\hat{\delta}$ using (8.8).

$$
\hat{\delta} = \begin{pmatrix} \hat{\delta}^{(1)} & \hat{\delta}^{(2)} & \hat{\delta}^{(3)} & \hat{\delta}^{(4)} \end{pmatrix}
\begin{pmatrix} 1 \\ -\rho_d \\ -\rho_o \\ -\rho_w \end{pmatrix}
\tag{8.8}
$$

We can use the expressions $\hat{\delta}^{(t)}$ for $t = 1, \ldots, 4$, to write these terms as a function of the sample data X and Y and the parameters ρ_d, ρ_o, ρ_w. This allows us to concentrate the log-likelihood with respect to the parameters $\hat{\delta}^{(t)}$, which contain the parameters $\alpha, \beta_d, \beta_o, \gamma$ associated with the model covariates.

Component residual matrices $\hat{E}^{(t)}, t = 1, \ldots, 4$ that take the form shown in (8.9) can be used to express the overall residual matrix $\hat{E} = \hat{E}^{(1)} - \rho_d \hat{E}^{(2)} - \rho_o \hat{E}^{(3)} - \rho_w \hat{E}^{(4)}$ related to the concentrated log-likelihood function.

---

[2]We note that this log-likelihood function can be used for the SDM model by replacing the covariate matrix Z with $\tilde{Z} = \begin{pmatrix} \iota_N & X_d & X_o & W_d X_d & W_o X_o & g \end{pmatrix}$ in the expressions set forth in the text.

$$\hat{E}^{(t)} = F^{(t)}(Y) - \hat{a}^{(t)} \iota_n \iota_n' - X \hat{\beta}_d^{(t)} \iota_n' - \iota_n (\hat{\beta}_o^{(t)})' X' - \hat{\gamma}^{(t)} G \qquad (8.9)$$

For the purpose of maximizing the log-likelihood we introduce the cross-product matrix Q that consists of various component residual matrices. Define $Q_{ij} = tr(\hat{E}^{(i)'} \hat{E}^{(j)})$, $i = 1, \ldots, 4$, $j = 1, \ldots, 4$, so the sum-of-squared residuals for our model become $S(\rho_d, \rho_o, \rho_w) = \tau(\rho_d, \rho_o, \rho_w)' Q \tau(\rho_d, \rho_o, \rho_w)$, where $\tau(\rho_d, \rho_o, \rho_w) = \begin{pmatrix} 1 & -\rho_d & -\rho_o & -\rho_w \end{pmatrix}'$.

Consequently, recomputing $S(\rho_d, \rho_o, \rho_w)$ for any set of values $\begin{pmatrix} \rho_d & \rho_o & \rho_w \end{pmatrix}$ requires a small number of operations that do not depend on n or k. This in conjunction with pre-computed values for the log-determinant term also expressed as a function of these parameters calculated using the efficient methods set forth in LeSage and Pace (2008) permits rapid optimization of the likelihood function with respect to these parameters. Using a 2.0 Ghz. Intel Core 2 Duo laptop computer it takes around 5 seconds to maximize the log-likelihood function for a problem involving migration flows between $n = 359$ ($N = 128,881$) US metropolitan areas.

The Bayesian Markov Chain Monte Carlo (MCMC) estimation procedures for standard spatial regression models set forth in Chapter 5 can be applied to these models. This allows us to extend the model in two useful ways. First, we can accommodate fat-tailed disturbance distributions using our spatial autoregressive model extension of the non-spatial model introduced by Geweke (1993) discussed in Chapter 5. Second, we can deal with a common problem that arises in modeling OD flows where many of the flows associated with OD pairs take on zero values. In Chapter 10, we discuss spatial Tobit models where zero observations of the dependent variable are viewed as arising from a sample truncation process. In the context of spatial econometric interaction models we could view zero flows as indicative of negative utility (or profits) associated with flows between these particular OD pairs. For example, the absence of migration flows between origin-destination pairs might be indicative of negative utility arising from moves between these locations.

Turning to Bayesian robust estimation of the spatial econometric interaction model, we introduce a set of latent variance scalars for each observation. That is, we replace $\varepsilon \sim N(0, \sigma^2 I_N)$, with:

$$\varepsilon \sim N[0, \sigma^2 \tilde{V}] \qquad (8.10)$$
$$\tilde{V}_{ii} = V_i, i = 1, \ldots, N$$
$$V = \text{vec}(R)$$
$$R = \begin{pmatrix} v_{11} & v_{12} & \cdots & v_{1n} \\ v_{21} & v_{22} & & v_{2n} \\ \vdots & & \ddots & \\ v_{n1} & & & v_{nn} \end{pmatrix}$$

Estimates for the N variance scalars in (8.10) are produced using an *iid* $\chi^2(\lambda)$ prior on each of the variance scalars $v_{ij}, i = 1, \ldots, n, j = 1, \ldots, n$ contained in the $n \times n$ matrix R, with a mean of unity and a mode and variance that depend on the hyperparameter λ of the prior. As discussed in Chapter 5, small values of λ (around 5) result in a prior that allows for the individual v_{ij} estimates to be centered on their prior mean of unity, but deviate greatly from the prior value of unity in cases where the model residuals are large. Large residuals are indicative of outliers or origin-destination combinations that are atypical or aberrant relative to the majority of the sample of origin-destination flows.

MCMC estimation requires that we sample sequentially from the complete set of conditional distributions for all parameters in the model. We present the conditional distributions (using our computationally efficient) moment matrix structure. The parameters of the model are: $\delta, \sigma, \rho_d, \rho_o, \rho_w$ and $\tilde{V}_{ii}, i = 1, \ldots, N$, where $\delta = \begin{bmatrix} \alpha & \beta_d & \beta_o & \gamma \end{bmatrix}'$. The $N \times N$ diagonal matrix $\tilde{V}$ contains the variance scalar parameters that distinguish this model from the homoscedastic model.

We present conditional distributions for the parameters δ and σ^2, when uninformative priors are assigned to the parameters δ, and an independent $IG(a, b)$ prior is assigned to σ^2. We rely on a uniform prior over the range $-1 < \rho_d, \rho_o, \rho_w < 1$ for these parameters, and impose stability restrictions that $\sum_i \rho_i > -1, \sum_i \rho_i < 1, i = d, o, w$, using rejection sampling. We rely on Geweke's *iid* chi-squared prior based on λ degrees of freedom for the variance scalars v_{ij}. We treat λ as a degenerate hyperparameter, but note that Koop (2003) provides an extension where an exponential prior distribution is placed on λ. Formally, our priors can be expressed as:

$$\pi(\delta) \propto N(c, T), \ T \to \infty \tag{8.11}$$
$$\pi(\lambda/v_{ij}) \sim iid \ \chi^2(\lambda) \tag{8.12}$$
$$\pi(\sigma^2) \sim IG(a, b) \tag{8.13}$$
$$\pi(\rho_i) \sim U(-1, 1), \quad i = d, o, w \tag{8.14}$$

The conditional posterior distribution for the δ parameters take the form of a multivariate normal:

$$p(\delta \mid \rho_d, \rho_o, \rho_w, \sigma^2, \tilde{V}) \propto N(\bar{\delta}, \sigma^2 \bar{D}) \tag{8.15}$$
$$\bar{\delta} = \beta^{(1)} - \rho_d \beta^{(2)} - \rho_o \beta^{(3)} - \rho_w \beta^{(4)}$$
$$\beta^{(i)} = (Z'\tilde{V}^{-1}Z)^{-1}Z'\tilde{V}^{-1}F^{(i)}(Y)$$
$$F^{(i)}(Y) = Y, \ WY, \ YW', \ WYW', \quad i = 1, \ldots, 4$$
$$\bar{D} = (Z'\tilde{V}^{-1}Z)^{-1}$$

The conditional posterior for the parameter σ^2 based on our prior $\sigma^2 \sim IG(a, b)$ is proportional to an inverse gamma distribution:

$$p(\sigma^2 \mid \rho_d, \rho_o, \rho_w, \delta, \tilde{V}) \propto IG[a + N/2, \tau'\tilde{Q}\tau/(2+b)]$$

$$\tau = \left(1 - \rho_d - \rho_o - \rho_w\right)'$$

$$\tilde{Q}_{ij} = tr[(E^{(i)'} \odot \tilde{R}')(\tilde{R} \odot E^{(j)})] \quad i, j = 1, \ldots, 4$$

$$\tilde{R} = \begin{pmatrix} v_{11}^{-\frac{1}{2}} & v_{12}^{-\frac{1}{2}} & \cdots & v_{1n}^{-\frac{1}{2}} \\ v_{21}^{-\frac{1}{2}} & v_{22}^{-\frac{1}{2}} & & v_{2n}^{-\frac{1}{2}} \\ \vdots & & \ddots & \\ v_{n1}^{-\frac{1}{2}} & & & v_{nn}^{-\frac{1}{2}} \end{pmatrix}$$

$$E^{(i)} = F^{(i)}(Y) - \alpha^{(i)} \iota_n \iota_n' - X\beta_d^{(i)} \iota_n' - \iota_n (\beta_o^{(i)})' X' - \gamma^{(i)} G$$

$$\beta^{(i)} \equiv \left(\alpha^{(i)} \; \beta_d^{(i)} \; \beta_o^{(i)} \; \gamma^{(i)}\right)' = (Z'\tilde{V}^{-1}Z)^{-1} Z'\tilde{V}^{-1} F^{(i)}(Y)$$

where $\tau'\tilde{Q}\tau$ represents the sum of squared residuals for any given values of the parameters ρ_d, ρ_o, ρ_w.

The conditional posterior for each variance scalar $v_{ij}, i, j = 1, \ldots, n$ can be expressed as in (8.16), where E_{ij} references the i, jth element of the matrix E.

$$p(\frac{E_{ij}^2 + \lambda}{v_{ij}} \mid \rho_d, \rho_o, \rho_w, \delta, \sigma^2) \propto \chi^2(\lambda + 1) \tag{8.16}$$

$$E = E^{(1)} - \rho_d E^{(2)} - \rho_o E^{(3)} - \rho_w E^{(4)}$$

In this model we must sample each of the three parameters ρ_d, ρ_o, ρ_w conditional on the two other dependence parameters and the remaining parameters (δ, σ^2, V). The log conditional posterior for ρ_d takes the form shown in (8.17), with analogous expressions for the other two spatial dependence parameters.

$$p(\rho_d | \rho_o, \rho_w, \delta, \sigma^2, \tilde{V}) \propto |I_N - \rho_d W_d - \rho_o W_o - \rho_w W_w| \tag{8.17}$$

$$\cdot \exp\left(-\frac{1}{2\sigma^2}\tau(\rho_d, \rho_o, \rho_w)'\tilde{Q}\tau(\rho_d, \rho_o, \rho_w)\right)$$

We note the presence of the determinant term which can be evaluated using the same algorithms for rapidly evaluating this expression as in maximum likelihood estimation (Chapter 4). Sampling for the parameters $\rho_i, i = d, o, w$ is accomplished using a Metropolis-Hastings algorithm based on a tuned normal random-walk proposal of the type discussed in Chapter 5.

8.3 Application of the spatial econometric interaction model

To illustrate the model, we used (logged) population migration flows between the 50 largest US metropolitan areas over the period from 1995 to 2000. The metropolitan area flows were constructed from population-weighted county-level migration flows and explanatory variables were taken from the 1990 Census. The population-weight (logged) level of 1990 county income and the metropolitan area (logged) population were used as explanatory variables to avoid potential endogeneity problems. A third explanatory variable was the proportion of 1990 metropolitan area residents who lived in the same house five years ago. A log transformation was applied to this proportion.

One problem that often arises with OD flows is the presence of zero flow magnitudes between origin-destination pairs, and for this sample there were 122 of the 2500 flows that were zero, making this a minor problem here.

A second problem is the presence of large flows on the diagonal of the OD flow matrix because of the large degree of intraregional migration relative to interregional migration reflected by smaller flows or zeros for the off-diagonal elements. One approach used in empirical studies is to set the diagonal elements of the flow matrix to zero (Tiefelsdorf, 2003; Fischer, Scherngell and Jansenberger, 2006). This reflects a view that intraregional flow elements represent a nuisance, since the focus of the model is on interregional flows. However, in our spatial econometric interaction model where spatial lags reflect local averages of the dependent variable, this would defeat the purpose of using local averages.

LeSage and Pace (2008) suggest an alternative approach to dealing with the large intraregional flow magnitudes which involves adding a separate intercept term for these observations as well as a set of explanatory variables. The intraregional explanatory variables contain non-zero observations for the intraregional observations extracted from the explanatory variables matrix X, and zeros elsewhere. We label this matrix X_i, and the associated intercept term $c = \text{vec}(I_n)$. This procedure introduces a separate model for the intraregional flows. This should allow the coefficients associated with the matrices X_d, X_o to reflect interregional variation in OD flows, and those associated with the matrix X_i to capture intraregional variation in flows. Implementing this approach requires that we adjust the moment matrix $Z'Z$ as well as the cross-product terms $Z' \, \text{vec}(F^{(t)}(Y))$ to reflect these changes.

We can write the adjusted model as in (8.18).

$$(I_N - \rho_d W_d)(I_N - \rho_o W_o)y = \iota_N \alpha + c\alpha_i + X_d\beta_d + X_o\beta_o + X_i\beta_i + \gamma g + \varepsilon \quad (8.18)$$

The modified moments matrix for this model is shown in (8.19), where $dg = \text{diag}(G)$.[3] The cross-product terms $Z'y$ required to produce least-squares estimates for the parameters, $(Z'Z)^{-1}Z'y$ would take the form shown in (8.19). We could use a similar set of definitions $Z' \text{ vec}(F^{(t)}(Y))$, where $(F^{(t)}(Y))$ equals Y, WY, YW', or WYW' for $t = 1, \ldots, 4$ to produce spatial autoregressive model estimates.

$$Z'Z = \begin{pmatrix} N & n & 0_k & 0_k & 0_k & \iota'_n G\iota_n \\ n & n & 0_k & 0_k & 0_k & 0 \\ 0_k & 0_k & nX'X & 0'_k 0_k & X'X & X'G\iota_n \\ 0_k & 0_k & 0'_k 0_k & nX'X & X'X & X'G'\iota_n \\ 0_k & 0_k & X'X & X'X & X'X & 0'_k \\ \iota'_n G'\iota_n & 0 & \iota'_n G'X & \iota'_n GX & 0_k & \text{tr}(G^2) \end{pmatrix}, Z'y = \begin{pmatrix} \iota'_n Y\iota_n \\ \text{tr}(Y) \\ X'Y\iota_n \\ X'Y'\iota_n \\ X'\text{diag}(Y) \\ \text{tr}(GY) \end{pmatrix}$$

$$(8.19)$$

TABLE 8.3: Spatial econometric interaction model estimates

Variables	Coefficient	t-statistic	t-probability
		Adjusted Model	
Constant	1.5073	8.19	0.0000
c	0.7170	2.63	0.0086
Destination Pop 90	0.2376	2.06	0.0387
Destination Inc 90	0.3269	5.56	0.0000
Destination Samehouse	-0.2371	-1.35	0.1757
Origin Pop 90	0.2129	1.82	0.0683
Origin Inc 90	0.2842	4.90	0.0000
Origin Samehouse	-0.4787	-2.69	0.0071
Intraregional Pop 90	0.3869	0.48	0.6269
Intraregional Inc 90	0.8018	2.05	0.0399
Intraregional Samehouse	2.0410	1.65	0.0991
Distance	-0.1255	-6.30	0.0000
$\hat{\rho}_d$	0.6428	39.78	0.0000
$\hat{\rho}_o$	0.6358	39.25	0.0000
$\hat{\rho}_w$	-0.5427	-16.39	0.0000
$\hat{\sigma}^2$	1.8119		
Log-Likelihood	$-3.6309 \cdot 10^3$		

Maximum likelihood estimates are presented in Table 8.3 for the model in (8.18). From the table, we see that all three spatial dependence param-

---

[3]In our discussion we have portrayed the diagonal of the distance matrix as containing zeros. However, a frequent practice in analysis of trade flows between countries is to use a non-zero intraregional distance which would make the main diagonal contain non-zero values.

eters are statistically significant. This suggests the presence of *origin-based*, *destination-based* and *origin-to-destination based* spatial dependence in the population migration flows between the largest 50 metropolitan areas.

Table 8.4 shows log-likelihoods from four models: the non-spatial model estimated using least-squares and models that we have labeled Model 1, Model 2, and Model 3 from the family of models enumerated in Section 8.1. These values along with likelihood ratio test statistics make it clear that the unrestricted version of the model (Model 3) is superior to the restricted variants of the model (Models 1 and 2). The non-spatial model has a much lower likelihood than all of the spatial models, and this model included the adjustments to the intercept as well as the variables X_i to account for large intraregional flows. The LR tests also reject Model 2 containing the restriction $\rho_w = -\rho_d \cdot \rho_o$ as inconsistent with this sample data. To draw inferences regarding the magnitude and significance of the coefficient estimates we need to calculate direct, indirect and total effects estimates for our model parameters.

Interpreting the parameter estimates requires that we implement our calculations for direct, indirect and total impact estimates. For this model the partial derivatives take a more complex form that can be derived from the expression in (8.20) for this extended type of SAR model.[4]

TABLE 8.4: Spatial econometric interaction model log-likelihoods

Model	Log-likelihood	LR test vs. Model 3	χ^2 (5%) Value
Model 3	-3630.9		
Model 2	-3645.5	29.2	$\chi^2(1) = 3.84$
Model 1	-3868.3	474.8	$\chi^2(1) = 3.84$
Non-spatial model	-4370.7	1479.6	$\chi^2(3) = 7.82$

To produce measures of dispersion for these estimates, the parameters δ and ρ_d, ρ_o, ρ_w were simulated using a multivariate normal distribution and the numerical Hessian estimate of the variance-covariance matrix. A sample of 1,000 simulated parameters were used in $S_r(W_d, W_o, W_w)$ with means and standard deviations used to construct t-statistics reported in Table 8.5.

The effects estimates indicate that the (cumulative) indirect impacts are larger than direct impacts, accounting for about two-thirds of the total effects magnitudes. Impact estimates for the 1990 population (*Pop 90*) and 1990 per capita income (*Inc 90*) can be interpreted as elasticities since these variables as

---

[4]Only traces are required to calculate the effects estimates so it would be computationally inefficient to use the large matrix inverse of $S(W_d, W_o, W_w)$. The trace-based methods described in Chapter 4 can be extended to this case.

well as the dependent variable are in log form. The effects estimates emphasize the relative importance of spatial spillovers when considering migration flows which have been ignored by empirical studies relying on non-spatial models.

$$(I_N - \rho_d W_d - \rho_o W_o - \rho_w W_w)y = Z\delta + \iota_N \alpha + c\alpha_i + \varepsilon \qquad (8.20)$$

$$E(y) = \sum_{r=1}^{k} S(W_d, W_o, W_w) Z_r \delta_r$$

$$\partial y / \partial Z_r' = S(W_d, W_o, W_w) I_N \delta_r$$

$$S(W_d, W_o, W_w) = (I_N - \rho_d W_d - \rho_o W_o - \rho_w W_w)^{-1}$$

A motivation for use of origin- and destination-specific variables such as per capita income and population, *Origin Pop 90, Origin Inc 90, Destination Pop 90, Destination Income 90*, in non-spatial models was the notion of *push* factors associated with origin regions and *pull* factors associated with destination regions. When we allow for spatial dependence taking the form of lagged dependent variables, this type of interpretation becomes problematical.

To illustrate, we consider the case of a positive pull factor, a ceteris paribus increase in per capita income of a single destination metropolitan area. A positive direct effect arising from this change would be relatively straightforward to interpret in terms of a pull factor, with some positive feedback loop effect. The existence of positive indirect effects or spatial spillovers in our model suggests that neighbors to the destination may also receive a positive pull from the increase in income. However, in addition to destination-based dependence, our model also includes origin-based dependence as well as origin-to-destination based dependence. Our measure of cumulative indirect effects captures spatial spillovers to all other regions as should be clear from the structure of $S_r(W_d, W_o, W_w)$. This means that if we partitioned these spillover impacts over space, we would expect to find spillovers falling on: 1) neighbors to the destination region and 2) neighbors to the regions where migration flows originate. This makes it more difficult to rely on the conventional pull factor interpretation, since spatial spillover effects that arise at origin regions would typically be associated with push factors, not pull factors.

Allowing for spatial dependence at origins, destinations, and between origins and destinations leads to a situation where changes at either the origin or destination will give rise to forces that set in motion a series of events. If we attempt to associate push with origin regions and pull with destination regions, any change gives rise to a series of *push* and *pull* events. A better way to view the forces set in motion by ceteris paribus changes in an explanatory variable associated with a single metropolitan area is in terms of multilateral effects that permeate the entire system of spatially interrelated regions. The point here is essentially that advanced by Anderson and van Wincoop (2004) and Behrens, Ertur and Koch (2007), who noted the difficulty of extending

TABLE 8.5: Spatial econometric interaction model effects estimates

Variables	Mean estimates	*t*-statistic	*t*-probability
		Direct effects	
Destination Pop 90	0.2982	2.13	0.0331
Destination Inc 90	0.4068	5.94	0.0000
Destination Samehouse	−0.2870	−1.31	0.1885
Origin Pop 90	0.2608	1.85	0.0640
Origin Inc 90	0.3517	4.69	0.0000
Origin Samehouse	−0.6007	−2.62	0.0087
Intraregional Pop 90	0.4543	0.46	0.6434
Intraregional Inc 90	1.0061	2.09	0.0362
Intraregional Samehouse	2.5476	1.68	0.0922
Distance	−0.1538	−6.36	0.0000
		Indirect effects	
Destination Pop 90	0.6293	1.95	0.0504
Destination Inc 90	0.8591	4.42	0.0000
Destination Samehouse	−0.5845	−1.27	0.2014
Origin Pop 90	0.5571	1.68	0.0918
Origin Inc 90	0.7448	3.65	0.0003
Origin Samehouse	−1.2491	−2.60	0.0093
Intraregional Pop 90	0.9895	0.45	0.6474
Intraregional Inc 90	2.1270	1.98	0.0474
Intraregional Samehouse	5.3675	1.60	0.1082
Distance	−0.3192	−9.01	0.0000
		Total effects	
Destination Pop 90	0.9275	2.04	0.0414
Destination Inc 90	1.2659	5.17	0.0000
Destination Samehouse	−0.8715	−1.30	0.1936
Origin Pop 90	0.8179	1.75	0.0787
Origin Inc 90	1.0965	4.13	0.0000
Origin Samehouse	−1.8498	−2.67	0.0076
Intraregional Pop 90	1.4438	0.46	0.6449
Intraregional Inc 90	3.1331	2.05	0.0404
Intraregional Samehouse	7.9151	1.64	0.0994
Distance	−0.4730	−9.82	0.0000

notions such as *push* and *pull* that arose in the context of bilateral flows to a multilateral world where indirect interactions link all metropolitan regions.

Given these caveats regarding interpretation of the effects estimates from our model of metropolitan migration flows, the most salient interpretation of these effects might be in terms of differences between the direct and indirect effects of origin and destination variables. For example, the positive difference between destination and origin direct effects associated with per capita income $(0.40 - 0.35)$ could be interpreted as meaning that positive

(1990) income gaps between metropolitan areas gave rise to increased migration flows to metropolitan areas having relatively higher per capita incomes. The positive difference between indirect effects associated with (1990) per capita income of destination and origin metropolitan areas (0.86 − 0.74) suggests that (1990) income gaps between metropolitan areas led to spillovers that increased migration flows to all other regions. Given what we know about the spatial structure of indirect or spatial spillover effects, these increased migration flows most likely impacted regions that were neighbors to origin and destination metropolitan areas exhibiting large relative income gaps.

There is a similar pattern of positive differences between the direct effects associated with the destination and origin (1990) population variable (0.29 − 0.26). This suggests positive migration flows arose as a result of (1990) population size differences that had a positive direct impact on destination metropolitan areas. The positive difference between indirect effects (0.63 − 0.55) suggests that (1990) population gaps between metropolitan areas led to spatial spillover effects that gave rise to higher levels of migration flows to all other regions, in a fashion similar to gaps in (1990) per capita income.

The origin and destination *Samehouse* variables exhibit negative direct and indirect effects, suggesting lower (1995-2000) migration flows for metropolitan areas where more people lived in the same house in 1990 as in 1985. This also resulted in negative spatial spillovers, meaning lower levels of migration flow for other metropolitan regions as well.

If our interest centered on interregional migration flows, we might view the intraregional variables as controls and the associated effects estimates as nuisance parameters. However, the positive direct effect elasticity of 1 and indirect effect elasticity of 2.12 leads to a total effect greater than 3. This suggests the direct or own-region effect of (1990) per capita income on migration flows was positive as was the indirect effect. Intraregional (1990) population also had positive but not significant direct, indirect and total effects. The *total* effect of the intraregional *Samehouse* variable is large and positive, but significant only at the 90 percent level.

The distance estimate has a total effect of −0.48 which is much larger than one would infer from the model coefficients reported in Table 8.3. The model coefficient is of course close to the *direct* effect estimate in Table 8.5, a typical result noted earlier in our discussion regarding interpretation of impact estimates from these models.

8.4   Extending the spatial econometric interaction model

There are a number of problems that are encountered in empirical modeling of OD flows. The next sections describe three extensions of these models that address some of these issues.

8.4.1 Adjusting spatial weights using prior knowledge

One way to extend the spatial econometric interaction model that is applicable to interregional commodity flows is to use a priori non-sample knowledge regarding the transportation network structure that connects regions. LeSage and Polasek (2008) point out that it is relatively simple to adjust the spatial weight matrix to reflect the presence or absence of interregional transport connectivity. They use truck and train commodity flows between 40 Austrian regions, where the mountainous terrain precludes the presence of major rail and highway infrastructure in all regions. Bayesian model comparison methods show that adjusting the spatial weight matrix to reflect transportation network structure results in an improved model.

To illustrate this type of adjustment, consider flows from origin A to destination Z depicted in Figure 8.1. Rook-type contiguity has been used to define neighbors to the origin region A and destination region Z. This reflects the type of movements that the Rook piece in the game of chess can make on our regular grid that resembles a checkerboard. Using the standard spatial weighting approach, *origin-based* dependence would rely on neighbors to origin region A (labeled c, e, f, h) to form the spatial lag vector $W_o y$. The spatial lag vector $W_d y$ reflecting *destination-based* dependence would rely on an average of neighbors to destination region Z (s, u, v, x). Finally, the spatial lag $W_w y$ that captures *origin-to-destination* dependence would be constructed using an average over all neighbors to both the origin and destination regions A and Z, (c, e, f, h, s, u, v, x).

LeSage and Polasek (2008) suggest using information on regions through which the transportation routes pass to modify the spatial weight structure that is used to form the matrices W_d, W_o and W_w. The figure provides an example where a highway extends from region A to Z, passing through regions h, A, c on the way to and from the origin region A, and through regions x, Z, s as it passes through the destination region Z. They make the plausible argument that if accessibility to the highway from regions such as e, f, or u, v is difficult or impossible, the matrices W_o, W_d and W_w should be adjusted to reflect this a priori information.

One example of a possible modification would be to construct $W_o y$ based on an average of neighboring regions h and c on the highway route near the origin region A. For the destination spatial lag, an average of regions x and s that are neighbors to Z and also on the highway route would be used to form $W_d y$. Finally, the spatial lag $W_w y$ could be constructed using an average involving regions h, c, x, s, those that neighbor both the origin and destination and are also on the highway route. Intuitively, they argue that we would expect to find higher levels of commodity flows between origin-destination pairs with a highway connection, than those that are not connected. Of course, one important role played by the spatial weight matrix in spatial

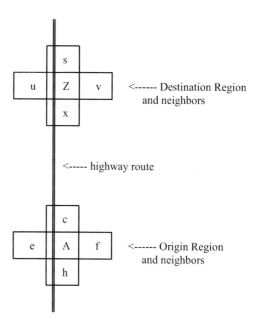

FIGURE 8.1:   Origin-Destination region contiguity relationships

econometric models is to capture interregional connectivity. It should not be surprising that LeSage and Polasek (2008) find evidence that improvements on this aspect of the spatial weight matrix lead to a superior fit of the model.

Of course, in most developed countries regions tend to be well-connected by transportation infrastructure, however there are exceptions where natural boundaries such as bodies of water or mountains lead to less connectivity. A map showing Austrian regions that lie on or off the major road and rail network is shown in Figure 8.2. From the map we can see clear evidence of regions that do not share a position along major transportation routes, and we would expect this to have an impact on interregional road/rail commodity flows.

8.4.2   Adjustments to address the zero flow problem

There are extensions that can be made to the spatial econometric interaction model to address the issue of zero flows. The presence of a large number of zero flows would invalidate the normality assumption needed for maximum likelihood estimation. At a finer spatial scale this problem becomes more

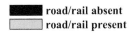

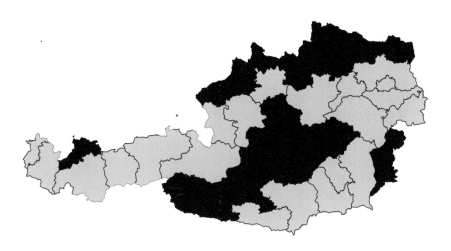

FIGURE 8.2:   Regions on and off of Austria's major road/rail network

acute, suggesting that aggregation to larger spatial units or cumulating flows over a longer time period is one approach to eliminating zeros.

For our sample of 1995-2000 population migration flows between the largest 50 metropolitan areas, 3.76% of the OD-pairs contained zero flows, whereas 9.38% of the OD-pairs were zero for the largest 100 metropolitan areas. In the case of the largest 300 metropolitan areas, 32.89% of the OD pairs exhibit zero flows. Since the largest 50 metropolitan areas contain around 49 percent of the population and 30 percent of persons migrating, one might not be willing to restrict the sample to only large metropolitan areas to avoid excessive zero flows.

In a non-spatial application to international trade flows, Ranjan and Tobias (2007) treat the zero flows using a threshold Tobit model. They note that commonly used $\ln(1 + y)$ as the dependent variable ignores the mixed discrete/continuous nature of flows and arbitrarily adds unity to the dependent variable to avoid taking the log of zero.

Their treatment interprets observed flows (y) as a latent indicator of desired trade $(y^\star)$, with zero trade volume viewed as resulting from a situation where the desired amount of trade is less than an amount that would be lost in transit (the threshold) using the conventional iceberg transport cost from trade theory. This results in a non-spatial model taking the form in (8.21), where the scalar τ represents the threshold, a parameter to be estimated.

$$\ln(y^\star + \iota_N \tau) = Z\delta + \varepsilon \tag{8.21}$$

$$y_i = \begin{cases} y_i^\star & \text{if } y_i^\star > 0, \\ 0 & \text{if } -\tau < y_i^\star \leq 0. \end{cases} \tag{8.22}$$

When the parameter $\tau = 0$, we have the standard Tobit model and the latent data interpretation for $y^\star$ is limited because we are simply modeling observed flows y. They use the fact that for $\tau > 0$, the model places a discrete mass over the zero flows that we noted are typically found in flow data. This approach allows the basic log-linear specification of the gravity equation to be used.

In Chapter 10 we introduce a spatial Tobit model that contains spatial lags of the dependent variable which could be used in conjunction with the threshold ideas from Ranjan and Tobias (2007) to treat the zero flows problem.

8.4.3   Spatially structured multilateral resistance effects

The monopolistic competition model in conjunction with a CES utility function was used by Anderson and van Wincoop (2004) to derive a gravity equation for trade flows that allows for transport costs and general barriers to trade. Conventional approaches to empirically incorporating multilateral trade resistance terms that arise in the form of "cost-in-freight" (c.i.f.) and "free-on-board" (f.o.b.) price differentials between regions suffered from the lack of reliable price information. A host of alternative approaches have been used to overcome this problem in empirical applications. One approach is to simply use published price indexes, but the validity of these has been questioned by Anderson and van Wincoop (2004). Ranjan and Tobias (2007) incorporate *random effects* parameters for the origin and destination regions in their threshold Tobit model, and it has become standard practice to rely on *fixed effects* parameters for origin and destination regions in non-spatial versions of the gravity model used in the empirical trade literature (Feenstra, 2002).

The fixed effects models take the form in (8.23), where Δ_o is an $N \times n$ matrix containing elements that equal 1 if region i is the origin region and zero otherwise, and θ_o is an $n \times 1$ vector of associated fixed effects estimates for regions treated as origins. Similarly, Δ_d is an $N \times n$ matrix containing elements that equal 1 if region j is the destination region and zero otherwise, and θ_d is an $n \times 1$ vector of associated fixed effects estimates for regions treated as destinations.

$$y = \alpha + \beta_o X_o + \beta_d X_d + \gamma g + \Delta_o \theta_o + \Delta_d \theta_d + \varepsilon \tag{8.23}$$

This approach interprets the parameters in the vector θ_o as latent "prices" or multilateral resistance indices for the n regions viewed as origins and similarly for the parameters θ_d for the destinations.

Spatially structured random effects seem more plausible since this approach introduces latent effects parameters that are structured to follow a spatial autoregressive process. We provided one motivation for this in Section 2.3 based on spatial heterogeneity. This type of model can be viewed as imposing a "stochastic restriction" on the origin-destination effects parameters so that multilateral resistance is similar for neighboring regions. This is motivated by the notion that latent or unobservable barriers to regional trade (which are usually motivated theoretically as price differential effects) should be similar for regions that are located nearby. As in the case of fixed effects, a set of $2n$ individual effects parameters is estimated using the sample of N OD flow observations. This approach was introduced by LeSage and Llano (2007) to model commodity flows between Spanish regions. The model takes the form in (8.24).

$$y = Z\delta + \Delta_d\theta_d + \Delta_o\theta_o + \varepsilon \qquad (8.24)$$
$$\theta_d = \rho_d W\theta_d + u_d$$
$$\theta_o = \rho_o W\theta_o + u_0$$
$$u_d \sim N(0, \sigma_d^2 I_n)$$
$$u_o \sim N(0, \sigma_o^2 I_n)$$

Given an *origin-centric* orientation of the flow matrix (columns as origins and rows as destinations), the matrices $\Delta_d = I_n \otimes \iota_n$ and $\Delta_o = \iota_n \otimes I_n$ produce N by n matrices. It should be noted that estimates for these two sets of random effects parameters are identified, since a set of n sample data observations are aggregated through the matrices Δ_d and Δ_o to produce each estimate in θ_d, θ_o.

The spatial autoregressive structure placed on the origin and destination effects reflects an implied prior for the spatial effects vector θ_d conditional on ρ_d, σ_d^2 and for θ_o conditional on ρ_o, σ_o^2 shown in (8.25) and (8.26).

$$\pi(\theta_d|\rho_d, \sigma_d^2) \sim (\sigma_d^2)^{n/2}|B_d|\exp\left(-\frac{1}{2\sigma_d^2}\theta_d' B_d' B_d\theta_d\right) \qquad (8.25)$$
$$\pi(\theta_o|\rho_o, \sigma_o^2) \sim (\sigma_o^2)^{n/2}|B_o|\exp\left(-\frac{1}{2\sigma_o^2}\theta_o' B_o' B_o\theta_o\right) \qquad (8.26)$$
$$B_d = I_n - \rho_d W$$
$$B_o = I_n - \rho_o W$$

Estimation of the spatially structured origin effects vector θ_o requires introduction of two additional parameters (ρ_o, σ_o^2) to the model. One of these controls the strength of spatial dependence between regions (treated as origins) and the other controls the variance/uncertainty of the prior spatial structure. Given these two scalar parameters along with the spatial structure, the n origin effects parameters are completely determined. One could view the spatial connectivity matrix W as introducing additional exogenous information that augments the sample data information. In contrast, the conventional fixed origin effects approach introduces n additional parameters to be estimated without (materially) augmenting the sample data information.

Another point about the spatially structured prior is that if the scalar spatial dependence parameters (ρ_o, ρ_d) are not significantly different from zero, the spatial structure of the effects vectors disappears, leaving us with normally distributed random effects parameters for the origins and destinations similar to the model of Ranjan and Tobias (2007). LeSage and Llano (2007) provide details regarding Bayesian MCMC estimation of this hierarchical linear model.

8.4.4 Flows as a rare event

For some OD flow matrices that contain an extremely large proportion of zeros, the argument for sample truncation seems questionable. LeSage, Fischer and Scherngell (2007) extend the model in (8.14) using results from Frühwirth-Schnatter and Wagner (2006). In their examination they treat interregional patent citations from a sample of European Union regions as representing knowledge flows. Counts of patents originating in region i that were cited by regions $j = 1, \ldots, n$ are used to form an OD *knowledge flow* matrix. Since cross-region patent citations are both counts and rare events, a Poisson distribution seems much more plausible than the normal distribution required for maximum likelihood estimation of the spatial econometric interaction model.

Frühwirth-Schnatter and Wagner (2006) argue that (non-spatial) Poisson regression models (including those with random-effects) can be treated as a partially Gaussian regression model by conditioning on two strategically chosen sequences of artificially missing data. Chapter 10 provides more details regarding Bayesian treatment of binary 0,1 observations as indicators of latent unobserved utility, an idea originated by Albert and Chib (1993). After conditioning on both of these latent sequences, Frühwirth-Schnatter and Wagner (2006) show that the resulting model can be sampled using Gibbs sampling of all regression parameters and the latent sequences. This requires random draws from only known distributions such as multivariate normal, inverse Gamma, exponential, and a discrete distribution with a limited number of categories, which eliminates the need for use of Metropolis-Hastings steps during sampling.

There is a large literature on Bayesian hierarchical spatial models (Banerjee, Carlin and Gelfand, 2004; Cressie, 1995), but this work cannot be applied to the case of non-linear Poisson regression models in a straightforward fashion. The tremendous advantage of the approach introduced by Frühwirth-Schnatter and Wagner (2006) is that this large suite of existing hierarchical linear spatial models can be directly applied after augmenting the existing sampling scheme with the two latent sequences. The addition of these two latent variable vectors does not affect the conditional distributions of existing hierarchical models, so existing algorithms and code can be used.

The one drawback to the approach is that one must sample two sets of latent parameters equal to $y_{ij} + 1$, where y_{ij} denotes the count for observation i. Specifically, if we have a sample $i = 1, \ldots, n$, and half of these observations exhibit magnitudes of $y_{ij} = 0$, while the other half take on non-zero count values of $y_{ij} = 10$, then we must sample two vectors of latent parameters equal to $11(n/2) + (n/2)$, where the $11(n/2)$ latent parameters are associated with the non-zero counts where $y_{ij} + 1 = 11$, and the $n/2$ parameters with the zero counts where $y_{ij} + 1 = 1$. For the sample of $n = 188$ regions in LeSage, Fischer and Scherngell (2007), there were 23,718 zero values and $\sum_i^j y_{ij} + 1 = 109,817$, for a total of 133,535 latent observations needed to sample each of the two latent variable vectors.

For the spatially structured random effects model from (8.14), let $y = (y_1, \ldots, y_N)$ denote our sample of $N = n^2$ counts for the OD pairs of regions. The assumption regarding y_i is that $y_i | \lambda_i$ follows a Poisson, $\mathcal{P}(\lambda_i)$ distribution, where λ_i depends on (standardized) covariates z_i with associated parameter vector γ as well as n vectors of latent spatial effects parameters. This model can be expressed as:

$$y_i | \lambda_i \sim \mathcal{P}(\lambda_i), \qquad (8.27)$$
$$\lambda_i = \exp(z_i \gamma + \delta_{di} \theta_d + \delta_{oi} \theta_o)$$

where δ_{di} represents the ith row from the matrix Δ_d that identifies region i as a destination region and δ_{oi} identifies origin regions using rows from the matrix Δ_o.

Frühwirth-Schnatter and Wagner (2006) note that the posterior density takes the form in (8.28), where $\mathcal{V}$ are parameters and ψ are the latent unobservables on which we are conditioning.

$$p(\mathcal{V}|\psi, y) \propto p(y|\mathcal{V})p(\mathcal{V}|\psi) \qquad (8.28)$$
$$p(y|\mathcal{V}) = \prod_{i=1}^{N} \frac{\exp(z_i \delta)^{y_i}}{y_i!} \exp(-\exp(z_i \delta))$$

The use of a normal distribution for the random effects in place of the conjugate gamma distribution results in a posterior density that does not belong to a density from a known distribution family. The contribution of Frühwirth-Schnatter and Wagner (2006) was to note that introduction of two sequences of artificially missing data (treated using data augmentation) can lead to a sequence of conditional posteriors for the parameters $\mathcal{V}$ that take the same form as those that would arise if our model was a normal linear model. One of the two sequences of artificially missing data eliminates the non-linearity of the Poisson model using ideas regarding unobserved inter-arrival times from a Poisson process. After eliminating the non-linearity, a linear regression model results, where the non-normal errors follow a log exponential distribution having a mean of unity. The second sequence of missing data is a component indicator associated with a normal mixture approximation to the log exponential distribution that is used to eliminate the non-normality.

8.5 Chapter summary

Standard spatial regression models rely on spatial autoregressive constructs that use spatial weight structures to specify dependence among n regions. Ways of parsimoniously modeling the connectivity among the sample of $N = n^2$ origin-destination (OD) pairs that arise in a closed system of interregional flows has remained a stumbling block. We demonstrated that the algebra of Kronecker products can be used to produce spatial weight structures that model dependence among the N OD pairs in a fashion consistent with standard spatial autoregressive processes. This allows us to extend spatial regression models that have served as the workhorse in applied spatial econometric analysis to model OD flows. The resulting models reflect a spatial filter for origin- and destination-based dependence, as well as an interaction term that we label origin-to-destination based dependence.

Important computational issues arise when working with OD flows since the sample is of dimension $N = n^2$, where n is the number of regions being modeled. We provide a moment-matrix approach that can be used for both maximum likelihood and Bayesian estimation of the models set forth here. The computational time and memory required by our approach to estimation does not depend on the sample size n, making it applicable to large problems.

Chapter 9

Matrix Exponential Spatial Models

This chapter discusses a *matrix exponential spatial specification* (MESS) introduced by LeSage and Pace (2007). As discussed in Section 7.6, one can view MESS as arising from a spatiotemporal process with exponential decay of influence from previous periods. The matrix exponential can provide an alternative to the spatial autoregressive process as a basis for building spatial regression models. It essentially replaces the geometric decay over space associated with the spatial autoregressive process with exponential decay over space.

This alternative has a computational advantage over conventional spatial autoregressive based regression models since it eliminates the need to calculate the log-determinant when producing maximum likelihood and Bayesian model estimates. There are also theoretical advantages associated with this type of spatial specification.

Section 9.1 presents the matrix exponential spatial specification, which we label the MESS model. This specification replaces the spatial autoregressive process with a matrix exponential spatial transformation. In Section 9.2 we introduce the idea of modeling spatial error variance-covariance matrices as a matrix exponential and show how this can be applied to produce a spatial regression specification. We describe a number of connections between various traditional spatial error model specifications that can be established using the matrix exponential spatial specification. A Bayesian version of the model is introduced in Section 9.3, and the basic MESS model is extended in Section 9.4 to include a parameterized spatial weight structure. Estimation of this extended model using MCMC is discussed and illustrated. Section 9.5 extends the MESS model to implement spatial fractional differencing.

9.1 The MESS model

In Section 9.1.1, we present a unique interior optimal spatial transformation of the dependent variable that leads to the MESS model. Section 9.1.2 discusses maximum likelihood estimation using this model, and Section 9.1.3 provides details on a closed form solution for this model using the approach discussed in Chapter 4.

9.1.1 The matrix exponential

Consider estimation of models where the dependent variable y undergoes a linear transformation Sy as in (9.1).

$$Sy = X\beta + \varepsilon \tag{9.1}$$

The vector y contains the n observations on the dependent variable, X represents the $n \times k$ matrix of observations on the independent variables, S is a positive definite $n \times n$ matrix, and the n-element vector ε is distributed $N(0, \sigma^2 I_n)$. Note that a conventional spatial autoregressive model can be written by setting $S = (I_n - \rho W)$ in (9.1). The concentrated log-likelihood for the model in (9.1) is shown in (9.2), where β and σ^2 have been concentrated out of the model.

$$L = \kappa + \ln|S| - (n/2)\ln(y'S'MSy) \tag{9.2}$$

The term κ represents a scalar constant and both $M = I_n - H$ and $H = X(X'X)^{-1}X'$ are idempotent matrices. The term $|S|$ is the Jacobian of the transformation from y to Sy. Without the Jacobian term, S containing all zeros would lead to a perfect, albeit pathological fit as noted in Chapter 4. The Jacobian term penalizes attempts to use singular or near singular transformations to artificially increase the regression fit.

The matrix exponential defined in (9.3) can be used as a model for S, where W represents an $n \times n$ non-negative matrix with zeros on the diagonal and α represents a scalar real parameter.

$$S = e^{\alpha W} = \sum_{i=0}^{\infty} \frac{\alpha^i W^i}{i!} \tag{9.3}$$

Of course, W is a spatial weight matrix, where $W_{ij} > 0$ indicates that observation j is a neighbor of observation i. As usual, $W_{ii} = 0$ to preclude an observation from directly predicting itself. Also, $(W^2)_{ij} > 0$ indicates that observation j is a neighbor to a neighbor of observation i. Similar relations hold for higher powers of W which identify higher-order neighbors. Thus the matrix exponential S, associated with matrix W, imposes a decay of influence for higher-order neighboring relationships. The MESS specification replaces the conventional geometric decay of influence from higher-order neighboring relationships implied by the spatial autoregressive process with an exponential pattern of decay in influence from higher-order neighboring relationships.

As in the case of conventional autoregressive processes, if W is row-stochastic, S will be proportional to a row-stochastic matrix, since products of row-stochastic matrices are row-stochastic (i.e., by definition $W\iota_n = \iota_n$ and therefore $W(W\iota_n) = \iota_n$, and so on, where ι_n denotes a vector of ones). Consequently, S is a linear combination of row-stochastic matrices and thus is proportional to a row-stochastic matrix.

In a non-spatial setting, Chiu, Leonard, and Tsui (1996) proposed use of the matrix exponential and discussed several of its salient properties, some of which are enumerated below:

1. S is positive definite,

2. any positive definite matrix is the matrix exponential of some matrix,

3. $S^{-1} = e^{-\alpha W}$,

4. $|e^{\alpha W}| = e^{\text{tr}(\alpha W)}$.

The last property greatly simplifies the MESS log-likelihood. Since $\text{tr}(W) = 0$ and by extension $|e^{\alpha W}| = e^{\text{tr}(\alpha W)} = e^0 = 1$, the concentrated log-likelihood takes the form: $L = \kappa - (n/2)\ln(y'S'MSy)$. Therefore, maximizing the log-likelihood is equivalent to minimizing $(y'S'MSy)$, the overall sum-of-squared errors.

The MESS model in (9.1) can be extended in a fashion similar to the SDM model. Let U represent a matrix of observations on p non-constant independent variables and let q be an integer large enough so that X approximately spans SU, but small enough so that X cannot span y. The design matrix X (assuming full rank) could have the form (9.4).

$$X = [\iota_n \; U \; WU \dots W^{q-1}U] \qquad (9.4)$$

In this case, X approximately spans SU and thus the MESS model based on (9.4) nests a spatial autoregression in the errors. Like the SDM model, this variant of the MESS model results in a situation where a set of linear restrictions on the parameters associated with the columns of X could yield the error autoregression. Hendry et al. (1984) advocates estimation of this type of general distributed lag model with subsequent imposition of restrictions that has been labeled the *general to specific approach* to model specification.

9.1.2 Maximum likelihood estimation

If elements of the powers of W represent magnitudes that do not rise with the power, the power series in (9.3) converges rapidly. Since row-stochastic, non-negative matrices W have a maximum of 1 in any row, the magnitude of the elements in the powers of W does not grow with powers of the matrix. Given a rapid decline in the coefficients of the power series, achieving a satisfactory progression towards convergence seems feasible with ten to twelve terms.

If the graph of W is strongly connected, meaning that a path exists between every pair of observations, then $\sum_{r=1}^{n} \omega_r W^r$ will be dense (all non-zeros) for positive ω_r (Horn and Johnson, 1993, p. 361-362), leading to a dense S. In this case, computing S separately would require prohibitive amounts of memory and time for large n. Fortunately, we do not need to compute S separately since S always appears in conjunction with y. This allows computation of Sy in $O((q-1)n^2)$ operations for dense W by sequential left-multiplication of y by W to form n−element vectors, (i.e., Wy, $W(Wy) = W^2 y$, and so on).

For sparse W the number of operations required to compute Sy declines to $O((q-1)n_{\neq 0})$, where $n_{\neq 0}$ denotes the number of non-zeros. For an m nearest neighbor spatial weight matrix that has m non-zero entries in each row, the operation count associated with computing Sy would decline to $O((q-1)mn)$. This results in an operation count for computing Sy in nearest neighbor specifications of W that is linear in n.

9.1.3   A closed form solution for the parameters

Section 4.10 presented a means of finding closed-form solutions for many single dependence parameter models. In this section we show how to find a closed-form solution to MESS using this framework. To illustrate this approach in detail, we define the $n \times q$ matrix Y comprised of powers of W times y in (9.5).

$$Y = [y \ Wy \ W^2 y \ldots W^{q-1} y] \tag{9.5}$$

We define a diagonal matrix G_1 containing some of the coefficients from the power series as shown in (9.6).

$$G_1 = \begin{pmatrix} 1/0! & & & \\ & 1/1! & & \\ & & \ddots & \\ & & & 1/(q-1)! \end{pmatrix} \tag{9.6}$$

In addition, we define the q-element column vector v shown in (9.7) that contains powers of the scalar real parameter α, $|\alpha| < \infty$.

$$v(\alpha) = [1 \ \alpha \ \alpha^2 \ldots \alpha^{q-1}]' \tag{9.7}$$

Using (9.5), (9.6) and (9.7), we can rewrite Sy as shown in (9.8).

$$Sy \approx Y G_1 v(\alpha) \tag{9.8}$$

Premultiplying Sy by the least-squares idempotent matrix M yields the residuals e, allowing us to express the overall sum-of-squared errors as in (9.9),

$$\begin{aligned} e'e &= v(\alpha)' G_1 (Y' M' M Y) G_1 v(\alpha) \\ &= v(\alpha)' G_1 (Y' M Y) G_1 v(\alpha) \\ &= v(\alpha)' Q v(\alpha) \end{aligned} \tag{9.9}$$

where $Q = G_1(Y'MY)G_1$. This allows us to rewrite $v(\alpha)'Qv(\alpha)$ as the $2q - 2$ degree polynomial $Z(\alpha)$, shown in (9.10).

$$Z(\alpha) = \sum_{i=1}^{2q-1} c_i \alpha^{i-1} = v(\alpha)'Qv(\alpha) \tag{9.10}$$

As discussed in Section 4.10, this is a polynomial in α and has a closed-form solution. With regard to second order conditions, LeSage and Pace (2007) show that the optimum α is unique for MESS.

The closed-form solution provides an optimal value of α and also the second derivative at this optimal value. Given the second derivative, a variant of the mixed analytical-numerical Hessian described in Section 3.2.1 provides standard errors. In addition, the analytic Hessian for MESS is more tractable than the SAR analytic Hessian.

9.1.4 An applied illustration

This section illustrates maximum likelihood estimation and demonstrates how the MESS model can be used to produce estimates for a data vector y generated using the more traditional spatial autoregressive specification: $y = \rho W y + X\beta + \varepsilon$. In many cases, the resulting estimates for the parameters β and the noise variance σ_ε^2 will be nearly identical to those from maximum likelihood estimation of the more traditional spatial autoregressive model. Given the computational advantages of the MESS model, this seems a desirable situation and provides a valuable tool for those working with large spatial data sets. The MESS model parameter α represents an analogue to the spatial dependence parameter ρ in the SAR model. While its value will not take on the same magnitudes, we establish a correspondence between α and ρ that can be used to provide a translation between these measures of spatial dependence from the two types of models.

As an illustration of the similarity in parameter magnitudes and inferences provided by conventional and MESS models, a dataset from Harrison and Rubinfeld (1978) containing information on housing values in 506 Boston area census tracts was used to produce SAR and MESS estimates.[1] The estimates shown in Table 9.1 are based on a first-order spatial contiguity matrix often used in conventional models. The estimation results indicate that identical inferences would be drawn regarding both the magnitude and significance of the 14 explanatory variables on housing values in the model. Both the point estimates as well as asymptotic t-values (based on a variance-covariance matrix obtained using a numerical Hessian to evaluate the log-likelihood function

---

[1] This data was augmented with with latitude-longitude coordinates described in Gilley and Pace (1996). These were used to create a first-order spatial contiguity weight matrix for the observations. The data is described in detail in Belsley, Kuh, and Welch (1980), with various transformations used presented in the table on pages 244-261.

at the maximum likelihood magnitudes) are presented in the table, where we see nearly identical values for both. Although we used the numerical Hessian in this application, the closed-form solution for α also yields the information needed to implement a mixed numerical-analytical Hessian as described in Chapter 4.

TABLE 9.1: Correspondence between SAR and MESS estimates

Variables	SAR model	t-statistic	MESS model	t-statistic
Constant	−0.00195	−0.1105	−0.00168	−0.0927
Crime	−0.16567	−6.8937	−0.16776	−6.8184
Zoning	0.08057	3.0047	0.07929	2.8732
Industry	0.04428	1.2543	0.04670	1.2847
Charlesr	0.01744	0.9327	0.01987	1.0406
Noxsqr	−0.13021	−3.4442	−0.13271	−3.4307
Roomsqr	0.16082	6.5430	0.16311	6.4428
Houseage	0.01850	0.5946	0.01661	0.5187
Distance	−0.21548	−6.1068	−0.21359	−5.8798
Access	0.27243	5.6273	0.27489	5.5188
Taxrate	−0.22146	−4.1688	−0.22639	−4.1435
Pupil/Teacher	−0.10304	−4.0992	−0.10815	−4.2724
Blackpop	0.07760	3.7746	0.07838	3.7058
Lowclass	−0.33871	−10.1290	−0.34155	−10.1768
ρ $\mid$ α	0.44799	11.892	−0.55136	−10.5852
R^2	0.8420		0.8372	
σ^2	0.1577		0.1671	

Regarding the relation between the spatial dependence parameters α and ρ, we can use the correspondence, $\rho = 1 - e^\alpha$ to transform the value of $\alpha = -0.55136$ reported in the table. This results in $\rho = 1 - e^{-0.55136} = 0.4238$, a value close to the reported SAR estimate of $\rho = 0.44799$. We derive this correspondence by equating the matrix norms from the two transformations $I_n - \rho W$ and $e^{\alpha W}$. The most convenient matrix norm to use is the maximum row sum norm which equals $1 - \rho$ for the autoregressive transformation and e^α for the matrix exponential transformation. Equating these leads to $\rho = 1 - e^\alpha$ or $\alpha = \ln(1 - \rho)$.

The correspondence between MESS and SAR models allows us to take advantage of the computational convenience arising from the matrix exponential spatial specification when analyzing spatial regression relationships traditionally explored using spatial autoregressive models.

9.2   Spatial error models using MESS

A host of regression models can accommodate spatial dependence in the disturbances and many of these rely upon particular implementations of multivariate normal regression models. Specifically, given n observations on the dependent variable y and k independent variables X, such models posit a linear (in the parameters) relation among the independent variables as shown in (9.11), where β represents a k element parameter vector and ε follows a multivariate normal distribution with variance-covariance matrix Ω.

$$y = X\beta + \varepsilon \tag{9.11}$$
$$\varepsilon \sim N(0, \Omega)$$

Models for spatially dependent errors rely on setting $\Omega_{ij} \neq 0$ when observation i is located near observation j. For example, an externality affecting both spatial locations i and j could result in ε_i and ε_j behaving similarly ($\Omega_{ij} > 0$), the case of positive spatial dependence.

Differences among multivariate normal spatial models arise from alternative specifications for the variance-covariance matrix Ω. Equation (9.12) shows the conditional (CSG), simultaneous (SSG), linear moving average (MAL), quadratic moving average (MAQ), and matrix exponential (MESS) specifications using the real scalar parameters $\phi, \lambda, \gamma, \theta, \alpha$ as well as the real $n \times n$ spatial weight matrix W.[2] We use a symmetric W in this section because CSG and MAL specifications require a symmetric spatial weight matrix. Of course, other spatial specifications can also accommodate symmetric spatial weights, so we rely on this to facilitate comparison of various spatial specifications.

$$
\begin{aligned}
\text{CSG} &: \Omega^{-1} = I_n - \phi W \\
\text{SSG} &: \Omega^{-1} = I_n - 2\rho W + \rho^2 W^2 \\
\text{MAL} &: \Omega = I_n + \gamma W \\
\text{MAQ} &: \Omega = I_n + 2\theta W + \theta^2 W^2 \\
\text{MESS} &: \Omega = e^{\alpha W} = \sum_{i=0}^{\infty} \frac{\alpha^i W^i}{i!}
\end{aligned}
\tag{9.12}
$$

To provide more insight into MESS, consider an expansion of the Taylor series in (9.12) shown in (9.13).

---

[2] Following Cressie (1993) we use SSG in this discussion of error models as a label for *simultaneous autoregressive models*. This is to avoid confusion because these are commonly referred to as SAR models in the spatial statistics literature. We also use CSG to reference what are commonly labeled CAR models in the spatial statistics literature.

$$\Omega = I_n + \alpha W + \frac{\alpha^2}{2}W^2 + \frac{\alpha^3}{6}W^3 + \frac{\alpha^4}{24}W^4 + \frac{\alpha^5}{120}W^5 + \cdots \quad (9.13)$$

$$\Omega^{-1} = I_n - \alpha W + \frac{\alpha^2}{2}W^2 - \frac{\alpha^3}{6}W^3 + \frac{\alpha^4}{24}W^4 - \frac{\alpha^5}{120}W^5 + \cdots$$

For positive α and non-negative W, MESS results in non-negative variance-covariance matrix elements, as in the linear (MAL) and quadratic (MAQ) moving average specifications. The inverse of the matrix exponential involves a simple switch in the sign of α, $\left((e^{\alpha W})^{-1} = e^{-\alpha W}\right)$, so negative α corresponds to specifying the inverse variance-covariance matrix. For negative values of α, MESS results in a potential mixture of negative and positive elements as in the SSG specification.

For all specifications either the variance-covariance matrix or the inverse variance-covariance matrix involve powers of the spatial weight matrix in their Taylor series expansions. These powers have a natural interpretation in the context of modeling spatial dependence. For example, positive elements in the squared weight matrix $(W^2)_{ij} > 0$ indicate that observation j is a neighbor of a neighbor to observation i (second-order neighbor). This relation holds for higher powers of W as well, where non-zero elements of W^h represent h-order neighbors. Over the relevant parameter domains, all the specifications place lower values on higher-order powers of the spatial weight matrices.

The CSG, SSG, MAL, and MAQ specifications have been mainstays of the spatial dependence literature (Ord, 1975; Anselin, 1988; Ripley, 1988; Haining, 1990). Chiu, Leonard, and Tsui (1996) introduced the matrix exponential function in the context of modeling general non-spatial covariance structures. However, the matrix exponential possesses many desirable properties and appears ideally suited for spatial applications.

First, $e^{\alpha W}$ is positive definite for all α and symmetric W and thus $e^{-\alpha W}$ is positive definite as well.[3] This freedom from singularities greatly simplifies both computational and theoretical work. In contrast to MESS, traditional specifications must obey various restrictions to ensure a positive-definite variance-covariance matrix (Cressie, 1993, p. 468).

Second, since any positive definite matrix is the matrix exponential of some matrix, this means the exponential of some matrix can yield the correct variance-covariance matrix. In fact, when $e^{\alpha W} = \Omega$, $W = \alpha^{-1}\ln(\Omega)$ where $\ln(\cdot)$ is the matrix logarithm function (Horn and Johnson, 1994, p. 448, 474).

We can again use the fact that the determinant is a simple function of the trace $|e^{\alpha W}| = e^{\alpha \cdot tr(W)}$, and this holds for any real, square matrix (Horn and Johnson, 1994, p. 474). Since the spatial weight matrix W has zeros on the diagonal $tr(W) = 0$, resulting in $|e^{\alpha W}| = e^0 = 1$ and $\ln|\Omega(\alpha)| = 0$ for all values of α. This produces a major simplification as shown by the profile or

---

[3] For non-symmetric W, $e^{-\frac{\alpha}{2}W'}e^{-\frac{\alpha}{2}W}$ is symmetric positive definite.

concentrated log-likelihood function in (9.14), where ($\cdot$) denotes the relevant spatial dependence parameter for each specification (Anselin, 1988, p. 110).

$$L\left(\cdot\right) = \kappa - (1/2)\ln\left|\Omega\left(\cdot\right)\right| - (n/2)\ln\left(\left(y - X\beta\left(\cdot\right)\right)'\Omega\left(\cdot\right)^{-1}\left(y - X\beta\left(\cdot\right)\right)\right)$$

$$\beta\left(\cdot\right) = \left(X'\Omega\left(\cdot\right)^{-1}X\right)^{-1}X'\Omega\left(\cdot\right)^{-1}y \tag{9.14}$$

As already emphasized, the presence of the $n \times n$ log-determinant of the variance-covariance matrix ($\ln\left|\Omega\left(\cdot\right)\right|$) ordinarily poses a computational challenge. For the MESS $\Omega\left(\cdot\right)$, maximum likelihood estimation can rely on non-linear least-squares.

Finally, any power of a matrix exponential remains a matrix exponential. Thus, $\left(e^{\alpha W}\right)^\tau = e^{\tau\alpha W}$ for some real scalar τ (Horn and Johnson, 1994, p. 435). This property leads to a simple expression of relations among the various modeling methods. As an alternative to (9.12), consider the weight matrices associated with the various specifications when these produce identical estimates. Let C, S, L, Q, and W represent symmetric spatial weight matrices associated with CSG, SSG, MAL, MAQ, and MESS. Equating the specifications yields the expression in (9.15), allowing us to express the other specifications in terms of the matrix exponential. This produces $C = I_n - e^{-\alpha W}$, $S = I_n - e^{-\frac{\alpha}{2}W}$, $Q = e^{\frac{\alpha}{2}W} - I_n$, and $L = e^{\alpha W} - I_n$.

$$\Omega = I_n + L = (I_n + Q)^2 = (I_n - S)^{-2} = (I_n - C)^{-1} = e^{\alpha W} \tag{9.15}$$

This formulation of the specifications reveals rather simple relations among the specifications in terms of the matrix exponential parameter α. Specifically, α varies by a factor of two between SSG and CSG, as well as between MAL and MAQ. This agrees with Ripley (1981, p. 97) who showed that the CSG parameter exceeded the SSG parameter by almost a factor of two for small CSG parameter values. Under MESS, the CSG parameter estimate of $-\alpha$ exactly doubles the SSG parameter estimate of $-(1/2)\alpha$.

As Cressie (1993, p. 409, 434) notes, any valid variance-covariance matrix can yield a conditional or simultaneous autoregression. Since the matrix exponential of a symmetric weight matrix always produces a valid variance-covariance matrix, this allows users to select their preferred interpretation (conditional or simultaneous). Many statisticians (Cressie, 1993, p. 408) prefer CSG, while SSG sees more usage in spatial econometrics (Anselin, 1988).

From a computational perspective, any algorithm for the rapid computation of the matrix exponential also yields rapid computation of other powers such as the inverse. Most applications only need $e^{\tau\alpha W}y$ or $e^{\tau\alpha W}X$ (not $e^{\tau\alpha W}$ itself), and particularly fast algorithms exist for these expressions. For dense matrices, these computations require $O(n^2)$ operations and sparse matrices may require as few as $O(n)$ operations (Pace and Barry, 1997). Consequently, the computational advantages of the matrix exponential spatial specification permit estimates, predictions, residuals, and other statistics for both simultaneous and conditional specifications at low computational cost.

9.2.1 Spatial model Monte Carlo experiments

We investigate the performance of MESS relative to the other common spatial models with a simple Monte Carlo experiment. The experiment involved generating sample data using each of the different spatial specifications as the data generating process (DGP) and estimating parameters for each model to examine the performance under reciprocal misspecification.

In these experiments, X contained a constant vector and a standard random normal vector (mean zero, variance unity). The first-order contiguity spatial weight matrix was constructed using Delaunay triangles based upon coordinates from a sample of 3,107 US counties in the lower 48 states from the 1990 Census.[4] The SSG specification most frequently employs a row-stochastic weight matrix while CSG and MAL require symmetric weight matrices, so we used a symmetric doubly stochastic weight matrix where each row and column sum to unity. The doubly stochastic weight matrix is constant preserving (e.g., $W\iota_n = \iota_n$, $W'\iota_n = \iota_n$), so each specification has an intercept variable proportional to a constant and thus the residuals in all specifications sum to zero, as in ordinary least-squares (OLS).

For the first two experiments, we used a relatively large dependence parameter of 0.8 for each DGP, and set the R^2 to 0.8 to hold signal-to-noise constant. These parameter choices mimic the situation found in housing data where models fit well, but also display residual spatial dependence (Pace et al., 2000). For all three experiments, we generated 100 trials per reported number and summarized these using the arithmetic mean.

Table 9.2 presents results from simulating each DGP, estimating all six models, and calculating spatial dependence estimates. Each estimator returned an average spatial dependence parameter estimate close to the true value of 0.8, when estimated under its own DGP. As previously discussed, the CSG and MAL estimates are almost twice the SSG and MAQ estimates for the CSG, MAL DGP. We note that when SSG, MAQ are the DGPs, the CSG and MAL specifications cannot produce an estimate that is twice 0.8 since they must be less than 1. The complement of the MESS parameter estimate lies between the SSG and MAQ parameter estimates for all but the SSG DGP (which produces the strongest spatial dependence).

Another experiment compared the mean and standard deviation of 100 regression parameter estimates ($\hat{\beta}$) under varying amounts of spatial dependence as well as the signal-to-noise. We used SSG parameter ρ values of 0.25, 0.50, 0.75, 0.90, and R^2 values of 0.20 and 0.80. We used MESS parameter α

---

[4]The lower 48 states in the US contain 3,111 counties. However, four counties did not have complete data in all fields and we deleted these. The deleted observations consisted of unusual counties such as Yellowstone within Yellowstone National Park.

TABLE 9.2: Dependence estimates across estimators and DGPs

	DGP	CSG	SSG	MESS	MAQ	MAL
Estimator						
CSG		0.801	0.989	0.911	0.870	0.596
SSG		0.483	0.800	0.639	0.597	0.343
MESS		−0.532	−1.073	−0.804	−0.762	−0.382
MAQ		0.533	0.988	0.813	0.804	0.403
MAL		0.971	0.990	0.990	0.990	0.805

values of -0.25, -0.50, -1.0, -2.0, and R^2 values of 0.20 and 0.80. Table 9.3 presents the average and standard deviation of parameter estimates ($\hat{\beta}$) on the non-constant variable across 100 trials for the SSG DGP (top panel) and the MESS DGP (bottom panel). Both SSG and MESS produced a distribution of estimates with nearly identical means and standard deviations under either DGP.

9.2.2 An applied illustration

To examine whether the findings from the Monte Carlo experiment hold for actual spatial data samples, we constructed models using 32 expenditure categories (e.g., alcohol, tobacco, furniture, etc.) from the 1998 Consumer Expenditure Survey. Regressions employed the double-log form with (logged) expenditure shares as the dependent variable and 12 (logged) explanatory variables measuring age (six separate variables based on age categories), race, gender, income, population, housing units, and land area. The data used for these variables came from the same sample of 3,107 US counties from the 1990 Census that was used in the previous section. The large number of estimated parameters (384) provides a natural setting to examine relations among the five spatial specifications (SSG, CSG, MESS, MAQ, MAL). We used the same doubly stochastic weight matrix based on Delaunay triangles as in the Monte Carlo experiment from the previous section.

For all five spatial specifications and for all 32 dependent variables, the estimated spatial dependence parameters were significant at the 1% level. Of 384 possible coefficients, the CSG, SSG, MESS, MAQ, MAL, and OLS estimators produced significant coefficients in 316, 313, 310, 311, 312, and 318 instances. OLS proved the most liberal (318) and MESS the most conservative (310) in finding significance.

Tables 9.4 and 9.5 present correlations among parameter estimates as well as signed square roots of the likelihood ratios. These results reinforce those from the Monte Carlo study and show a close relation between SSG and MESS as well as between MAQ and MESS. Of course, as n becomes large we would expect the correlation between estimates to approach 1.0 given the

TABLE 9.3:	Mean and dispersion
of estimates across estimators and DGPs

R^2	Autoregressive DGP			
	ρ	OLS	SSG	MESS
0.80	0.25	0.9994	0.9992	0.9992
0.80	0.25	0.0087	0.0086	0.0086
0.80	0.50	0.9996	0.9991	0.9991
0.80	0.50	0.0094	0.0086	0.0086
0.80	0.75	0.9999	0.9991	0.9990
0.80	0.75	0.0118	0.0085	0.0087
0.80	0.90	1.0006	0.9990	0.9989
0.80	0.90	0.0167	0.0084	0.0088
0.20	0.25	0.9977	0.9969	0.9970
0.20	0.25	0.0348	0.0343	0.0344
0.20	0.50	0.9983	0.9965	0.9965
0.20	0.50	0.0378	0.0343	0.0345
0.20	0.75	0.9996	0.9963	0.9960
0.20	0.75	0.0472	0.0339	0.0346
0.20	0.90	1.0026	0.9961	0.9955
0.20	0.90	0.0667	0.0335	0.0353

R^2	Matrix exponential DGP			
	α	OLS	SSG	MESS
0.80	−0.25	0.9997	1.0004	1.0003
0.80	−0.25	0.0076	0.0074	0.0074
0.80	−0.50	0.9994	1.0005	1.0006
0.80	−0.50	0.0080	0.0072	0.0072
0.80	−1.00	0.9987	1.0006	1.0009
0.80	−1.00	0.0098	0.0067	0.0066
0.80	−2.00	0.9967	1.0003	1.0010
0.80	−2.00	0.0191	0.0055	0.0050
0.20	−0.25	0.9990	1.0014	1.0014
0.20	−0.25	0.0306	0.0296	0.0296
0.20	−0.50	0.9977	1.0022	1.0024
0.20	−0.50	0.0322	0.0289	0.0289
0.20	−1.00	0.9948	1.0026	1.0036
0.20	−1.00	0.0393	0.0267	0.0265
0.20	−2.00	0.9866	1.0011	1.0040
0.20	−2.00	0.0764	0.0221	0.0202

relation among error models discussed in Section 3.3.1 where we developed
the Hausman test.

TABLE 9.4: Correlations among β estimates

	CSG	SSG	MESS	MAQ	MAL	OLS
CSG	1.000	0.997	0.991	0.986	0.980	0.942
SSG	0.997	1.000	0.998	0.996	0.992	0.963
MESS	0.991	0.998	1.000	0.999	0.997	0.976
MAQ	0.986	0.996	0.999	1.000	0.999	0.983
MAL	0.980	0.992	0.997	0.999	1.000	0.988
OLS	0.942	0.963	0.976	0.983	0.988	1.000

TABLE 9.5: Correlations among signed root deviances

	CSG	SSG	MESS	MAQ	MAL	OLS
CSG	1.000	0.993	0.980	0.969	0.958	0.912
SSG	0.993	1.000	0.996	0.990	0.984	0.949
MESS	0.980	0.996	1.000	0.999	0.995	0.971
MAQ	0.969	0.990	0.999	1.000	0.999	0.981
MAL	0.958	0.984	0.995	0.999	1.000	0.988
OLS	0.912	0.949	0.971	0.981	0.988	1.000

Table 9.6 shows the number of times a specification produced a significant result of opposite sign relative to another specification. The spatial specifications differ in various cases. For example, CSG and the moving average specifications disagreed in 3 cases regarding the sign of a significant coefficient. Interestingly, MESS never produced an opposite inference when compared to other spatial specifications.

TABLE 9.6: Number of significant coefficients with opposite signs

	CSG	SSG	MESS	MAQ	MAL	OLS
CSG	0.000	0.000	0.000	3.000	3.000	7.000
SSG	0.000	0.000	0.000	2.000	2.000	3.000
MESS	0.000	0.000	0.000	0.000	0.000	1.000
MAQ	3.000	2.000	0.000	0.000	0.000	1.000
MAL	3.000	2.000	0.000	0.000	0.000	0.000
OLS	7.000	3.000	1.000	1.000	0.000	0.000

9.3   A Bayesian version of the model

A Bayesian approach to the MESS model would include specification of prior distributions for the parameters in the model, α, β, σ. Prior information regarding the parameters β and σ^2 is unlikely to exert much influence on the posterior distribution of these estimates in the case of very large samples where application of MESS models holds an advantage over spatial autoregressive specifications. However, the parameter α could exert an influence even in large samples, because of the important role played by the spatial dependence parameter in these models. Given this motivation, we begin with an uninformative prior $\pi(\beta, \sigma^2|\alpha) \propto \kappa$, and let $\pi(\alpha)$ denote an arbitrary prior for α.

Using Bayes' theorem to combine the likelihood and prior, we obtain the kernel posterior distribution:

$$p(\beta, \sigma^2, \alpha|\mathcal{D}) \propto \sigma^{-(n+1)}\exp\left[-(1/2\sigma^2)(Sy - X\beta)'(Sy - X\beta)\right]\pi(\alpha) \quad (9.16)$$

Using the properties of the gamma distribution (Judge et al., 1982, p. 86), we can integrate out the parameter σ^2 to obtain:

$$p(\beta, \alpha|\mathcal{D}) \propto ([y'S(\alpha)'MS(\alpha)y] + [\beta - \beta(\alpha)]'\,X'X\,[\beta - \beta(\alpha)])^{-n/2}\pi(\alpha)$$
$$\beta(\alpha) = (X'X)^{-1}X'S(\alpha)y$$
$$S(\alpha) = e^{\alpha W} \qquad\qquad\qquad (9.17)$$

where we write $S(\alpha)$ and $\beta(\alpha)$ to reflect the dependence of these expressions on the spatial dependence parameter α.

9.3.1   The posterior for α

The joint distribution in (9.17) is a multivariate t-distribution (conditional on α) that can be integrated with respect to β to arrive at the posterior distribution for the spatial dependence parameter α.

$$p(\alpha|\mathcal{D}) \propto [y'S(\alpha)'MS(\alpha)y]^{-(n-k)/2}\pi(\alpha) \qquad (9.18)$$

The expression in (9.18) represents the marginal posterior for α. The $2q - 2$ degree polynomial expression in (9.10) for $Z(\alpha) = y'S(\alpha)'MS(\alpha)y$ proves particularly convenient for integration of this marginal posterior. The posterior expectation of the parameter α is:

$$E(\alpha|\mathcal{D}) = \alpha^* = \frac{\int_{-\infty}^{+\infty} \alpha \cdot p(\alpha|\mathcal{D})d\alpha}{\int_{-\infty}^{+\infty} p(\alpha|\mathcal{D})d\alpha} \qquad (9.19)$$

A few points to note regarding the limits of integration in (9.19). First, restriction of the upper limit of integration to zero imposes positive spatial dependence, an approach often taken in applied practice. Without loss of generality we could extend the limit of integration to allow for negative spatial dependence estimates. Second, we can use the correspondence between ρ in conventional spatial autoregressive (SAR) models and α from the MESS model to show that $\alpha = -5$ implies $\rho = 0.9933$. Since the upper bound on ρ is unity, and values of 0.99 are seldom encountered during empirical application of SAR models, we can set the lower integration limit to -5, rather than rely on $-\infty$. The correspondence also indicates that we can accommodate negative spatial autocorrelation ranging down to -1 by extending the upper limit of integration to 0.7.

The integrand in the normalizing constant of the denominator in (9.19) can be expressed using our polynomial $Z(\alpha)$ from (9.10) as:

$$p(\alpha|\mathcal{D}) \propto \left(\sum_{i=1}^{2q-1} c_i \alpha^{i-1} \right)^{-(n-k)/2} \pi(\alpha) \tag{9.20}$$

which makes univariate integration a simple scalar problem. This is true irrespective of the number of observations in the problem.

Turning attention to the posterior variance for α, this takes the form shown in (9.21), where the limits of integration are those noted in the discussion surrounding (9.19).

$$\text{var}(\alpha|\mathcal{D}) = \frac{\int [\alpha - \alpha^*]^2 \cdot p(\alpha|\mathcal{D})d\alpha}{\int p(\alpha|\mathcal{D})d\alpha} \tag{9.21}$$

This numerical integration problem would also benefit from the scalar form of $Z(\alpha)$ which is embedded in $p(\alpha|\mathcal{D})$ in (9.21).

It is informative to contrast this result with that arising in more traditional spatial autoregressive models such as: $y = \rho W y + X\beta + \varepsilon$. Using the development from Chapter 5, we have a marginal posterior for ρ, the spatial dependence parameter in these models shown in (9.22), where $\pi(\rho)$ denotes a prior for the parameter ρ (see Chapter 5).

$$p(\rho|\mathcal{D}) \propto |I_n - \rho W|[(n-k)^{-1}u(\rho)'u(\rho)]^{-(n-k)/2}\pi(\rho)$$
$$u(\rho) = (I_n - \rho W)y - X\beta(\rho)$$
$$\beta(\rho) = (X'X)^{-1}X'(I_n - \rho W)y \tag{9.22}$$

To compute the posterior expectation of ρ in this model one would need to perform univariate numerical integration on the expression in (9.23), where the limits of integration involve those for the parameter ρ set forth in Chapter 4 involving eigenvalues of the spatial weight matrix W.

$$E(\rho|\mathcal{D}) = \rho^* = \frac{\int \rho \cdot p(\rho|\mathcal{D})d\rho}{\int p(\rho|\mathcal{D})d\rho} \qquad (9.23)$$

Note that this involves calculating the $n \times n$ determinant $|I_n - \rho W|$ over a grid of values for the parameter ρ as well as finding the eigenvalue limits. It should be clear that this is a more difficult problem to solve, especially for large spatial data samples. In addition to computing the log determinant, we would also need to compute minimum and maximum eigenvalues of W, or restrict the range of the spatial dependence to an interval such as $(-1, 1)$. The eigenvalue information is not needed for the case of the MESS model where the lower and upper limits of integration can be set for all estimation problems (see the discussion surrounding (9.19)).

In summary, solution of the Bayesian MESS model for the posterior mean and variance of the spatial dependence parameter α, as well as the entire posterior distribution of α requires simple univariate integration involving the scalar polynomial $Z(\alpha)$.

9.3.2   The posterior for β

Turning attention to the posterior distribution for β in the Bayesian MESS model, we can use the multivariate t-density centered at $\beta(\alpha^*)$, suggesting that the posterior mean can be computed analytically using:

$$E(\beta|\mathcal{D}) = (X'X)^{-1}X'S(\alpha^*)y \qquad (9.24)$$

where α^* denotes the posterior mean from (9.19). The posterior variance-covariance matrix unconditional on α takes the form:

$$\text{var-cov}(\beta) = \frac{1}{n-k-2}\left(\int (Z(\alpha)p(\alpha|\mathcal{D})d\alpha\right)(X'X)^{-1} \qquad (9.25)$$

This requires univariate integration of the posterior expectation: $E(Z(\alpha)|\mathcal{D}) = \int Z(\alpha)p(\alpha|\mathcal{D})d\alpha$. As we have already seen, the scalar polynomial expression for $Z(\alpha)$ makes this a simple computation. One might also rely on the approximation $Z(\alpha^*)/(n-k)$, which would involve simply evaluating the expression $Z(\alpha)$ at the posterior mean α^*.

Given the multivariate t-density for β, we can express this joint distribution as the product of a marginal and conditional distribution. We can use standard expressions from Zellner (1971, p. 67) to analyze the posterior distributions for individual elements of β. Here as in the case of the posterior distribution for α, the scalar polynomial expression $Z(\alpha)$ plays an important role in simplifying the computational tasks involved.

9.3.3   Applied illustrations

Three applied illustrations of the Bayesian version of the MESS model are provided. The first illustration uses a dataset from Pace and Barry (1997) and examines voter turnout in the 1980 presidential election by county for a sample of 3,107 US counties and four explanatory variables. A second illustration involves 30,987 house sales in Lucas county, Ohio and 10 explanatory variables, and the third relies on expenditure budget shares for gasoline in 59,025 census tracts with 4 explanatory variables. A standardized first-order contiguity matrix was used for the spatial weight matrix W in all illustrations. For this application, the matrices W were constructed using Delaunay triangle algorithms described in Chapter 4 applied to the location coordinates measuring relative position in the map plane.

Estimation results based on maximum likelihood and the Bayesian MESS models are presented in Table 9.7. We discuss each of these applications in turn.

For the presidential election example, explanatory variables were: a constant term, education (high school graduates), homeownership, and median household income. The dependent variable is the population voting as a proportion of population 19 years or older (those eligible to vote). This proportion was logged to induce normality. All explanatory variables were expressed as logs of the population proportion, e.g., the log of homeowners in the county as a proportion of the county population. A diffuse prior on all parameters, α, β, σ was employed in the Bayesian model, which should produce estimates nearly identical to those from maximum likelihood estimation.

In the table, we see that Bayesian and maximum likelihood estimates are identical to at least 3 decimal places in all cases. This similarity of the two sets of estimates also extends to the inferential parameter estimates shown in the table. We used a numerical Hessian evaluation of the log-likelihood at the ML estimates to produce the standard errors, although one could produce the ML standard errors using the analytic or mixed numerical-analytic Hessian described in Chapter 4. The time required to produce estimates for this 3,107 observation example was 0.110 seconds based on lower and upper integration limits of -4 and 0 respectively. An important point to note is that a priori knowledge regarding the magnitude of spatial dependence reflected in the parameter α can be used to further improve the speed of solution. For example, setting the lower integration limit to -2 and the upper limit to 0 reduced the time needed to solve the problem to 0.08 seconds. Increasing the limits of integration to -5 and 0 resulted in 0.14 seconds. Varying the limits of integration produced estimates that were nearly identical. In the table, we report times based on integration limits of -4 to 0 for timing compatibility in all three examples.

The second illustration involves a fairly typical housing price model, based on houses sold over the period from 1993 to 1997 in a single Ohio county. The dependent variable was the log of selling price. Explanatory variables

TABLE 9.7:   Bayesian estimation results for three applied examples

Presidential election, 3,107 Observations				
Variables	Bayes mean	Bayes std	ML mean	ML std
constant	0.696283	0.042381	0.696371	0.042360
education	0.272566	0.013923	0.272640	0.013917
homeowners	0.505877	0.015182	0.505883	0.015174
median income	−0.128554	0.016474	−0.128601	0.016466
α	−0.675480	0.023520	−0.675204	0.023174
σ^2	0.015336	0.000389	0.015331	0.000395
time (secs)	0.110			

House sales, 30,987 Observations				
Variables	Bayes mean	Bayes std	ML mean	ML std
House age	0.464807	0.018170	0.464774	0.018224
$(\text{House age})^2$	−0.967741	0.038228	−0.967641	0.038461
$(\text{House age})^3$	0.308795	0.022616	0.308757	0.022682
log(living area)	0.299507	0.002835	0.299488	0.002940
log(lotsize)	0.068662	0.002785	0.068647	0.002855
1993 dummy	−0.092811	0.002816	−0.092811	0.002817
1994 dummy	−0.077587	0.002864	−0.077587	0.002865
1995 dummy	−0.059417	0.002902	−0.059417	0.002902
1996 dummy	−0.052613	0.002975	−0.052613	0.002975
1997 dummy	−0.030402	0.002986	−0.030402	0.002986
α	−0.785965	0.006363	−0.786043	0.006347
σ^2	0.161119	0.001294	0.161109	0.001301
time (secs)	0.591			

Census tracts, 59,025 Observations				
Variables	Bayes mean	Bayes std	ML mean	ML std
constant	0.328584	0.015496	0.328840	0.015495
log(vehicles/spending)	0.563411	0.006083	0.563363	0.006082
log(median income)	−0.036234	0.000289	−0.036231	0.000289
log(employment)	0.000969	0.000248	0.000969	0.000248
α	−0.844754	0.004958	−0.844892	0.001480
σ^2	0.001206	0.000007	0.001206	0.000008
time (secs)	0.641			

consisted of housing characteristics such as: house age, as well as house age-squared and cubed, living area and lotsize measured in square feet and dummy variables for each of the 5 years covered by the sample. Here again, we see estimates that are identical to a least 3 decimal digits in all cases, including the standard deviation estimates. The time required for this data sample was 0.5910 seconds when using integration limits of −4 to 0. Although the sample contained nearly 10 times as many observations as the presidential election example, we see only a six-fold increase in time required to produce estimates. Here again, the time required to solve the problem was reduced to

0.54 seconds when the integration limits were set to -2 and 0, with identical estimation results.

For the third example using 59,025 census tracts, the relationship explored involved the log share of all expenditures devoted to gasoline on average in each census tract. One might expect spatial dependence in these observations as similarly located census tracts would exhibit similar commuting patterns for work and shopping. A constant term and three explanatory variables were used: the log budget share of expenditures on vehicles, log median income and log of employment in the census tract. Here we see a time of 0.641 seconds based on the integration limits of -4 to 0. In this example, changing the limits of integration to -2 and 0 had a very modest impact on the time required, reducing it to 0.621 seconds.

9.4   Extensions of the model

We can extend the MESS model to include a more flexible spatial weight specification that is governed by the introduction of hyperparameters used in the weight specification. Bayesian MCMC estimation methods can be used to produce estimates of the hyperparameters that provide information regarding the nature and extent of spatial influence.

In Section 9.4.1 we introduce this extended version of the model, and in Section 9.4.2 we describe estimation using Markov Chain Monte Carlo methods. In Section 9.4.3 we illustrate the method in an application.

9.4.1   More flexible weights

Additional flexibility can be introduced by specifying a spatial weight that includes a decay parameter ϕ that lies between 0 and 1, along with a variable number of nearest neighbor spatial weight matrices N_i, where the subscript i is used to refer to a weight matrix containing non-zero elements for the ith closest neighbor. The weight structure specification is shown in (9.26), where m denotes the maximum number of neighbors considered.

$$W = \sum_{i=1}^{m} \left(\frac{\phi^i N_i}{\sum_{i=1}^{m} \phi^i} \right) \tag{9.26}$$

In (9.26), ϕ^i weights the relative effect of the ith individual neighbor matrix, so that S depends on the parameters ϕ as well as m in both its construction and the metric used. By construction, each row in W sums to 1 and has zeros on the diagonal. To see the role of the spatial decay hyperparameter ϕ, consider that a value of $\phi = 0.87$ implies a decay profile where the 6th nearest neighbor exerts less than $1/2$ the influence of the nearest neighbor. We might think of this value of ϕ as having a *half-life* of six neighbors. On the other hand, a value of $\phi = 0.95$ has a half-life between 14 and 15 neighbors.

The flexibility arising from this type of weight specification adds to the burden of estimation requiring that we draw an inference on the parameters ϕ and m. Together these hyperparameters determine the nature of the spatial weight structure. To the extent that the weight structure specification in (9.26) is flexible enough to adequately approximate more traditional weight matrices based on contiguity, the model introduced here can replicate results from models that assume the matrix W is fixed and known. However, all inferences regarding β and σ^2 drawn from a model based on a fixed matrix W are conditional on the particular W matrix employed. The model we introduce here produces inferences regarding β and σ^2 that are conditional only on a family of spatial weight transformations that we denote Sy, where $S = e^{\alpha W}$, with the matrices W taking the form in (9.26). Of course, this raises the issue of inference regarding these hyperparameters, and we show that the Bayesian MESS model introduced here can produce a posterior distribution for the joint distribution of the parameters α, ϕ and m as well as the other model parameters of interest, β and σ^2.

9.4.2 MCMC estimation

The extended variant of the Bayesian MESS is presented in (9.27), where the prior distributions for the parameters are also listed.

$$Sy = X\beta + \varepsilon$$
$$S = e^{\alpha W}$$
$$W = \sum_{i=1}^{m} \left(\phi^i N_i / \sum_{i=1}^{m} \phi^i \right)$$
$$\varepsilon \sim N(0, \sigma^2 V), \quad V_{ii} = (v_1, \ldots, v_n), \quad V_{ij} = 0 \ (i \neq j)$$
$$\pi(\beta) \sim N(c, T)$$
$$\pi(r/v_i) \sim iid \ \chi^2(r)$$
$$\pi(\sigma^2) \sim IG(a, b)$$
$$\pi(\alpha) \sim U(-\infty, 0]$$
$$\pi(\phi) \sim U(0, 1)$$
$$\pi(m) \sim U^D[1, m_{\max}] \tag{9.27}$$

We rely on a normal prior for $\beta \sim N(c, T)$, and inverse gamma prior for σ^2, with prior parameters a, b, where the normal and inverse gamma priors are independent. The prior assigned for α can be a relatively non-informative uniform prior that allows for the case of no spatial effects when $\alpha = 0$.

The relative variance terms $(v_1, v_2, \ldots, v_n)$ represent our variance scalars to accommodate outliers and heteroscedasticity as motivated in Chapter 5. We rely on the same *iid* $\chi^2(r)/r$ distribution as a prior for these variance scalars.

A relatively non-informative approach was taken for the hyperparameters ϕ and m where we rely on a uniform prior distribution for ϕ and a discrete uniform distribution for m, the number of nearest neighbors. The term $m_{\max}$ denotes a maximum number of nearest neighbors to be considered in the spatial weight structure, and U^D denotes the discrete uniform distribution that imposes an integer restriction on values taken by m. Note that practitioners may often have prior knowledge regarding the number of neighboring observations that are important in specific problems, or the extent to which spatial influence decays over neighboring units. Informative priors could be developed and used here as well, but in problems where interest centers on inference regarding the spatial structure, relatively non-informative priors would be used for these hyperparameters.

Given these distributional assumptions, it follows that the prior densities for $\beta, \sigma^2, \alpha, \phi, m, v_i$ are given up to constants of proportionality by (9.28), (where we rely on a uniform prior for α).

$$\pi(\beta) \propto \exp[-\frac{1}{2}(\beta - c)'T^{-1}(\beta - c)] \tag{9.28}$$

$$\pi(\sigma^2) \propto (\sigma^2)^{-(a+1)}\exp\left(-\frac{b}{\sigma^2}\right)$$

$$\pi(\phi) \propto 1$$

$$\pi(\alpha) \propto 1$$

$$\pi(m) \propto 1$$

$$\pi(v_i) \propto v_i^{-(\frac{r}{2}+1)}\exp\left(-\frac{r}{2v_i}\right)$$

9.4.3 MCMC estimation of the model

Given the prior densities from section 9.4.2, the Bayesian identity,

$$p(\beta, \sigma^2, V, \phi, \alpha, m|\mathcal{D}) = p(\mathcal{D}|\beta, \sigma^2, V, \phi, \alpha, m) \cdot \pi(\beta, \sigma^2, V, \phi, \alpha, m) \tag{9.29}$$

together with the assumed prior independence of the prior distributions for the parameters allows us to establish the joint posterior density for the parameters, $p(\beta, \sigma^2, V, \phi, \alpha, m|\mathcal{D})$. This posterior is not amenable to analysis of the type described previously, because we would need to integrate over the hyperparameters m and ϕ. We can however use Markov Chain Monte Carlo (MCMC) to sample from the posterior distribution for the parameters in our model.

We rely on Metropolis-Hastings to sample from the posterior distributions for the parameters α, ϕ and m in the MESS model. A normal distribution is used as the proposal density for α and rejection sampling can be used to constrain α to a range such as $[-5, 0.7]$ discussed in Section 9.3.1. A uniform proposal distribution for ϕ over the interval $(0, 1)$ was used along with a discrete uniform for m over the interval $[1, m_{\max}]$. The parameters β, V and σ in the MESS model can be estimated using draws from the conditional distributions of these parameters that take a known form.

Summarizing, we will rely on Metropolis sampling for the parameters α, ϕ and m within a sequence of Gibbs sampling steps to obtain β, σ and V.

9.4.4   The conditional distributions for β, σ and V

To implement our Metropolis within Gibbs sampling approach to estimation we need the conditional distributions for β, σ and V which are presented here.

For the case of the parameter vector β conditional on the other parameters in the model, $\alpha, \sigma, V, \phi, m$ we find that:

$$p(\beta|\alpha, \sigma, V, \phi, m) \sim N(c^*, T^*)$$
$$c^* = (X'V^{-1}X + \sigma^2 T^{-1})^{-1}(X'V^{-1}Sy + \sigma^2 T^{-1}c)$$
$$T^* = \sigma^2(X'V^{-1}X + \sigma^2 T^{-1})^{-1} \tag{9.30}$$

Note that given the parameters V, α, ϕ, σ and m, the vector Sy and $X'V^{-1}X$ can be treated as known, making this conditional distribution easy to sample. This is often the case in MCMC estimation, which makes the method attractive.

The conditional distribution of σ^2 is shown in (9.31), (Gelman et al., 1995).

$$p(\sigma^2|\beta, \alpha, V, \phi, m) \propto (\sigma^2)^{-(\frac{n}{2}+a)}\exp\left[-\frac{e'V^{-1}e + 2b}{2\sigma^2}\right] \tag{9.31}$$

where $e = Sy - X\beta$, which is proportional to an inverse gamma distribution with parameters $(n/2) + a$ and $e'V^{-1}e + 2b$.

The conditional distribution of V given the other parameters is proportional to a chi-square density with $r + 1$ degrees of freedom (Geweke, 1993). Specifically, we can express the conditional posterior of each v_i as:

$$p(\frac{e_i^2 + r}{v_i}|\beta, \alpha, \sigma^2, v_{-i}, \phi, m) \sim \chi^2(r + 1) \tag{9.32}$$

where $v_{-i} = (v_1, \ldots, v_{i-1}, v_{i+1}, \ldots, v_n)$ for each i.

As noted above, the conditional distributions for α, ϕ and m take unknown distributional forms that require Metropolis-Hastings sampling. By way of summary, the MCMC estimation scheme involves starting with arbitrary initial values for the parameters which we denote $\beta^0, \sigma^0, V^0, \alpha^0, \phi^0, m^0$. We then sample sequentially from the set of conditional distributions for the parameters in our model.

1. $p(\beta|\sigma^0, V^0, \alpha^0, \phi^0, m^0)$, which is a normal distribution with mean and variance-covariance defined in (9.30). This updated value for the parameter vector β we label β^1.

2. $p(\sigma^2|\beta^1, V^0, \alpha^0, \phi^0, m^0)$, which is inverse gamma distributed as shown in (9.31). Note that we rely on the updated value of the parameter vector $\beta = \beta^1$ when evaluating this conditional density. We label the updated parameter $\sigma = \sigma^1$ and note that we will continue to employ the updated values of previously sampled parameters when evaluating the next conditional densities in the sequence.

3. $p(v_i|\beta^1, \sigma^1, v_{-i}, \alpha^0, \phi^0, m^0)$ which can be obtained from the chi-squared distribution shown in (9.32). Note that this draw can be accomplished as a vector, providing greater speed.

4. $p(\alpha|\beta^1, \sigma^1, V^1, \phi^0, m^0)$, which we sample using a Metropolis step with a normal proposal density, along with rejection sampling to constrain α to the desired interval. The likelihood is proportional to the desired conditional distribution of α.

5. $p(\phi|\beta^1, \sigma^1, V^1, \alpha^1, m^0)$, which we sample using a Metropolis step based on a uniform distribution that constrains ϕ to the interval (0,1). Here again, we rely on the likelihood (which is proportional to the conditional distribution) to evaluate the candidate value of ϕ. As in the case of the parameter α it would be easy to implement a normal or some alternative prior distributional form for this hyperparameter.

6. $p(m|\beta^1, \sigma^1, V^1, \alpha^1, \phi^1)$, which we sample using a Metropolis step based on a discrete uniform distribution that constrains m to be an integer from the interval $[1, m_{\max}]$. As in the case of α and ϕ, we rely on the likelihood to evaluate the candidate value of m.

Sampling proceeds sequentially through steps 1) to 6) and on each pass through the sampler we employ the updated parameter values in place of the initial values $\beta^0, \sigma^0, V^0, \alpha^0, \phi^0, m^0$. On each pass through the sequence we collect the parameter draws which are used to construct a joint posterior distribution for the parameters in our model.

9.4.5   Computational considerations

Use of the likelihood when evaluating candidate values of α, ϕ and m in the MCMC sampling scheme requires that we form the matrix exponential $S = e^{\alpha W}$, which in turn requires computation of $W = \sum_{i=1}^{m}(\phi^i N_i / \sum_{i=1}^{m} \phi^i)$ based on the current values for the other two parameters. For example, in the case of update $\alpha = \alpha^1$, we use $\phi = \phi^0$ and $m = m^0$ to find W. The nearest neighbor matrices N_i can be computed outside the sampling loop to save time, but the remaining calculations can still be computationally demanding if the number of observations in the problem is large.

Further aggravating this problem is the need to evaluate both the existing value of the parameters α, ϕ and m, given the updated values for β, σ and V as well as the candidate values. In all, we need to form the matrix product Sy, along with the matrix W six times on each pass through the sampling loop.

To enhance the speed of the sampler, we compute the part of Sy that depends only on ϕ and m, for a grid of values over these two parameters prior to beginning the sampler. During evaluation of the conditionals and the Metropolis-Hastings steps, a simple table look-up recovers the stored component of Sy and applies the remaining calculations needed to fully form Sy.

The ranges for these grids can be specified by the user, with a trade-off between selecting a large grid that ensures coverage of the region of posterior support and a narrow grid that requires less time. In a typical spatial problem, the ranges might be $0.5 \leq \phi \leq 1$, and $4 < m < 30$. If the grid range is too small, the posterior distributions for these parameters should take the form of a censured distribution, indicating inadequate coverage of the region of support.

Simpler models than that presented in (9.27) could be considered. For example either ϕ or m, or both ϕ and m could be fixed a priori. This would enhance the speed of the sampler because eliminating one of the two hyper-parameters from the model reduces the computational time needed by almost one-third since it eliminates two of the six computationally intensive steps involving formation of Sy. For example, labels for the various MESS models used in the experiments presented in the next section are enumerated below from simplest to most complex.

MESS1 – a model with both ρ and m fixed, and no v_i parameters.

MESS2 – a model with ρ fixed, m estimated and no v_i parameters.

MESS3 – a model with m fixed, ρ estimated and no v_i parameters.

MESS4 – a model with both ρ and m estimated and no v_i parameters.

MESS5 – a model with both ρ and m estimated as well as estimates for the v_i parameters.

The use of nearest neighbors also accelerates computation. As described in Section 4.11, nearest neighbor calculations using index arithmetic in place of matrix multiplication can greatly reduce computation time as indexing into a matrix is one of the fastest digital operations.

9.4.6　An illustration of the extended model

We provide illustrations of the extended Bayesian MESS model in using a generated model with only 49 observations taken from Anselin (1988). Use of a generated example where the true model and parameters are known allows us to illustrate the ability of the model to find the true spatial weight structure used in generating the model.

A traditional spatial autoregressive (SAR) model: $y = \rho W y + X\beta + \varepsilon$ was used to generate the vector y based on 49 spatial observations from Columbus neighborhoods presented in Anselin (1988). The spatial weight matrix, $W = \sum_{i=1}^{m} \phi^i N_i / \sum_{i=1}^{m} \phi^i$, was based on $m = 5$ nearest neighbors and distance decay determined by $\phi = 0.9$. The two explanatory variables from Anselin's data set (in studentized form) along with a constant term and ρW were used to generate a vector $y = (I_n - \rho W)^{-1} X\beta + (I_n - \rho W)^{-1}\varepsilon$. The parameters β and the noise variance, σ_ε^2 were set to unity and the spatial correlation coefficient ρ was set to 0.65.

This generated data was used to produce maximum likelihood estimates of the parameters based on a SAR and MESS model specification as well as Bayesian MCMC estimates. Of course, traditional implementation of the SAR model would likely rely on a first-order contiguity matrix treated as exogenous information, which we label W_1. Maximum likelihood estimation of this MESS specification would attempt to determine values for the hyperparameters ϕ, m using a concentrated likelihood grid search over these values. The Bayesian model would produce posterior estimates for the hyperparameters as part of the MCMC estimation as in the models labeled MESS4 and MESS5 in the previous section. Of course, it would be possible to rely on MCMC estimation and the simpler models labeled MESS1 to MESS3 in the previous section, but the computational requirements for this small sample are minimal.

We illustrate the difference in estimates and inferences that arise from using these three approaches. Note that two variants of the SAR model were estimated, one based on a first-order contiguity matrix, W_1 and another based on the true W matrix used to generate the model. In practice of course, one would not know the true form of the W matrix. One point to note is that the first-order contiguity matrix for this data set contains an average number of neighbors equal to 4.73 with a standard deviation of 1.96. Of the total $49 \times 49 = 2{,}401$ elements there are 232 non-zero entries. We might expect that the differences between SAR models based on W_1 and the true W containing five nearest neighbors and a small amount of distance decay should be small.

The non-Bayesian MESS model implemented maximum likelihood estimation by searching over a grid of ϕ values from 0.01 to 1 in 0.01 increments and neighbors m ranging from 1 to 10. Estimates were produced based on the values of ϕ and m that maximized the concentrated log likelihood function. The Bayesian MESS model was run to produce 5500 draws with the first 500 discarded to allow the MCMC chain to converge to a steady state.[5] Diffuse

---

[5]This is actually an excessive number of draws, since the estimates were the same to one or two decimal places as those from a sample of 1250 draws with the first 250 discarded.

priors were used for β and σ and two variants of the model were estimated: one that included the parameters V and another that did not. The latter Bayesian model assumes that $\varepsilon \sim N(0, \sigma^2 I_n)$, which is consistent with the assumption made by the non-Bayesian SAR and MESS models.

The estimation results are presented in Table 9.8. Measures of precision for the parameter estimates are not reported in the table because all coefficients were significant at the 0.01 level. In the table we see that the SAR model based on the true spatial weight matrix W performed better than the model based on W_1, as we would expect. (True values used to generate the data are reported in the first column next to the parameter labels). Both the concentrated likelihood approach and the posterior distribution from the Bayesian MESS models identified the correct number of neighbors used to generate the data. The Bayesian MESS models produced posterior estimates for ϕ based on the mean of the draws equal to 0.91 and 0.89 compared to the true value of 0.90, whereas the concentrated likelihood search resulted in an estimate of $\phi = 1.0$. Nonetheless, the MESS models produced very similar β estimates as well as estimates for the spatial dependence parameter in this model, α. The estimate of σ^2 from one Bayesian MESS model was close to the true value of unity, while the other Bayesian model produced an estimate closer to the maximum likelihood estimates for the SAR model based on the true W matrix.

TABLE 9.8:   A comparison of models from experiment 1

Parameters	SAR W_1	SAR W	ML MESS	MESS4	MESS5
$\beta_0 = 1$†	1.3144	1.1328	1.1848	1.1967	1.1690
$\beta_1 = 1$	1.1994	0.9852	1.0444	1.0607	1.0071
$\beta_2 = 1$	1.0110	1.0015	1.0144	1.0102	0.9861
$\sigma^2 = 1$	1.4781	0.7886	0.8616	0.9558	0.7819
$\rho = 0.65$	0.5148	0.6372			
α			-0.8879	-0.8871	-0.9197
R^2	0.8464	0.9181	0.9160	0.9141	0.9134
$m = 5$			5	5.0466	5.0720
$\phi = 0.90$			1.0	0.9171	0.8982

† true values used to generate the data.

The concentrated likelihood approach identified the correct number of neighbors used to generate the data and points to a value of $\phi = 1$, versus the true value of 0.9. The posterior distribution of ϕ was skewed, having a mean of 0.9171, a median of 0.9393 and a mode of 0.9793. This partially explains the difference between the maximum likelihood estimate of unity and the Bayesian estimate reported in Table 9.8. The posterior distributions for the hyperparameters ϕ and m provide a convenient summary that allows the user to rely on mean, median or modes in cases where the resulting distributions are skewed.

Note that the parameter α in the MESS model plays the role of ρ in the traditional spatial autoregressive models capturing the extent of spatial dependence. Inferences about spatial dependence are based on a test of the magnitude of α versus zero. Figure 9.1 shows the posterior distribution of α from the MESS4 model, which should make it clear that this estimate would lead to an inference of spatial dependence, that is, $\alpha \neq 0$.

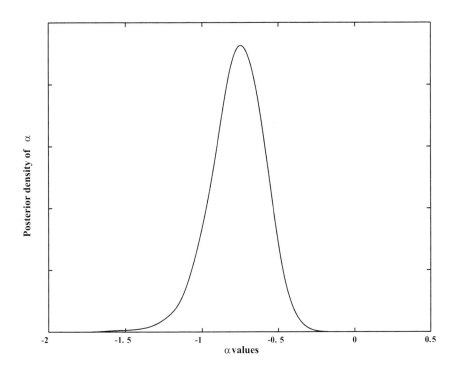

FIGURE 9.1: Posterior distribution of α parameter

As an illustration of the ability of the MESS model to find the correct model specification, we produced estimates for models based on a first-order contiguity matrix used to generate the data in this experiment as well as models based on the two through six nearest neighbors. Note that use of spatial weight matrices based on nearest neighbors represents a misspecification since the first-order contiguity matrix was used to generate the dependent variable vector y. No hyperparameters were used in this experiment, so the specification: $W_i = N_i/i, i = 2, \ldots, 6$, was used, where the binary nearest neighbor matrix N_i contains ones for the i nearest neighbors to each observation.

The question of interest here is whether the MESS models can distinguish the first-order contiguity matrix used to generate the data from the nearest neighbor matrices. Posterior probabilities for these six models are shown in Table 9.9 for the Bayesian model and the log likelihood function values are shown for the non-Bayesian MESS model.[6] From the table we see that the MESS models correctly identified the model associated with the true weight matrix. Almost all of the posterior probability weight was placed on this model, indicating that the flexibility associated with a specification that allows varying the number of neighbors did not lead the model to pick an inferior spatial weight structure when confronted with the true structure.

TABLE 9.9: Specification search example involving six models

Neighbors	ML MESS Log likelihood	MCMC MESS Posterior probability
Correct W matrix	-75.7670	0.9539
2 neighbors	-85.0179	0.0001
3 neighbors	-80.2273	0.0094
4 neighbors	-81.9299	0.0017
5 neighbors	-79.3733	0.0247
6 neighbors	-80.3274	0.0102

Relatively diffuse priors along with a prior reflecting a belief in constant variance across space were used in the experiments above to illustrate that the Bayesian MESS model can replicate maximum likelihood estimates. This is however a computationally expensive approach to producing MESS estimates. A practical motivation for the Bayesian model would be cases involving outliers or non-constant variance across space. To illustrate the Bayesian approach to non-constant variance over space we compare six models based on alternative values for the hyperparameter r that specifies our prior on heterogeneity versus homogeneity in the disturbance variances. These tests are carried out using two data sets, one with homoscedastic and another with heteroscedastic disturbances. Non-constant variances were created by scaling up the noise variance for the last 20 observations during generation of the y vector. This might occur in practice if a neighborhood in space reflects more inherent noise in the regression relationship being examined. The last 20 observations might represent one region of the spatial sample.

We test a sequence of declining values for r with large values reflecting a prior belief in homogeneity and smaller values indicating heterogeneity. Posterior probabilities for these alternative values of r are shown in Table 9.10 for

---

[6] Posterior probabilities can be computed using the log marginal likelihood which is described in LeSage and Pace (2007) for this model (see Chapter 6).

both sets of generated data. For the case of constant variances, the posterior model probabilities correctly point to a model based on large r values of 50. In the case of heteroscedastic disturbances, the models based on r values of 10, 7 and 4 receive high posterior probability weights, reflecting the non-constant variance.

TABLE 9.10:   Homogeneity test results for two data sets

r-value	Homoscedastic data Posterior probabilities	Heteroscedastic data Posterior probabilities
50	0.9435	0.0001
20	0.0554	0.0039
10	0.0011	0.2347
7	0.0001	0.6200
4	0.0000	0.1413
1	0.0000	0.0000

In addition to correctly identifying the existence of heterogeneity in the disturbance variances, a plot of the posterior means of the v_i estimates can be a useful diagnostic regarding the nature and extent of the heterogeneity. Figure 9.2 shows a plot of these estimates for the heteroscedastic Bayesian MESS model as well as the heteroscedastic Bayesian SAR model. From the figure we see that the pattern of inflated variances over the last 20 observations is correctly identified by the v_i estimates from both models.

9.5   Fractional differencing

In this chapter we developed a spatial model based on the matrix exponential and in Chapter 4 we considered matrix logarithms when examining alternative ways to calculate the log-determinant. Having the ability to work with matrix exponentials and logarithms suggests possible model extensions such as,

$$e^{a\ln(A)}y = X\beta + \varepsilon \tag{9.33}$$
$$A^a y = X\beta + \varepsilon \tag{9.34}$$

where we assume that A is positive definite and a is real. This is a fractional transformation of A. An attractive computational feature of this specification

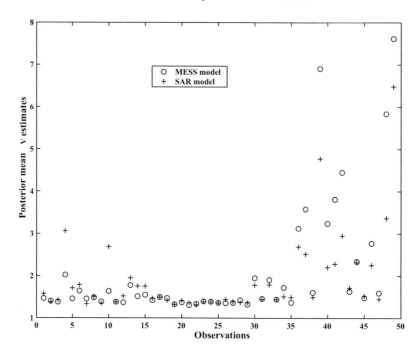

FIGURE 9.2:   Posterior means of the v_i estimates for a heteroscedastic model

is that $\ln|A^a| = a\ln|A|$. Therefore, updating $\ln|A^a|$ over a range of changing values for the parameter a requires simple multiplication of two scalars, a and $\ln|A|$. The log-determinant term would be computed once for any particular A.

In the time series literature fractional transformations are usually associated with differencing so that $A = I_n - L$ where L is a triangular temporal lag matrix. Fractional differencing has proven useful for situations where dependence slowly declines with time (Hosking, 1981). A variety of mechanisms can yield this type of dependence pattern. For example, Granger (1980) showed that fractional differencing could arise from aggregation. In finite time series, fractional differencing can be used to represent some high order ARMA processes (Haubrich, 1993, p. 767).

If we view the spatial equilibrium as the long-run outcome of a spatiotemporal process as motivated in Chapter 7, a fractional differencing spatiotemporal process could lead to a fractional differencing spatial equilibrium. Therefore, some of the motivations used in the time series literature may also apply to the spatial analogs.

Various aspects of spatial systems may be more likely to produce higher order dependence than in time. First, in space there are a large number of paths to each observation from every other observation. This means that changes in one location could take a very indirect path to influence another location leading to small amounts of high order dependence. Although individual paths may have a very small influence, the cumulative effect may be non-trivial. Second, boundaries such as borders and natural features may lead to a reflection of influences, and these influences may not die away as quickly as they would in an infinite, featureless plain. Borders, oceans, rivers, jurisdictions, and other boundaries are integral aspects of the spatial system and not data artifacts. Complicated geographic features may thus lead to multiple reflections and high order dependence.

In time series analysis $I_n - L$ is non-singular, whereas in a spatial setting $A = I_n - W$ is singular for stochastic W. Although spatial differencing has appeared in the literature (Ord, 1975), the singular nature of the transformation and its poor empirical performance have not led to much use. However, a small modification to W can allow for an operation that acts like differencing, but still results in positive definite A.

To see this, consider the usual SAR model in (9.35) where one of the columns of X equals the unit constant vector ι_n. The residuals e in (9.36) result from multiplying the transformed y by the idempotent matrix M_X.

$$(I_n - \rho W)y = X\beta + \varepsilon \tag{9.35}$$
$$M_X(I_n - \rho W)y = e \tag{9.36}$$

As a brief review of idempotent matrices, for some matrix Z, the matrix $H_Z = Z(Z'Z)^{-1}Z'$ is the idempotent projection matrix or *Hat matrix*, with its complement being the idempotent matrix $M_Z = I_n - H_Z$. Properties of idempotent matrices include: $M_Z^2 = M_Z$, $H_Z^2 = H_Z$, and $M_Z H_Z = 0_n$. If Z_1 is a subset of Z_2, $M_{Z_1} M_{Z_2} = M_{Z_2} M_{Z_1} = M_{Z_2}$.

Since ι_n is a column of X, $M_X M_\iota = M_X$. In terms of the Z notation, $Z_1 = \iota_n$ and $Z_2 = X$. The matrix M_ι acts to mean-center columns of matrices or vectors. For example, a regression using an intercept on a variable v yields residuals u with zero mean ($u = M_X v$). This same result could be obtained by calculating $u = M_X M_\iota v$, since mean-centering a second time still yields residuals with zero mean. The equality in (9.37) arises since $M_X M_\iota = M_X$,

$$M_X(I_n - \rho W)y = M_X(I_n - \rho M_\iota W)y \tag{9.37}$$

Although $M_\iota W$ in (9.37) acts the same as W, $I_n - M_\iota W$ could be positive definite for some forms of W.

Specifically, we introduce a spatial weight matrix W so that $I_n - M_\iota W$ is positive definite. One form of W that will produce a positive definite matrix $I_n - M_\iota W$ is a symmetric, doubly stochastic weight matrix W_{ds} such that

$W_{ds}^r > 0$ for some positive integer r.[7] This weight matrix has a number of convenient properties that lead to positive definite $I_n - M_\iota W_{ds}$. First, because W_{ds} is real and symmetric, $W_{ds} = U\Lambda U'$ where U is an $n \times n$ matrix of orthogonal eigenvectors and Λ is a real matrix with associated eigenvalues on the diagonal (Horn and Johnson, 1993, 4.1.5). We use u_i to denote column i of U or the ith eigenvector. Second, doubly stochastic matrices have a maximum eigenvalue of 1 with a corresponding constant eigenvector (Marcus and Minc, 1992, 5.13.2). Third, since $W_{ds}^r > 0$ for some positive integer r, this implies that the largest eigenvalue of 1 is unique (Horn and Johnson, 1993, Theorem 8.5.2). This condition will be satisfied if there are paths of order r or less between any two entries involved in higher order neighboring relations specified by W_{ds}^r.

As a result, the doubly stochastic weight matrix W_{ds} has the eigenvalue expansion shown in (9.38) and (9.39), where the outer product of the first eigenvector equals H_ι and is associated with the maximum eigenvalue, $\lambda_1 = 1$.

$$W_{ds} = n^{-1}\iota_n\iota_n'\lambda_1 + u_2u_2'\lambda_2 + \ldots + u_nu_n'\lambda_n \tag{9.38}$$
$$W_{ds} = H_\iota\lambda_1 + u_2u_2'\lambda_2 + \ldots + u_nu_n'\lambda_n \tag{9.39}$$
$$\lambda_1 = \max(\lambda) = 1, \quad \text{abs}(\lambda_i) < 1 \quad (i = 2, \ldots n) \tag{9.40}$$

Multiplication of W_{ds} by M_ι in (9.41) strips away the first eigenvector term associated with the largest eigenvalue of 1. This defines $W_{-\iota}$ where the largest magnitude eigenvalue is now strictly less than 1 as stated in (9.42).

$$M_\iota W_{ds} = W_{-\iota} = u_2u_2'\lambda_2 + \ldots + u_nu_n'\lambda_n \tag{9.41}$$
$$\max(\text{abs}(\lambda_{W_{-\iota}})) < 1 \tag{9.42}$$

A brief example may make this clearer. Expression (9.43) presents a symmetric doubly stochastic matrix $W_{ds}^{(o)}$.

$$W_{ds}^{(o)} = \begin{bmatrix} 0.0000 & 0.3389 & 0.1895 & 0.2627 & 0.2089 \\ 0.3389 & 0.0000 & 0.1933 & 0.1618 & 0.3060 \\ 0.1895 & 0.1933 & 0.0000 & 0.3538 & 0.2634 \\ 0.2627 & 0.1618 & 0.3538 & 0.0000 & 0.2217 \\ 0.2089 & 0.3060 & 0.2634 & 0.2217 & 0.0000 \end{bmatrix} \tag{9.43}$$

The eigenvectors $U^{(o)}$ and the eigenvalues on the diagonal of $\Lambda^{(o)}$ associated with $W_{ds}^{(o)}$ appear in (9.44) and (9.45).

---

[7]Row stochastic matrices could also be used, but these require a more involved development based on the Schur decomposition.

$$U^{(o)} = \begin{bmatrix} 0.4472 & 0.3512 & -0.6123 & 0.5265 & 0.1568 \\ 0.4472 & 0.5657 & 0.0769 & -0.5409 & -0.4261 \\ 0.4472 & -0.5247 & 0.1534 & 0.3343 & -0.6240 \\ 0.4472 & -0.5158 & -0.3202 & -0.5255 & 0.3940 \\ 0.4472 & 0.1237 & 0.7022 & 0.2056 & 0.4993 \end{bmatrix} \qquad (9.44)$$

$$\mathrm{diag}(\Lambda^{(o)}) = \begin{bmatrix} 1.00 & -0.05 & -0.19 & -0.41 & -0.35 \end{bmatrix} \qquad (9.45)$$

The first column of $U^{(o)}$ contains the constant eigenvector, where each of the five elements equal $0.4472 = \sqrt{(1/5)}$, and this eigenvector has an associated eigenvalue of 1. Multiplication of the constant eigenvector by M_ι, which mean-centers vectors, essentially eliminates the first eigenvector of $U^{(o)}$, but does not change the other eigenvectors. Since the eigenvectors are orthogonal and one of the eigenvectors was a constant vector, the other eigenvectors have a zero mean. Multiplication by M_ι does not change the other eigenvectors, so the multiplication $M_\iota U^{(o)}$ effectively removes the largest eigenvalue of 1 from $W_{ds}^{(o)}$.

$$M_\iota U^{(o)} = \begin{bmatrix} 0.0000 & 0.3512 & -0.6123 & 0.5265 & 0.1568 \\ -0.0000 & 0.5657 & 0.0769 & -0.5409 & -0.4261 \\ -0.0000 & -0.5247 & 0.1534 & 0.3343 & -0.6240 \\ -0.0000 & -0.5158 & -0.3202 & -0.5255 & 0.3940 \\ 0.0000 & 0.1237 & 0.7022 & 0.2056 & 0.4993 \end{bmatrix} \qquad (9.46)$$

This allows us to use the matrix $W_{-\iota}$ that has a largest eigenvalue less than 1 to define a positive definite spatial differencing transformation, $\Delta_{-\iota}$, shown in (9.47). We label this term a *feasible spatial differencing* transformation. The transformation has a log-determinant equal to ψ as indicated in (9.49).

$$\Delta_{-\iota} = I_n - W_{-\iota} \qquad (9.47)$$

$$|\Delta_{-\iota}| = |I_n - W_{-\iota}| > 0 \qquad (9.48)$$

$$\ln|\Delta_{-\iota}| = \psi \qquad (9.49)$$

Although it is possible to use (9.47) as a feasible spatial differencing transformation, it is more flexible to rely on a transformation that introduces a real fractional parameter, δ, as shown in (9.50). Following Hosking (1981), we assume $\delta \in (-0.5, 0.5)$. An outstanding advantage of the fractional transformation is that this leads to a linear log-determinant term in (9.51).

$$\Delta_{-\iota}^{\delta} = e^{\delta \ln(\Delta_{-\iota})} \qquad (9.50)$$

$$\ln|\Delta_{-\iota}^{\delta}| = \delta\psi \qquad (9.51)$$

In turn, the linear log-determinant term leads to a simple concentrated log likelihood shown in (9.52).

$$\ln L(\delta) = \kappa + \delta\psi - \frac{n}{2}\ln\left(e(\delta)'e(\delta)\right) \tag{9.52}$$

$$e(\delta) = M_X e^{\delta\ln(\Delta_{-\iota})}y \tag{9.53}$$

To provide an idea of the performance of spatial fractional feasible differencing, we examine two sets of sample data using spatial fractional feasible differencing in the next section.

9.5.1   Empirical illustrations

This section provides two illustrations of spatial fractional differencing, one based on a census tract sample involving housing and the other based on a sample of US counties and election data. Use of the smaller sample of US counties versus the larger sample of US Census tracts should allow variation in the level and relative importance of higher-order spatial dependence.

We compare the fractional differencing method to a variety of other estimators. Specifically, for each data set we fitted the model using ordinary least-squares (OLS) as well as moving average (MA), matrix exponential (ME), autoregressive (AR), and fractional differencing (FD) estimators.

$$(I_n - W_{-\iota})^\delta y = X\beta_{FD} + \varepsilon_1 \tag{9.54}$$

$$(I_n - \rho W)y = X\beta_{AR} + \varepsilon_2 \tag{9.55}$$

$$e^{\alpha W}y = X\beta_{ME} + \varepsilon_3 \tag{9.56}$$

$$(I_n - \theta W)^{-1}y = X\beta_{MA} + \varepsilon_4 \tag{9.57}$$

$$y = X\beta_{OLS} + \varepsilon_5 \tag{9.58}$$

The motivation for using alternative specifications in the series of experiments set forth here is that one approach to accommodating higher-order spatial dependence would be to rely on alternative model specifications such as the FD, AR, ME, and MA shown in (9.54)–(9.57).

Another means of capturing neighboring relations is through the weight matrix, and we use two specifications for W, one based on contiguity with doubly stochastic scaling that we label (W_c). The second is a weight matrix based on 30 nearest neighbors with a geometric decay parameter of 0.9 for each order of neighbor as shown in (9.26). Let $W_s = N_1 + N_2 0.9 + N_3 0.81 + \ldots$ where N_i are individual neighbor matrices described in Chapter 4. By itself W_s is non-symmetric. Forming $(W_s + W_s')$ and scaling it to make the rows and columns sum to 1 yields the symmetric doubly stochastic nearest neighbor matrix that we label W_{nn}.

The experiments will examine whether use of the relatively more sophisticated FD model in conjunction with the simpler contiguity weight matrix can produce results comparable to those from simpler model specifications such as the AR, ME, and MA based on the richer 30 nearest neighbor weight matrix. It should be clear that the 30 nearest neighbor weight matrix has more connections among neighbors than does a contiguity-based weight matrix.

The experiments will examine the trade-off between specifying dependence via methods which differ in the emphasis placed on higher-order neighboring relations versus specifying dependence via weight matrix choice. As will be shown later, ranking methods in terms of the role of high order dependence yields (from low to high) OLS, MA, MESS, AR, and FD. Intuitively, methods such as fractional differencing which allow a role for high order dependence may prefer a less connected weight matrix such as one based on contiguity. Methods such as moving averages which allow for almost no role for high order dependence may prefer a more connected weight matrix such as one based on nearest neighbors. Of course, a third modeling strategy is to use *both* a more sophisticated model and weight matrix in an effort to model higher order spatial dependence. This is also considered in our experiments.

The first application uses sample data on the votes cast in the 1980 presidential election across U.S. counties taken from Pace and Barry (1997). To determine the contiguous US counties, we relied on the geographic centroids of all counties (or their equivalents) from the Census. The dependent variable reflects the total number of recorded votes cast for all parties in the 1980 presidential election as a proportion of the voting age population, ln(Votes/Pop) or ln(Votes) − ln(Pop). Explanatory variables used were: the population 18 years of age or older (Pop) in each county, the population in each county with a 12th grade or higher education (Education), the number of owner-occupied housing units (Houses), and aggregate county-level income (Income). These were used to form the $3,107 \times 5$ matrix X shown in (9.59), where we have added a constant term vector ι_n.

$$X = \begin{bmatrix} \iota_n & \ln(\text{Pop}) & \ln(\text{Education}) & \ln(\text{Houses}) & \ln(\text{Income}) \end{bmatrix} \qquad (9.59)$$

Table 9.11 contains the coefficient and dependence parameter estimates along with signed root deviances and log-likelihoods for the various specifications based on the doubly stochastic symmetric contiguity weight matrix, W_c. A clear pattern emerges. Specifically, the log-likelihoods rise as the methods place greater emphasis on the role of higher-order neighboring relations with fractional differencing producing the highest log likelihood and OLS the lowest log likelihood. In addition, the fractionally differenced method produced a material improvement in log-likelihood function values, achieving a value that is 92.9 higher than its nearest competitor, the AR model.

Table 9.12 contains the coefficient estimates along with signed root deviances and log-likelihoods for the various specifications based on the doubly stochastic symmetric nearest neighbor weight matrix, W_{nn}. Table 9.12 shows that every spatial specification except fractional differencing displayed a higher likelihood for the nearest neighbor weight matrix W_{nn} relative to a contiguity based weight matrix W_c. This result is consistent with the notion that a richer weight matrix specification should improve the model estimates from methods that place less emphasis on higher order neighboring relations. These results produced a higher concentrated-log likelihood function value for the AR specification than the fractional differencing specification. However, the geometric parameter used to create the weight matrix was set to maximize performance of the AR specification. Relative to the fractional differencing results based on the contiguity-based weight matrix, here we find that an AR specification based on the nearest neighbor weight matrix produced the highest likelihood. However, the small difference of 6.21 between log likelihoods may be partially due to fitting the geometric parameter governing the weights assigned to individual neighbors.

Another interesting pattern emerges from comparing the coefficient estimates from Table 9.11 and Table 9.12. If we compare the AR estimates as we move from contiguity (W_c) to nearest neighbor weights (W_{nn}) , these move toward the FD estimates based on contiguity (W_c). This suggests that the FD model is better capable of using the simpler contiguity weight matrix (W_c) to capture patterns of higher-order spatial dependence relative to the AR model using W_c.

The second application uses housing data. Housing provides a classic example of spatially dependent data, and we examine (logged) housing values as a function of (logged) households, median household income, median years of education, and land area. The sample data represent 62, 226 census-tract level observations from the 2000 Census. This application uses the definition of X from (9.60).

$$X = \begin{bmatrix} \iota_n & \ln(\text{Households}) & \ln(\text{Income}) & \ln(\text{Education}) & \ln(\text{Land Area}) \end{bmatrix} \quad (9.60)$$

Table 9.13 contains the coefficient and dependence estimates along with signed root deviances and log-likelihoods for the various specifications based on the doubly stochastic symmetric contiguity weight matrix, W_c. Like the county-level election data results, Table 9.13 again shows a pattern where the log-likelihood rises as each method places more emphasis on higher order dependence (FD>AR>ME>MA>OLS). The difference between log likelihoods is material with FD exceeding the likelihood of the AR specification by 2, 588.60.

Table 9.14 contains results for the various specifications based on the doubly stochastic symmetric nearest neighbor weight matrix, W_{nn} in the same format as Table 9.13. We see a similar pattern to those from the election data, with the AR specification exhibiting the highest likelihood. In this case, the log-likelihood from the ME specification also exceeded the FD log likelihood, but unlike the election data example, FD using the simpler contiguity weight matrix W_c outperformed an AR specification based on the richer 30 nearest neighbors weight matrix. This makes the point that the FD model specification can exploit a simpler weight structure to successfully capture higher-order patterns of dependence.

It is also noteworthy that the fractional differencing specifications in all four tables resulted in similar estimates for the parameter δ, which ranged from 0.22 to 0.29. These dependence estimates also displayed less variation over the alternative data samples and weight matrices than the spatial dependence parameters from other model specifications.

TABLE 9.11: Maximum likelihood estimates for election data using W_c

Variables	FD	AR	ME	MA	OLS
Intercept	0.7157	0.8952	1.0100	1.1711	1.5576
	16.9542	20.4268	22.5129	25.6052	30.7777
Voting Pop	−0.5675	−0.6493	−0.6948	−0.7451	−0.8464
	−29.1112	−31.7391	−32.6442	−33.8985	−34.7211
Education	0.1281	0.2274	0.2739	0.3431	0.5167
	8.6762	14.9267	17.9317	22.3849	30.8928
Home Ownership	0.3991	0.3986	0.4134	0.4232	0.4291
	26.3627	25.3995	25.2175	24.9422	23.0066
Income	0.0151	−0.0079	−0.0265	−0.0585	−0.1439
	0.9211	−0.4727	−1.5058	−3.2359	−7.2373
Parameter	0.2189	0.5320	−0.5845	−0.4600	0.0000
	33.6086	30.7143	29.2650	−25.2714	0.0000
$n^{-1}\ln L$	−1.8482	−1.8781	−1.8921	−1.9272	−2.0300

To summarize the empirical results, both the election and housing data showed a consistent pattern of FD having the highest likelihood when using contiguity W_c. We also found a pattern of improvement in the likelihood function values from the other spatial specifications when the model switched from the simpler contiguity weights, W_c to the richer 30 nearest neighbor weights W_{nn}. Examining the alternative estimators in terms of the weight assigned to low-order versus high-order neighbors provides some insight into these patterns. Ranking the alternative specifications on the basis of low-versus high-order neighbor emphasis leads to: OLS, MA, ME, AR, and FD.

We now provide an empirical examination of the emphasis placed on high-order neighboring relations by the various dependence specifications. The results we present were constructed using estimates from fitting the election data example with the contiguity weight matrix W_c shown in (Table 9.11). We can express $E(y)$ using (9.61)–(9.65) as a function of X and the empirical estimates from the various dependence specifications. Each of the dependence specifications has a series approximation based on powers of the weight matrix. The emphasis each specification gives to the various orders of neighbors distinguishes these specifications.

TABLE 9.12: Maximum likelihood estimates for election data using W_{nn}

Variables	FD	AR	ME	MA	OLS
Intercept	0.6953	0.7659	0.8264	0.9859	1.5576
	15.7086	17.5550	18.8537	22.8608	30.7777
Voting Pop	−0.5618	−0.6108	−0.6397	−0.6913	−0.8464
	−28.1070	−30.2889	−31.2022	−32.9737	−34.7211
Education	0.1083	0.1604	0.1882	0.2583	0.5167
	6.8455	10.3655	12.3896	18.0939	30.8928
Home Ownership	0.4099	0.4025	0.4097	0.4190	0.4291
	26.3939	26.0726	25.9185	25.7206	23.0066
Income	0.0191	0.0194	0.0115	−0.0195	−0.1439
	1.1257	1.1535	0.6753	−1.1408	−7.2373
Parameter	0.2899	0.6700	−0.9392	−0.9100	0.0000
	33.2438	33.7897	33.2053	−30.1890	0.0000
$n^{-1}\ln L$	−1.8521	−1.8462	−1.8525	−1.8833	−2.0300

TABLE 9.13: Maximum likelihood estimates for housing data using W_c

Variables	FD	AR	ME	MA	OLS
Intercept	−5.4482	−4.8989	−5.8024	−7.4016	−11.0700
	−154.5890	−148.0487	−161.3703	−188.2670	−208.6987
Households	0.0185	0.0142	0.0221	0.0374	0.0767
	10.6710	8.0028	11.1578	16.9300	25.8473
Income	0.3464	0.3643	0.4470	0.5867	0.9105
	111.0039	116.6410	132.0263	158.0656	181.5669
Education	0.6208	0.4360	0.4808	0.5751	0.7828
	65.8672	45.9048	45.4117	48.6392	49.3896
Land Area	0.0036	−0.0112	−0.0203	−0.0349	−0.0711
	8.0856	−24.7413	−41.2206	−64.3828	−97.6894
Parameter	0.2652	0.7260	−0.9551	−0.7700	0.0000
	246.6843	235.9442	224.6063	−190.9716	0.0000
$n^{-1}\ln L$	−4.0564	−4.0980	−4.1400	−4.2523	−4.5453

TABLE 9.14: Maximum likelihood estimates for housing data using W_{nn}

Variables	FD	AR	ME	MA	OLS
Intercept	−6.4223	−5.4099	−5.4572	−7.2926	−11.0700
	−160.8616	−157.6558	−155.2703	−200.2547	−208.6987
Households	0.0263	0.0143	0.0147	0.0347	0.0767
	14.0014	7.8368	7.7959	16.7868	25.8473
Income	0.4011	0.3799	0.3960	0.5639	0.9105
	113.8333	117.0507	120.5236	165.2767	181.5669
Education	0.7350	0.5157	0.4906	0.5869	0.7828
	72.0133	52.6201	48.5588	53.0282	49.3896
Land Area	0.0065	−0.0024	−0.0063	−0.0272	−0.0711
	13.0689	−5.0206	−13.0949	−55.0203	−97.6894
Parameter	0.2721	0.7770	−1.3320	−0.9900	0.0000
	236.2361	238.8063	237.9460	−211.8613	0.0000
$n^{-1}\ln L$	−4.0969	−4.0871	−4.0904	−4.1847	−4.5453

$$E(y_{FD}) = (I_n - W_{-\iota})^{-0.2189} X \beta_{FD} \tag{9.61}$$

$$E(y_{AR}) = (I_n - 0.5320\,W)^{-1} X \beta_{AR} \tag{9.62}$$

$$E(y_{ME})y = e^{0.5845\,W} X \beta_{ME} \tag{9.63}$$

$$E(y_{MA})y = (I_n + 0.4600\,W) X \beta_{MA} \tag{9.64}$$

$$E(y_{OLS}) = X \beta_{OLS} \tag{9.65}$$

To make this less abstract, Table 9.15 presents the weights assigned to various powers of W based on the estimates shown in (9.61)–(9.64). Inspection of Table 9.15 shows that relative to the AR specification, FD assigns lower weight to the first three orders of neighbors, about the same weight to fourth order neighbors, and larger weights for fifth and higher order neighbors. Relative to the other spatial specifications, the FD weights decline more slowly with order.

9.5.2 Computational considerations

From a computational standpoint, one can use many of the same calculations set forth in the case of the matrix exponential spatial specification to produce estimates for the FD specification. For example, we can rely on the closed-form solution method from Chapter 4, where the expression for G_1 remains the same. However, Y has a different definition.

$$Y = \begin{bmatrix} y & \ln(\Delta_{-\iota})y & \ln(\Delta_{-\iota})^2 y & \dots & \ln(\Delta_{-\iota})^{q-1} y \end{bmatrix} \tag{9.66}$$

TABLE 9.15: Weights by order of
neighbors

Order	FD	AR	ME	MA
0	0.0000	1.0000	1.0000	1.0000
1	0.2189	0.5320	0.5845	0.4600
2	0.1334	0.2830	0.1708	0.0000
3	0.0987	0.1506	0.0333	0.0000
4	0.0794	0.0801	0.0049	0.0000
5	0.0670	0.0426	0.0006	0.0000
6	0.0583	0.0227	0.0001	0.0000
7	0.0518	0.0121	0.0000	0.0000
8	0.0467	0.0064	0.0000	0.0000
9	0.0427	0.0034	0.0000	0.0000
10	0.0393	0.0018	0.0000	0.0000
11	0.0365	0.0010	0.0000	0.0000
12	0.0342	0.0005	0.0000	0.0000
13	0.0321	0.0003	0.0000	0.0000
14	0.0303	0.0001	0.0000	0.0000
15	0.0287	0.0001	0.0000	0.0000
16	0.0273	0.0000	0.0000	0.0000
17	0.0261	0.0000	0.0000	0.0000
18	0.0249	0.0000	0.0000	0.0000
19	0.0239	0.0000	0.0000	0.0000
20	0.0230	0.0000	0.0000	0.0000

In the matrix exponential case, $W^2 y$ is calculated as $W(Wy)$ as opposed to forming W^2 and multiplying it by y. In the fractional differencing case, we calculate $\ln(\Delta_{-\iota})^2 y$ by finding $v = \ln(\Delta_{-\iota})y$ and then by forming $\ln(\Delta_{-\iota})v$. In turn, $v = -\sum_{i=1}^{p} i^{-1} W^i M_\iota y$, where p is the highest-order power used. Since this converges slowly, p should be large (e.g., 1000). Calculating Y represents the most time consuming part of fractional differencing estimation. However, this only needs to be done once for a given W, making estimation feasible for large n.

The matrix $W_{-\iota}$ is dense by itself even though W is sparse. Therefore, calculation of ψ by direct evaluation of $\ln|I_n - W_{-\iota}|$ is not practical. Also, $\ln|I_n - W|$ is singular. However, the constant $\psi = \ln|I_n - W_{-\iota}| = \lim_{\omega \to 1}(\ln|I_n - \omega W| - \ln(1-\omega))$. This is the overall log-determinant $\ln|I_n - W|$ with the part $(\ln(1-\omega))$ associated with the eigenvalue of 1 subtracted out. A practical computational approach is to calculate $\ln|I_n - \omega W| - \ln(1-\omega)$ for a sequence of values of ω approaching (but not including) 1. This sequence can be used to extrapolate $\ln|I_n - \omega W| - \ln(1-\omega)$ for $\omega = 1$. This method permits use of non-symmetric or symmetric matrices, takes advantage of sparseness in W, and avoids the singularity at $\omega = 1$.

The computational time required by the various procedures is quite moderate. For the $n = 3,107$ data set it took 0.2 seconds to find the contiguity weight matrix, 0.06 seconds to calculate ψ, 9.61 seconds to compute Y (for $p = 2,500$, $q = 16$) and only 0.13 seconds to produce the fractional differencing estimates. The 30 nearest neighbor case required 0.47 seconds to find the weight matrix, 0.28 seconds to compute ψ, 19.84 seconds to compute Y, and 0.14 seconds to find the estimates.[8] We need only form ψ and Y when changing W, and thus exploring models based on alternative independent variables requires very little computational time.

As a check on this approach to calculating ψ, we found the eigenvalues of W_c for the election data and calculated the log-determinant of $W_{-\iota}$ directly. The difference between the log-determinant from the eigenvalue calculation and the proposed approach was 0.0054, a very small number. We also checked the accuracy of the fractional differencing approximation. This was done by changing p, the degree of the matrix logarithm approximation, and q, the degree of the matrix exponential approximation. Changing p and q had some effect on the accuracy of $\tilde{\delta}$, and therefore on the regression coefficients. For the election data using W_c, changing from $p = 1,000$ and $q = 8$ to $p = 2,500$ and $q = 16$ resulted in a change in the δ estimate from 0.2197 to 0.2189, a difference of 0.0008.

For the larger $n = 62,226$ data set, it took 2.4 seconds to form the spatial weight matrix, 0.80 seconds to determine ψ, 4.65 minutes to compute Y using $q = 16$ and $p = 2,500$, and only 0.17 seconds to find the fractional differencing estimates. The 30 nearest neighbor case required 6.03 seconds for ψ, and 13.47 minutes to compute Y. Again, when changing X the marginal computational cost would just be the time to find the fractional differencing estimates. The key computational burden is computing Y, and this could be reduced by going to a Chebyshev approximation of the type described in Chapter 4.

9.6   Chapter summary

We have introduced the matrix exponential spatial specification (MESS) as an alternative to the spatial autoregressive process. MESS can be used to construct spatial regression models that replace geometric decay from the spatial autoregressive process with exponential decay.

This type of specification has both computational as well as theoretical advantages over the spatial autoregressive specification. These arise from the ease of inversion, differentiation, and integration of the matrix exponential. Moreover, the covariance matrix associated with the matrix exponential is always positive definite. Finally, the matrix exponential has a simple matrix

---

[8] All the times were for a machine using an AMD 3.2 Ghz Athlon.

determinant which vanishes for the common case of a spatial weight matrix with a trace of zero. This simplification was used to produce a closed-form solution for maximum likelihood estimates, and to provide Bayesian estimates based on univariate numerical integration of a scalar polynomial expression. In addition, some of the benefits of MESS extend to the case of spatial fractional differencing.

LeSage and Pace (2007) provide a further illustration that demonstrates how the analytical and computational advantages of MESS can be exploited in Bayesian model comparison MC^3 methods of the type described in Chapter 6. The MC^3 method was implemented by drawing on straightforward extensions of the existing results in the regression model literature.

The chapter also set forth a spatial specification based on a fractional differencing transformation. In the time series literature fractional differencing has proven useful in situations where dependence slowly declines with time. We argued that spatial systems may be more likely to produce patterns of higher order dependence than in the case of time series analysis. This type of pattern seems likely to arise when the number of connection paths between each observation and all others is large, or when boundaries or borders produce multiple reflections as a result of changes to nodes in the system.

Chapter 10

Limited Dependent Variable Spatial Models

This chapter introduces approaches to modeling dependent variables that reflect binary choice outcomes generated by spatially dependent processes. Spatial dependence in choice outcomes result in a situation where observed choices at one location are similar to choices made at nearby locations. There are a number of scenarios where we might see this type of outcome in observed choices. For example, in the aftermath of Hurricane Katrina the decision of a business owner in New Orleans to rebuild and reopen a store might depend on the decision of neighboring businesses to reopen. When considering origin-destination flows of commuters traveling to work, the choice between mass transit and automobile mode of travel might exhibit spatial dependence because commuters located at nearby origins would be faced with the same presence or absence of mass transit opportunities. Holloway, Shankara, and Rahman (2002) show that binary choices regarding adoption of an agricultural program by Bangladeshi rice producers exhibited spatial dependence. Applications to land-use decisions regarding conversion from agricultural to non-agricultural uses, where land-use decisions of neighboring property owners exert an influence on the decision outcome have also been popular (Zhou and Kockelman, 2008; Irwin and Bockstael, 2004). Probit variants of the SAR model were considered by McMillen (1992), who proposed an EM algorithm as a way to produce consistent (maximum likelihood) measures of dispersion for estimates β from these models. A major contribution to the non-spatial probit literature was the work of Albert and Chib (1993) who proposed treating the binary dependent variable observations as indicators that relate to underlying unobservable or latent levels of utility. They introduce these latent levels as parameters that can be estimated using a Bayesian MCMC framework. We discuss this type of approach in Section 10.1 which we extend to the case of the spatial probit SAR model (LeSage, 2000). We consider a related Tobit (or censored regression) model variant of the SAR model in Section 10.3.

The first type of spatial probit model that we discuss takes the SAR form shown in (10.1), where the $n \times 1$ vector y contains a set of 0,1 binary values that reflect choice outcomes, or they might reflect presence or absence of a tax or other feature in each region/observation. We could also have a measure of negative or positive change in (average) land values for a sample of regions, and so on.

$$y = \rho W y + X \beta + \varepsilon, \qquad \varepsilon \sim N(0, \sigma_\varepsilon^2 I_n) \tag{10.1}$$

This probit variant of the SAR model could of course be extended to the case of the SDM model by adding spatial lags of the explanatory variables. The motivations for a spatial lag of the dependent variable already described in the initial chapters should apply here as well. For example, if we begin with a time-lagged model that relates y_t to $W y_{t-1}$, then we are stating that decision outcomes (or the presence or absence of some feature in each region) exert an impact on future decisions (or features) of neighboring regions.

Similarly, beginning with a non-spatial relationship: $y = X\beta + u$, our motivation for a spatial lag of the dependent variable as arising from the presence of an omitted variable that is correlated with an included variable and exhibits spatial dependence would also provide a motivation for the SAR or SDM model. As an example, if we have a binary measure of the presence or absence of patenting activity for a sample of regions, the existence of tacit unmeasurable knowledge that is excluded from the set of explanatory variables in the model should lead to spatial dependence in the observed measures that record the presence or absence of regional patenting activity.

In addition to the SAR probit model we also discuss SAR ordered probit, SAR Tobit, and SAR multinomial probit variants of the SAR probit model. We approach the estimation task from a Bayesian MCMC sampling viewpoint. For an extensive discussion of alternative approaches to estimating these models see Flemming (2004). Maximum likelihood estimation seems quite difficult as pointed out by Beron and Vijverberg (2000), who report estimation times for a SAR probit model requiring many hours for a 49 observation problem. Early use of Bayesian MCMC sampling for spatial probit models can be found in Bolduc, Fortin and Gordon (1997), who model a spatial error covariance structure.

A second type of model that we explore in Section 10.6 was introduced by Smith and LeSage (2004). This model relies on an error structure that involves an additive error specification first introduced by Besag, York and Mollie (1991) and subsequently employed by many authors (Gelman et al., 1995). The approach of Smith and LeSage (2004) allows both spatial dependencies and general spatial heteroscedasticity to be treated simultaneously and has been popular in marketing applications (Allenby et al., 2002; Yang and Allenby, 2003; Ter Hofstede, Wedel and Steenkamp, 2002).

Smith and LeSage (2004) illustrate the method using county-level voting outcomes for a presidential election. The model relies on spatially structured effects parameters as well as common variance scalars for broader regions such as states in the county-level voting application. This allows for state-level differences in the effects parameters as well as the variance.

Section 10.6 also discusses a dynamic spatial ordered probit extension of this model described in Wang and Kockelman (2008a,b). This dynamic variant of the model can capture patterns of spatial and temporal autocorrelation in ordered categorical response data.

The next section begins with a discussion of Bayesian treatment of unobserved latent utilities, which is a key feature of MCMC estimation of probit, tobit and multinomial probit models.

10.1   Bayesian latent variable treatment

The Bayesian approach to modeling binary limited dependent variables treats the binary 0,1 observations in y as indicators of latent, unobserved (net) utility. The unobservable utility underlies the observed choice outcomes. For example, if the binary dependent variable reflects the decision to buy or not buy a product, the observed 0,1 indicator variable y represents observed decision outcomes in our sample. These are viewed as merely a proxy for the fact that when net utility is negative, a decision not to buy ($y = 0$) is made, and when net utility associated with the purchase is positive, a buy decision ($y = 1$) is made.[1] The Bayesian estimation approach to these models is to replace the unobserved latent utility with *parameters* that are estimated. For the case of a SAR probit model, given estimates of the $n \times 1$ vector of missing or unobserved (parameter) values that we denote as y^*, we can proceed to estimate the remaining model parameters β, ρ by sampling from the same conditional distributions that we used in the continuous dependent variable Bayesian SAR models from Chapter 5.

More formally, the choice depends on the difference in utilities: $(U_{1i} - U_{0i}), i = 1, \ldots, n$ associated with observed 0,1 choice indicators. The probit model assumes this difference, $y_i^* = U_{1i} - U_{0i}$, follows a normal distribution. We do not observe y_i^*, only the choices made, which are reflected in:

$$
\begin{aligned}
y_i = 1, & \quad \text{if} \quad y_i^* \geq 0 \\
y_i = 0, & \quad \text{if} \quad y_i^* < 0
\end{aligned}
$$

There are strict interpretations of this relationship that rely on utility maximization to argue that an individual located in region i choosing alternative 1 implies: $\Pr(y_i = 1) = \Pr(U_{1i} \geq U_{0i}) = \Pr(y_i^* \geq 0)$. Smith and LeSage (2004) provide a more detailed discussion of these issues. Albert and Chib (1993) adopt a less formal economic interpretation and view the y_i^* as simply unobserved values associated with observed choice events. These are modeled using the non-spatial regression relation: $y_i^* = X_i\beta + \varepsilon_i, \ \varepsilon_i \sim N(0, \sigma_\varepsilon^2)$.

---

[1]The utility derived from owning the good could be considered minus that from retaining income equal to the purchase price when the consumer does not buy the good.

If the vector of latent utilities y^* were known, we would also know y, which led Albert and Chib (1993) to conclude: $p(\beta, \sigma_\varepsilon^2 | y^*) = p(\beta, \sigma_\varepsilon^2 | y^*, y)$. The insight here is that if we view y^* as an additional set of parameters to be estimated, then the (joint) conditional posterior distribution for the model parameters $\beta, \sigma_\varepsilon^2$ (conditioning on both y^*, y) takes the same form as a Bayesian regression problem involving a continuous dependent variable rather than the problem involving the discrete-valued vector y. We extend this approach to the case of a SAR model where the model parameters are $\beta, \rho, \sigma_\varepsilon^2$. If an additional set of n parameters $y_i^*, i = 1, \ldots, n$ were introduced to the model, estimation via MCMC sampling would require that we sequentially sample each of these parameters from their conditional distributions. Recall that our MCMC estimation scheme for the SAR model simply cycles through the sequence of conditional distributions for all model parameters taking samples from each of these. A large number of passes through the sampler produces a sequence of draws for the model parameters that converge to the unconditional joint posterior distribution.

Albert and Chib (1993) argued that if we could introduce a vector of parameters y^* and obtain a sample from the conditional posterior distribution of each element (parameter) in this vector, then estimation of the remaining parameters of interest $\beta, \sigma_\varepsilon^2$, would be relatively simple. The simplicity arises from the fact that given y^* values in place of the binary y values, we can use the same conditional posterior distributions that arise for the case of a continuous dependent variable regression model. We follow this approach for our Bayesian SAR model from Chapter 5 rather than the conventional regression model used by Albert and Chib (1993). In this case, given the vector of n parameter values in y^* in place of the binary y values, we can use the same conditional posterior distributions set forth in Chapter 5 to sample the remaining model parameters $\beta, \rho, \sigma_\varepsilon^2$, where y^* is used to replace the vector y containing binary values. This approach is quite simple since the form of the distributions from which we need to sample the parameters $\beta, \rho, \sigma_\varepsilon^2$ *conditional* on the parameters y^* are the same as those from Chapter 5 for the continuous dependent variable model.

Albert and Chib (1993) go on to derive the form of the joint posterior distribution $p(\beta, \sigma_\varepsilon^2 | y^*, y)$ and associated conditional posterior distributions that allow MCMC estimation for their non-spatial probit regression model. However, their results concerning the conditional posterior distributions are not applicable to our case where the dependent variable follows a spatial dependence process.

For the case of independent observations considered by Albert and Chib (1993), combining the normality assumption from the non-spatial regression model with the sample data information contained in y, leads to conditional distributions for the important parameters y_i^* that take the form of *univariate* truncated normal distributions shown in (10.2) and (10.3).

$$y_i^*|y_i, \beta, \sigma_\varepsilon^2 \sim N(X_i\beta, \sigma_\varepsilon^2) \, \delta(y_i^* \geq 0) \quad \text{if} \quad y_i = 1 \qquad (10.2)$$
$$y_i^*|y_i, \beta, \sigma_\varepsilon^2 \sim N(X_i\beta, \sigma_\varepsilon^2) \, \delta(y_i^* < 0) \quad \text{if} \quad y_i = 0 \qquad (10.3)$$

We use $\delta(A)$ as an indicator function for each event A (in the appropriate underlying probability space), so $\delta(A) = 1$ for outcomes where A occurs and $\delta(A) = 0$ otherwise. Expression (10.2) represents a univariate normal distribution truncated to the left at 0 if $y_i = 1$, where $X_i\beta$ is the mean of the distribution and σ_ε^2 is the variance. Similarly, expression (10.3) is a univariate normal distribution truncated to the right at zero.

There is an identification problem with the non-spatial probit model since multiple values for the model parameters $\beta, \sigma_\varepsilon^2$ give rise to the same likelihood function values. This arises because $Pr(X_i\beta + \varepsilon_i \geq 0|\beta, \sigma_\varepsilon^2) = Pr(cX_i\beta + c\varepsilon_i \geq 0|\beta, \sigma_\varepsilon^2)$. That is, multiplying the mean $X_i\beta$ and variance σ_ε^2 by the scalar $c > 0$ leads to a distribution for the disturbances: $c\varepsilon_i \sim N(0, c^2\sigma_\varepsilon^2)$, which is the same model with different coefficients and error variance. This means that the probit model cannot identify both β and σ_ε^2, which is conventionally solved by setting $\sigma_\varepsilon^2 = 1$.

10.1.1 The SAR probit model

An important difference between the non-spatial regression model and the SAR model is that the dependence leads to a *multivariate* truncated normal distribution (TMVN) for the latent y^* parameters from which we need to sample these parameters. Specifically, for the SAR model we have a mean vector and variance-covariance matrix shown in (10.4), where we have set $\sigma_\varepsilon^2 = 1$ for identification.

$$y^* \sim TMVN\{(I_n - \rho W)^{-1}X\beta, [(I_n - \rho W)'(I_n - \rho W)]^{-1}\} \qquad (10.4)$$
$$y^* \sim TMVN(\mu, \Omega)$$

We introduce $\mu = (I_n - \rho W)^{-1}X\beta$ as the mean and $\Omega = [(I_n - \rho W)'(I_n - \rho W)]^{-1}$ as the variance-covariance matrix. As in the case of independent observations, the insight of Albert and Chib (1993) holds so the (joint) conditional distribution for the model parameters $p(\beta, \rho|y^*) = p(\beta, \rho|y^*, y)$ takes the same form as in the case of a continuous dependent variable SAR model. It also leads to individual conditional posterior distributions for the parameters $p(\beta|\rho, y^*)$ and $p(\rho|\beta, y^*)$ that are the same as in the case where we have a continuous dependent variable y in place of y^*. The key conditional posterior distribution that we require to implement this scheme is the n-variate truncated normal for $p(y^*|\beta, \rho, y)$.

10.1.2 An MCMC sampler for the SAR probit model

For clarity we refer to estimation of the SAR probit model as an MCMC sampling scheme that samples sequentially from the conditional posterior distributions for the model parameters β, ρ, y^*. Within this sequence, we need to sample a set of n values to fill-in the vector y^*. Details regarding this are described in the next section. For clarity we describe the MCMC sampling scheme here without details regarding this step.

If we use the same independent prior distributions $\pi(\beta, \rho) = \pi(\beta)\pi(\rho)$ as in Chapter 5, where we assign a normal prior $\beta \sim N(c, T)$ and a uniform (or $\mathcal{B}(a, a)$) prior for the parameter ρ, these two conditional distributions given the parameters y^* should be the same. Specifically, we can sample:

$$p(\beta|\rho, y^*) \propto N(c^*, T^*) \tag{10.5}$$
$$c^* = (X'X + T^{-1})^{-1}(X'Sy^* + T^{-1}c)$$
$$T^* = (X'X + T^{-1})^{-1} \tag{10.6}$$
$$S = (I_n - \rho W)$$

To see the insight of Albert and Chib (1993), suppose we had three scalar parameters $\theta_1, \theta_2, \theta_3$ and sample data y. The joint distribution: $p(\theta_1, \theta_2|\theta_3, y)$ would be used as the basis for deriving a conditional distribution for the parameter θ_1, and another conditional distribution for the parameter θ_2. These two conditionals would take the form: $p(\theta_1|\theta_2, \theta_3, y)$ and $p(\theta_2|\theta_1, \theta_3, y)$. Of course, to complete the sampler we would need to also have a conditional distribution: $p(\theta_3|\theta_1, \theta_2, y)$.

Applying the result from Albert and Chib (1993), the (joint) conditional distribution for the parameters $p(\theta_1, \theta_2|\theta_3) = p(\theta_1, \theta_2|\theta_3, y)$. Further, using $\beta = \theta_1, \rho = \theta_2$ and $y^* = \theta_3$, we have the result in (10.5), after noting that the *parameters* y^* play the role of the continuous data vector y from Chapter 5.

Following this same line of reasoning, the parameter ρ can be sampled from $p(\rho|\beta, y^*)$. This can be accomplished using either the Metropolis-Hastings approach or integration and draw by inversion set forth in Section 5.3.2. This requires evaluating the expression in (10.7)

$$p(\rho|\beta, y^*) \propto |I_n - \rho W| \exp\left(-\frac{1}{2}[Sy^* - X\beta]'[Sy^* - X\beta]\right) \tag{10.7}$$

Finally, we need to sample each value of y^* from its conditional distribution. In the work of Albert and Chib (1993), each value y_i^* in the vector had a univariate truncated normal conditional distribution. This univariate truncated normal distribution with a mean and variance that was easy to calculate provided the basis for sampling these n parameters.

In the case of our SAR probit model, the conditional distribution of the parameter vector y^* takes the form of a truncated multivariate distribution. For the case of a *non-truncated* n-dimensional multivariate normal distribution, one can sample from a sequence of n conditional univariate normal distributions to obtain the n parameters. Smith and LeSage (2004) provide an example of this type of procedure. However, Geweke (1991) points out that this *cannot* be done for the case of a truncated multivariate distribution. That is, the individual elements from a vector such as y^* cannot be obtained by sampling from a sequence of univariate truncated normal distributions. This has been a source of misunderstanding in work on the SAR probit model such as that by LeSage (2000).

It is possible to sample the n parameters in the vector y^* from the truncated multivariate normal distribution using a method proposed by Geweke (1991). Details regarding this are provided in the next section. For now, we simply assume that these parameters can be sampled.

Given the ability to sample from the complete sequence of conditional distributions for all of the model parameters, MCMC estimation can be applied to the SAR probit model. A single sequence of samples from $p(\beta|\rho, y^*), p(\rho|\beta, y^*)$ and $p(y^*|\beta, \rho)$ constitutes only a single pass through the MCMC sampler. We must make a large number of passes to produce a large sample of draws from the joint posterior distribution of the model parameters.

The sample of draws can be used to construct parameter estimates based on posterior means and standard deviations as described in Chapter 5.

10.1.3 Gibbs sampling the conditional distribution for y^*

The key conditional distribution required to implement our MCMC sampling scheme for the SAR probit model is $p(y^*|\beta, \rho)$. This section focuses on details regarding how to obtain samples from this conditional distribution. As already noted, we cannot sample the individual elements y_i^* by sampling from a sequence of univariate truncated normal distributions. That is, the marginal distributions for individual elements of the $n \times 1$ vector y^* are not univariate truncated normal.

Geweke (1991) sets forth one approach to sample from a multivariate truncated normal distribution. We will rely on the Geweke (1991) approach, but note that this is an active area of research and other approaches have also appeared in the literature. Geweke (1991) uses a Gibbs sampling algorithm to carry out draws from a multivariate truncated normal distribution. Recall that we use the term Gibbs sampler to refer to situations where the conditional distributions from which we need to sample take known forms.

The approach involves Gibbs sampling that produces draws for individual elements y_i^* from the $n \times 1$ vector y^* based on the (conditional) distribution of each element y_i^* conditional on all other $n - 1$ elements, which we denote using y_{-i}^*. For our SAR probit model, we wish to sample from a truncated n-variate normal distribution: $y^* \sim TMVN(\mu, \Omega)$ subject to vector of linear inequality restrictions $a \le y^* \le b$, where the truncation bounds a and b depend on the observed values 0, 1 for elements of y, with details provided later. Geweke (1991) establishes that sampling from a truncated n-variate normal distribution: $y^* \sim TMVN(\mu, \Omega)$ subject to linear inequality restrictions $a \leq y^* \leq b$, is equivalent to constructing samples from the n-variate normal distribution $z \sim N(0, \Omega)$ subject to the linear restrictions: $\underline{b} \leq z \leq \overline{b}$. Where $\underline{b} = a - \mu, \overline{b} = b - \mu$. We then obtain the sample for y^* using: $y^* = \mu + z$.

The method of Geweke (1991) works with the *precision* matrix, or inverse of the variance-covariance matrix of the truncated multivariate normal distribution from which we wish to sample. We label this for our SAR probit model using: $\Psi = \Omega^{-1} = (1/\sigma_\varepsilon^2)(I_n - \rho W')(I_n - \rho W)$. As in the case of non-spatial probit, we impose the identification restriction that $\sigma_\varepsilon^2 = 1$.

Geweke's procedure takes into account the fact that the marginal distributions of the elements of z *are not* univariate truncated normal. He exploits the fact that the (conditional) distribution of each element of z_i, conditional on all other elements z_{-i} can be expressed as univariate distributions with (conditional) mean and (conditional) variance that are easy to calculate. These expressions for the mean and variance can be used to produce a draw from a univariate truncated normal distribution subject to appropriate constraints. This allows us to use Gibbs sampling to *build up* a sample from the (joint) multivariate truncated normal distribution in which we are interested. We emphasize that we use the term Gibbs sampler because Geweke's approach takes advantage of the fact that the (conditional) distribution of each element i, $z_i|z_{-i}$, conditional on all of the other elements $-i$ takes a known form from which random deviates can be easily generated.

Geweke uses expressions for the partitioned (symmetric) matrix inverse to establish that $E(z_i|z_{-i}) = \gamma_{-i}z_{-i}$ for the case of a non-truncated multivariate normal distribution $N(0, \Omega)$, where $\gamma_{-i} = -\Psi_{-i}/\Psi_{i,i}$, and Ψ_{-i} is the ith row of Ψ excluding the ith element. This implies that for the truncated distribution we have normal conditional distributions taking the form in (10.8).

$$z_i|z_{-i} = \gamma_{-i}z_{-i} + h_iv_i \qquad (10.8)$$
$$h_i = (\Psi_{i,i})^{-1/2}$$

Samples for $v_i \sim N(0, 1)$ are subject to the truncation constraints:

$$(\underline{b_i} - \gamma_{-i}z_{-i})/h_i < v_i < (\overline{b_i} - \gamma_{-i}z_{-i})/h_i$$
$$\underline{b_i} = -\infty \text{ and } \overline{b_i} = -\mu_i \text{ for } y_i = 0$$
$$\underline{b_i} = -\mu_i \text{ and } \overline{b_i} = +\infty \text{ for } y_i = 1$$

These can be used to produce a vector z of $z_i, i = 1, \ldots, n$, where previously sampled values $z_1, z_2, \ldots, z_{i-1}$ are used during sampling of element z_i. In addition, we use $z_{i+1}, z_{i+2}, \ldots, z_n$ from the previous pass through the Gibbs sampler when updating z_i. More formally, let $z_i^{(0)}$ denote the initial values of zero, and $z_i^{(m)}$ the values after pass m through the Gibbs sampler. On the first pass we use: $z_1^{(1)}, z_2^{(1)}, \ldots, z_{i-1}^{(1)}$ when filling in the initial zero value for $z_i^{(0)}$. Having reached the final element $i = n$, we have a set of values $z_i^{(1)}, i = 1, \ldots, n$. For the second pass, $m = 2$, we sample the first element $z_1^{(2)}$, using previously generated $z_i^{(1)}, i = 2, \ldots, n$ values, leading to a new value $z_i^{(2)}, i = 1$. For the second element we use $z_1^{(2)}$ in conjunction with $z_i^{(1)}, i = 3, \ldots, n$. We follow a similar process for the third and subsequent elements. At the end of the second pass through the sampler, we have an updated vector of values $z_i^{(2)}, i = 1, \ldots, n$. This procedure is continued on each of the m passes.

This procedure represents a Gibbs sampling scheme that takes into account the dependence between observations using the basic idea of Gibbs sampling. That is, the joint distribution for the vector z can be constructed by sampling from the complete sequence of conditional distributions for each element, $z_i | z_{-i}$.

Having made a series of m passes through the n-observation vector z, we generate $y^* = \mu + z^{(m)}$. The vector y^* can then be used to produce draws from the conditional distributions for the remaining model parameters β, ρ as described in the previous section.

10.1.4 Some observations regarding implementation

For clarity we refer to estimation of the SAR probit model as an MCMC sampling scheme that samples sequentially from the conditional posterior distributions for the model parameters β, ρ, y^*. Within this sequence, we rely on an m-step Gibbs sampler to produce the vector of parameters y^*. The m-step Gibbs sampler is the procedure set forth in the previous section.

The typical implementation of the m-step Gibbs sampler sets the vector z to zero values on the initial step and uses the m-steps to *build up* a sample of values for the n-vector z. Taking this approach to sampling y^* is a computationally intensive operation. To see this, consider an example where we are working with a sample of 3,000 US counties, and wish to produce 5,000 draws by making passes through the MCMC sampler. Let the m-step Gibbs sampler for y^* be based on $m = 10$, so *on each* of the 5,000 passes we need to make $10 \times 3,000 = 30,000$ passes over the sample of counties to produce a single vector z that can be used to construct a single draw for the vector y^*. Of course, we must do this 5,000 times, so this amounts to a total of $5,000 \times 30,000 = 150,000,000$ evaluations of the inner-most m-step Gibbs sampler.

Using a value of $m = 10$ is fairly standard in applied code used in Bayesian multinomial probit applications by Koop (2003). It might seem surprising that only 10 passes are required to build up an adequate sample from the truncated multivariate normal distribution. However, keep in mind that we will make many passes through the MCMC sampler. On each MCMC pass we only need a sample of y^* values that are reasonably accurate, since our procedure will involve thousands of samples drawn for the y^* values.

Fortunately, it is often possible to rely on a single step, that is we can set $m = 1$. When doing this, we rely on the values z from the previous trip through the MCMC sampler rather than initializing the vector z to zero on each pass. To see why this is possible, consider a case where the parameters β and ρ exhibit a great deal of precision. This is typically the case in estimation problems involving large spatial samples, say the 3,000 US counties. In this situation, these parameters will not change greatly on each pass through the MCMC sampler. If these parameter values are approximately equal on each MCMC pass, then each pass is equivalent to taking another step m when we rely on the vector z from the previous MCMC pass. Recall that for step $m = 2$, we would rely on the same parameters β, ρ and the values in the vector z from the $m = 1$ step.

We note that the true criterion for whether this scheme of reusing the vector of values in z and $m = 1$ will work well is the amount of precision in the model parameters. This is unfortunately not known a priori. In applied practice, one could produce estimates for increasing values of m to see if the same estimates arise.

Obviously, reducing $m = 10$ to $m = 1$ would decrease the time required to produce MCMC estimates for the SAR probit model. Theoretically, even if the parameters β and ρ change on each pass through the MCMC sampler, sampling from the complete sequence of conditional distributions will still lead to the joint posterior distribution for the parameters in which we are interested. However, this theory is asymptotic, so with small samples it would be important to check for sensitivity of the estimates to values of m used. In exploratory work involving data generated experiments where the true parameters are known, $m = 1$ seems to produce estimates that were similar to those from using $m = 10$ or $m = 20$ for larger sample sizes (those where $n > 500$). For smaller samples (those where $n < 500$), reliance on $m = 1$ produces estimates with larger standard deviations than $m = 10$ or $m = 20$, and some bias in estimates for the parameters β. It seems intuitively plausible that increasing the sample size should produce more precise estimates, so smaller values of m could be used. We also note that carrying out the full $m-$step procedure for smaller samples is not a problem.

There may be a trade-off between increased speed and the need to carry out more draws, an issue that needs to be studied further. There are a number of hybrid approaches that could also be explored. For example, one could begin using $m = 10$ during the *burn-in* period and then switch to $m = 1$ for the remaining MCMC draws. The rationale for this would be to allow the sampler to work harder during the initial exploratory phase to obtain high-quality estimates for the parameters y^*. After this, we rely on passes through the MCMC sampler to capture the dependence rather than the m-step Gibbs sampler.

When we set $m = 1$, it takes around 45 minutes to produce 1,000 draws for a sample of 3,100 US counties using a relatively slow laptop computer and MATLAB. Using compiled code to carry out the innermost Gibbs sampling task should provide a six times speed improvement, and there are other ways to optimize the coding implementation. Some timing experiments indicate that doubling the number of sample observations results in doubling the time required to produce the same number of MCMC draws.

There are some fortunate computational aspects to this procedure. One positive aspect is that we can work with the precision matrix Ψ, so we avoid the need to calculate the inverse: $[(I_n - \rho W)'(I_n - \rho W)]^{-1}$.

Computing an inverse for $S^{-1} = (I_n - \rho W)^{-1}$ is problematical. Despite use of a sparse matrix W, the inverse is a dense matrix containing all non-zero elements. Computer memory requirements for storing elements of a dense matrix when n is large place severe constraints on the size of the problem that can be handled. LeSage and Pace (2004) point out that the memory requirements for the inverse matrix S^{-1} increased 50 fold over those for the matrix $S = (I_n - \rho W)$ in their application involving the prices of sold and unsold homes. See Chapter 4 for a discussion of these issues. Chapter 4 also discusses computing $\mu = (I_n - \rho W)^{-1} X \beta$ by solving the equation $(I_n - \rho W)\mu = X\beta$ for μ, to avoid forming an explicit inverse.

A final positive aspect of the m−step Gibbs sampling scheme within the MCMC sampler is that problems involving large n can be solved without resort to a large amount of computer memory. This is because the larger problem is broken into a series of n draws from univariate conditionals. This in conjunction with use of the precision matrix in place of the variance covariance matrix (which avoids the need to compute the inverse of the variance-covariance matrix), limits the memory required.

10.1.5 Applied illustrations of the spatial probit model

As an initial illustration, two samples of $n = 400$ and $n = 1,000$ continuous values y^* were generated. These were used to determine y_i values of zero if $y_i^* < 0$ and one when $y_i^* \geq 0$. The SAR model DGP used was:

$$y^* = (I_n - \rho W)^{-1} X \beta + (I_n - \rho W)^{-1} \varepsilon$$
$$\varepsilon \sim N(0, I_n)$$

The matrix X consisted of an intercept term and two standard random normal deviates, and the coefficients $\beta = \begin{pmatrix} 0 & 1 & -1 \end{pmatrix}'$. A value of $\rho = 0.75$ was used and a spatial weight matrix based on six nearest neighbors was constructed, using vectors of standard normal deviates as locational coordinates.

Results are shown in Table 10.1 alongside those for the Albert and Chib (1993) non-spatial probit model, and maximum likelihood estimates based on the continuous y^* values that were used to produce the binary dependent variable. Of course, in applied practice one would not know these values, but they serve as a benchmark for the accuracy of our m-step Gibbs sampling procedure which simulates these values as model parameters. The correlation coefficient between these actual latent utilities and the posterior mean of the simulated values was 0.92, for both the 400 and 1,000 observation samples, indicating accurate sampling.

A value of $m = 10$ was used for the Gibbs steps within the MCMC sampling procedure to produce estimation results for the $n = 400$ sample and $m = 1$ was used for the $n = 1,000$ sample. The estimates reported in the table represent posterior means based on 1,200 draws with the first 200 omitted to account for burn-in of the MCMC sampler. For the case of $n = 400$ and $m = 10$ on each MCMC draw, the latent variable values z_i were initialized to zero on each MCMC draw, so the sample of latent z values were built up anew using the $m = 10$ step Gibbs sampler. We contrast the results in Table 10.1 based on this procedure with results presented in Table 10.2, based on reusing z values from previous MCMC draws and values for $m = 1, 2$ and 10. For the $n = 1,000$ sample with $m = 1$ presented in Table 10.1, we relied on reuse of z values from previous draws.

The posterior means for the SAR probit model are all within one standard deviation of the coefficients that resulted from maximum likelihood estimation based on the actual y^* values. Using the non-spatial probit model produced biased estimates that are more than three standard deviations away from the true coefficient values as well as the maximum likelihood estimates. For the larger sample of $n = 1,000$, we see SAR probit model posterior mean estimates for the parameters β that are nearly indistinguishable from those based on the actual y^* values. Interestingly, the standard deviations from the SAR probit model are around twice those from maximum likelihood estimation. Working with a binary rather than continuous dependent variable imposes some costs, which include more uncertainty in the coefficient estimates. This is apparent in the larger standard deviations relative to those from a model based on the actual *utilities*.

As an illustration of the impact of changing the number of Gibbs steps, a comparison of estimates from $m = 1, m = 2$ and $m = 10$ are shown in Table 10.2 for a sample of $n = 400$ observations.

The times required to produce 1,200 draws (with the estimates based on the last 1,000 draws) are reported in the table. Convergence tests indicated that the same posterior means and standard deviations were associated with 1,200 draws as with the last 1,000 draws from a sample of 5,000 draws, suggesting no problems with convergence of the MCMC sampler. As indicated earlier, doubling the sample size results in a doubling of the time required to produce the sample number of MCMC draws.

In contrast, the speed improvement from reducing m from 10 to 1 is not a ten-fold decrease in time required as we might suppose, but rather a decrease around 6.5 in time required. Caching and other loop optimizing features of the MATLAB software used to produce the estimates are such that the time cost of looping within the inner m-step Gibbs sampler is not linear in m.

TABLE 10.1: SAR probit model estimates

Estimates	SAR Mean	SAR Std dev	SAR probit Mean	SAR probit Std dev	Probit Mean	Probit Std dev
			$m = 10,\ n = 400$			
$\alpha = 0$	−0.1196	0.0549	−0.1844	0.0686	−0.5715	0.0763
$\beta_1 = 1$	1.0187	0.0493	0.9654	0.1179	0.7531	0.0946
$\beta_2 = -1$	−1.0078	0.0495	−0.8816	0.1142	−0.7133	0.0855
$\rho = 0.75$	0.7189	0.0290	0.6653	0.0564		
Time(sec.)	0.2		1,276		2.8	
			$m = 1,\ n = 1,000$			
$\alpha = 0$	0.0436	0.0316	0.05924	0.0438	0.0980	0.0466
$\beta_1 = 1$	0.9538	0.0318	0.96105	0.0729	0.7409	0.0528
$\beta_2 = -1$	−1.0357	0.0315	−1.04398	0.0749	−0.8003	0.0586
$\rho = 0.75$	0.7019	0.0116	0.69476	0.0382		
Time(sec.)	0.4		586		4.7	

As the results show, using only $m = 1$ leads to similar estimates and standard deviations as $m = 2$ and $m = 10$. Recall also that setting $m = 1$ involves reuse of z-values from previous draws of the MCMC sampler versus initializing z to zero values. This appears to have no impact on estimation outcomes. We note that reuse of z-values from previous draws may create dependence in the sample of draws for y^*, but typically these are not an object of inference. This sampling dependence could carry over to other model parameters, but this is a question to be addressed by a detailed study of alternative implementation schemes for this procedure. In our data generated experiments where the true model parameters are known, the simple scheme based on $m = 1$ produced estimates very close to truth based on only a small sample of 1,200 draws with the first 200 draws excluded from the sample used to calculate posterior means. This suggests dependence in the sequence of MCMC draws was not a problem.

TABLE 10.2: The impact of changing m on SAR probit model estimates

Coefficients	$m = 1$ Mean	$m = 1$ Std dev	$m = 2$ Mean	$m = 2$ Std dev	$m = 10$ Mean	$m = 10$ Std dev
constant= 0	0.0241	0.0649	0.0236	0.0585	0.0223	0.0551
$\beta_1 = 1$	1.0637	0.1394	1.0457	0.1172	1.0364	0.1176
$\beta_2 = -1$	−1.0081	0.1323	−0.9788	0.1087	−0.9844	0.1067
$\rho = 0.75$	0.7454	0.0538	0.7374	0.0515	0.7468	0.0488
time (seconds)	195		314		1,270	

An applied example involved a county-level model of presidential voting from the 2000 US presidential election where $y = 1$ for the 2,438 counties won by George Bush and $y = 0$ for 669 counties won by Al Gore. Explanatory variables used were a constant term, a binary vector of 0,1 values with 1 for counties won by the Republican party presidential candidate in the 1996 election and 0 for counties won by the democratic party candidate (*Repub96*), the log of other party votes going to candidates other than Republican or Democrat candidates (*Oparty*), the log of county population over age 25 having college degrees (*College*), the log of median household income in the county (*Income*), the log of the number of persons who lived in the same house five years ago in 1995 (*Shouse*), the log of persons who were foreign born (*Fborn*), the log of persons in poverty (*Poverty*), and the log of homes built within the last year (1999) prior to the year 2000 Census (*Nhomes*).

Conventional non-spatial probit estimates are reported in Table 10.3 alongside the SAR probit model estimates. These were based on 1,200 MCMC draws and $m = 1$. The time required to produce 1,200 draws for the sample of $n = 3,107$ was around 45 minutes. Another run with $m = 5$ and the z-values initialized to zero on each pass through the MCMC sampler produced nearly identical results.

TABLE 10.3:   SAR probit model estimates

Coefficients	SAR probit model			Probit model		
	Mean	Std	p-level	Mean	Std	p-level
Constant	8.1454	3.0757	0.004	13.7828	3.2507	0.000
Repub96	2.0604	0.1305	0.000	2.4367	0.1240	0.000
Oparty	0.0864	0.0469	0.035	0.0751	0.0565	0.180
College	−0.5730	0.1107	0.000	−0.6477	0.1186	0.000
Income	−1.1003	0.3154	0.001	−1.7270	0.3334	0.000
Shouse	0.2433	0.2994	0.213	−0.7385	0.3027	0.014
Fborn	−2.0122	0.8532	0.008	−3.6200	0.9378	0.000
Poverty	−0.8945	0.1304	0.000	−1.1215	0.1412	0.000
Nhomes	2.0381	0.5968	0.000	2.3017	0.6255	0.000
ρ	0.4978	0.0307	0.000			

The table reports posterior means and standard deviations for the SAR probit model along with a *Bayesian p-level* that measures whether the coefficient is sufficiently different from zero. This statistic should be comparable to the conventional p-level associated with the asymptotic t-statistic from the non-spatial probit model reported in the table (Gelman et al., 1995).

From the reported estimates we see non-spatial probit estimates that are generally larger in absolute magnitude than those from the spatial model and around two standard deviations away from the spatial estimates. This is consistent with significant bias in the non-spatial estimates. There is one interesting reversal in sign associated with the *Shouse* variable, which is negative and significant in the non-spatial model, but positive and not different from zero in the spatial model. However, as in the case of continuous spatial regression models, we cannot directly compare these coefficient magnitudes. As is well-known for conventional probit models we need to evaluate the non-linear probit relationship by calculating *marginal effects* estimates. This is the subject to which we turn attention in the next section.

10.1.6 Marginal effects for the spatial probit model

In non-spatial probit models, the parameter magnitudes associated with the estimated coefficients $\hat{\beta}$ do not have the same marginal effects interpretation as in standard regression models. This arises due to non-linearity in the normal probability distribution. The magnitude of impact on the expected probability of the event y occurring varies with the level of say the rth explanatory variable, x_r. The nature of this non-linear relationship between changes in the dependent variable (the expected probability of the event) and changes in x_r is determined by the standard normal density such that:

$$\partial E[y|x_r]/\partial x_r = \phi(x_r \beta_r)\beta_r \qquad (10.9)$$

where β_r is a non-spatial probit model estimate, and the expression $\phi(\cdot)$ is the standard normal density. Because the magnitude of impact on changes in expected probability varies with the level of x_r, model estimates are often interpreted using mean values of a regressor such as $\bar{x}_r$. The marginal effects are then interpreted as the change in the event probability associated with a change in the average or typical sample observation for variable x_r. In addition to the non-linear nature of the mean response of expected probability to changes in x_r, there is also the need to consider a measure of dispersion for this to provide a basis for statistical inference regarding the significance of these changes.

In spatial regression models that involve spatial lags of the dependent variable (such as our SAR probit model), a change in the ith observation of the explanatory variable vector x_{ir} will impact the own region y_i plus other-regions $y_j, j \neq i$ in the sample. This of course suggests that changes in the level of a single observation x_{ir} will have an impact on the expected probability of the event being analyzed in both own- and other-regions.

We have already established that $E(\partial y/\partial x'_r) = (I_n - \rho W)^{-1} I_n \beta_r$ for the SAR model, which is an $n \times n$ matrix. The diagonal of this matrix captures what we have labeled the direct impact of a change in x_{ir} on the own-observation y_i, with the off-diagonal elements representing indirect or spatial spillover impacts. This matrix replaces the non-spatial model coefficient β_r, so we can calculate the marginal effects for our spatial SAR probit model by replacing β_r in (10.9) with this matrix. This leads to the expression in (10.10), where: $S = (I_n - \rho W)$, and $\bar{x}_r$ denotes the mean value of the rth variable.

$$\partial E[y|x_r]/\partial x'_r = \phi(S^{-1}I_n\bar{x}_r\beta_r) \odot S^{-1}I_n\beta_r \qquad (10.10)$$

In place of the expression $\phi(\bar{x}_r\beta_r)$ that scales the parameter estimate β_r in the non-spatial probit model, we have a matrix, $\phi(S^{-1}I_n\bar{x}_r\beta_r)$. And, in place of the (scalar) coefficient β_r from the non-spatial probit model, we have another matrix, $S^{-1}I_n\beta_r$.

We can adopt the same approach taken earlier to develop scalar summary measures for the continuous dependent variable SAR model. The main diagonal elements of $\phi\left[(I_n - \rho W)^{-1}I_n\bar{x}_r\beta_r\right] \odot (I_n - \rho W)^{-1}I_n\beta_r$ represent the direct impacts, which we average over. Similarly, the average of the row (or column) sums can be used to produce a total impact scalar summary measure, and the indirect impacts are the difference between these two measures. We interpret the average of the row-sums as the *average total impact from* changing an observation, and the average of the column-sums as the *average total impact to* an observation.

As an example, consider the decision to install a home security system, where the binary dependent variable indicates the absence ($y = 0$) or presence ($y = 1$) of security systems in a sample of n homes. Suppose we are interested in the marginal effects of a variable x_r recording the number of burglaries that have occurred for each home in the sample. From the viewpoint of a single homeowner i who is contemplating spending on an alarm system, the conventional probit model would indicate that a burglary at a neighboring home j would have *no effect* on homeowner i's decision to purchase a security system. Only an increase in burglaries at home i represented by a change in x_{ir} would impact the probability that homeowner i makes a security system purchase. The conventional use of the average or mean burglary rate $\bar{x}_r$ to calculate marginal effects has the implication that an increase in the mean burglary rate $\bar{x}_r$ across the sample of homes would increase the probability of all homeowners in the sample purchasing security systems. For this reason, the marginal effects are usually analyzed in the conventional model by varying the level of x_r over the range of values taken by this variable to assess how the level of burglaries at various homes in the sample would impact the decision to purchase a security system. Doing this creates a type of spatial variation in the effects estimates since the level of x_r values vary over space.[2] However, this does not imply any spatial interaction between observations, as this is not possible in a non-spatial probit model.

In contrast, the SAR probit model implies that a change in burglaries of neighboring homes j would have an *effect* on the probability that homeowner i purchases a security system. The effect would depend on spatial proximity

---

[2]This evaluation is of course conditional on all explanatory variables and associated coefficient estimates) in the model, which are typically held fixed at their respective means.

of homeowner i to j, captured by the spatial weight matrix W, as well as the strength of spatial dependence measured by the parameter ρ. A burglary at home j would have both a direct impact on the probability that homeowner j purchases a security system, as well as an indirect or spatial spillover impact on neighbors. The total effect is the sum of these two impacts.

Conventional practice in the non-spatial probit model calculates marginal effects using the point estimates (or posterior means in the case of Bayesian analysis) and average variable values. One advantage of MCMC estimation is that the sample draws arising from estimation can be used to produce separate marginal effects for every observation at each iteration (or using the sample of draws from the MCMC procedure in a post-estimation procedure). Averaging over these results recognizes the global nature of spatial spillovers that cumulate across interrelated observations in the sample, and reflect the joint posterior distribution of the model parameters.

Average marginal effects calculated in this fashion based on our scalar summary measures of the direct and indirect impacts would produce impact estimates that could be interpreted in the following way. The direct impact would show how a rise in burglaries (burglaries on average across the sample of homes) would affect the decision of (the average) individual homeowner being burglarized to purchase a security system, where the impact is of course measured in expected probability terms. The indirect effect would represent the probabilistic impact of these increased burglaries on (the average) neighboring homeowners' purchase decisions. We could of course carry out a partitioning of these impacts to determine the spatial extent of the effects, that is, the rate at which the impacts decay as we move to more distant neighbors. We further note that if interest centered on a sub-sample representing a particular neighborhood, these individual observations could be analyzed in the same fashion.

In terms of the *from an observation* and *to an observation* interpretation for our scalar summary measures of the impacts, we might consider the following interpretation. From a seller of security systems perspective, the model estimates would be useful in determining *to an observation* impacts. These measure how changes in the mean burglaries would influence the probability of all homeowners in the sample purchasing systems (the total impact), which would be useful for projecting sales of security systems. In terms of the *from an observation* viewpoint, a security system salesperson could use the model estimates to determine how likely individual homeowners who have been burglarized (the direct impact) and their neighbors (the indirect impact) are to purchase a system.

In addition to calculating mean summary measures for the *spatial marginal effects*, there is also a need to calculate measures of dispersion for these estimates. This could be done using the MCMC draws in expression (10.10) to construct a posterior distribution for the *spatial marginal effects* summary measures. A computationally efficient approach to doing this remains a subject for future research.

Table 10.4 shows the marginal effects estimates for both the non-spatial and SAR probit model for our year 2000 presidential election example. The table illustrates that the non-spatial model has only a single marginal effect that would be interpreted as a direct impact. This would equal the total impact since there are no indirect (spatial spillover) impacts. In contrast, the SAR probit model has an indirect or spatial spillover impact, which is added to the direct impact to produce a summary measure of the total impact associated with changes in each explanatory variable.

From the table we see some similarity between the direct effects estimates from the spatial model and the marginal effects estimates of the non-spatial model. This result is consistent with previous comparisons of spatial and non-spatial models for the case of a continuous dependent variable. However, there are some striking differences which arise from differences in the estimated parameter magnitudes $\hat{\beta}$ representing the posterior means. For example, the marginal effects and direct effects for the *Shouse* variable noted earlier are widely divergent, as are those for the *Fborn* variable.

TABLE 10.4: Probit and SAR probit marginal effects estimates

Variables	probit model $\hat{\beta}$	marginal effects	SAR probit model $\hat{\beta}$	direct impacts	indirect impacts	total impacts
Repub96	2.4367	0.4520	2.0605	0.4729	0.7807	1.2536
Oparty	0.0751	0.0289	0.0865	0.0340	0.0329	0.0668
College	−0.6477	−0.0851	−0.5730	−0.0929	−0.2167	−0.3096
Income	−1.7270	−0.0000	−1.1003	−0.0000	−0.2985	−0.2985
Shouse	−0.7386	−0.2669	0.2433	0.0997	0.0925	0.1921
Fborn	−3.6201	−1.4341	−2.0122	−0.8319	−0.7647	−1.5966
Poverty	−1.1216	−0.0136	−0.8945	−0.0337	−0.3358	−0.3695
Nhomes	2.3018	0.8862	2.0382	0.8196	0.7744	1.5940

The indirect effects are larger than the direct effects for 4 of the 8 explanatory variables, and nearly equal to the direct effects in the other 4 cases. Keep in mind that the indirect effects are cumulated over all neighboring observations, so the impact on individual neighboring counties is likely smaller than the direct effects. This example illustrates a substantial role played by spatial spillovers, so that changes in the level of an explanatory variable such as *College* graduates will exert an indirect or spillover impact (cumulated over all other counties) on the probability of voting for George Bush that is twice the size of the direct own-county effect. Because the explanatory variables have all been transformed using logs, these effects estimates have an elasticity interpretation.

The largest total effects are associated with the *Nhomes* and *Fborn* variables, having elasticities around 1.6 and -1.6 respectively. The binary variable *Repub96* representing a win by the 1996 Republican party presidential candidate is also large, but does not have the same elasticity interpretation as other explanatory variables. Votes for candidates such as Nader or Buchanan (*Oparty*) had positive direct and indirect impacts around 0.033 leading to a total effect around 0.066. This suggests that a 10 percent increase in these votes increased the probability of Bush winning the county by around 2/3 of one percent. We note that the coefficient estimate for this variable is not significantly different from zero in the conventional probit model, but has a Bayesian p-level of 0.035, suggesting a non-zero impact. To further explore the importance of other party candidates on the probability of Bush winning a county, one would need to calculate a measure of dispersion for the mean effects estimates reported in the table.

10.2 The ordered spatial probit model

One extension to the spatial probit model is an *ordered probit model*. This type of model describes a situation where we can observe more than two choice outcomes, but the alternatives must take a particular form. They must be ordered, which may arise in certain modeling situations where the alternatives exhibit a natural or logical ordering. For example, if we had survey information where participants are asked to choose from alternatives such as: Strongly Agree, Agree, Uncertain, Disagree, Strongly Disagree, then the choice set exhibits a natural ordering.

The ordered spatial probit model would generalize the basic model from Section 10.1, but still rely on the relationship between y^* and y. If the observed choice outcomes y_i can take ordered values $\{j = 1, \ldots, J\}$, where J is the number of ordered alternatives, then we posit the relationship in (10.11),

$$y_i = j, \quad \text{if} \quad \phi_{j-1} < y_i^* \le \phi_j \tag{10.11}$$

where $\phi_0 \le \phi_1 \le \ldots \le \phi_J$ are parameters to be estimated. The probit model we already examined is a special case of this model where $J = 2$ and $\phi_0 = -\infty, \phi_1 = 0$, and $\phi_2 = +\infty$. This turns out to be an identification restriction that is typically placed on the ordered probit model, where the restriction for a J alternative model takes the form: $\phi_0 = -\infty, \phi_1 = 0$, and $\phi_J = +\infty$. The other $\phi_j, j = 2, \ldots, J - 1$ are parameters to be estimated.

The parameters $\phi_j, j = 2, \ldots, J - 1$ can be added to a Bayesian MCMC estimation scheme by sampling from their conditional posterior distributions. In the non-spatial model, Koop (2003) shows that the form of the conditional posterior can be deduced for the case of a flat or uniform prior by arguing:

1. Conditional on knowing the other parameters which we denote ϕ_{-j}, we know that ϕ_j must lie in the interval $[\phi_{j-1}, \phi_{j+1}]$.

2. Conditional on both y^*, y we know which values of the latent data y^* correspond to the observed choices y, the insight of Albert and Chib (1993).

3. The conditional posterior distribution for the parameters ϕ is based on no other information from the model.

Items 1) to 3) above imply a uniform conditional posterior distribution for $\phi_j, j = 2, \ldots, J-1$ taking the form:

$$p(\phi_j | \phi_{-j}, y^*, y, \beta) \sim U(\bar{\phi}_{j-1}, \bar{\phi}_{j+1}), \quad j = 2, \ldots, J-1 \tag{10.12}$$

The bounding or cut-point values ϕ are determined by examining the maximum (and minimum) values of the latent data y_i^* over all individuals i who have chosen alternative j, that is where $y_i = j$. Since individuals' choices are independent from those of other individuals in the non-spatial model, this leads to:

$$\bar{\phi}_{j-1} = \max\{\max\{y_i^* : y_i = j\}, \phi_{j-1}\}$$
$$\bar{\phi}_{j+1} = \min\{\min\{y_i^* : y_i = j+1\}, \phi_{j+1}\}$$

For the case of our ordered spatial probit model we do not have independence between choices of individuals, so we need to consider if this same approach can be applied. In Section 10.6 we discuss a space-time dynamic ordered probit model introduced by Wang and Kockelman (2008a,b) that allows for spatially structured random effects. In their model where individual choices are dependent across both time and space, the cut-points exhibit dependence invalidating the non-spatial approach.

In our cross-sectional model, we can sample from the conditional distribution for each y_i^*, making these unconditional on other y_j^*. This allows us to use an argument that there is *conditional independence* in the sampled y_i^*, and apply the same approach to sampling the cut-points $\bar{\phi}_{j-1}$ and $\bar{\phi}_{j+1}$. This involves making a claim that for all i:

$$\max\{\max\{y_i^* | z^{(m)} : y_i = j\}, \phi_{j-1}\} = \max\{\max\{y_i^* : y_i = j\}, \phi_{j-1}\}$$
$$\min\{\min\{y_i^* | z^{(m)} : y_i = j\}, \phi_{j+1}\} = \min\{\min\{y_i^* : y_i = j\}, \phi_{j+1}\}$$

where the equality arises from using the Gibbs sampler to produce a distribution for the vector y^* using $z_i | z_{-i}$ that takes spatial dependence into account. If this argument is plausible, then we can resort to an m-step Gibbs sampler:

$$z_i | z_{-i} = \gamma_{-i} z_{-i} + h_i v_i \tag{10.13}$$
$$h_i = (\Psi_{i,i})^{-1/2}$$

where we sample $v_i \sim N(0,1)$ and use the truncation constraints:

$$(\underline{b_1} - \gamma_{-i} z_{-i})/h_i < v_i \leq (\overline{b_1} - \gamma_{-i} z_{-i})/h_i, \text{ for } y_i = 1$$
$$(\underline{b_2} - \gamma_{-i} z_{-i})/h_i < v_i \leq (\overline{b_2} - \gamma_{-i} z_{-i})/h_i, \text{ for } y_i = 2$$
$$\vdots$$
$$(\underline{b_J} - \gamma_{-i} z_{-i})/h_i < v_i \leq (\overline{b_J} - \gamma_{-i} z_{-i})/h_i, \text{ for } y_i = J$$

Where:

$$
\begin{array}{llll}
\underline{b_1} = & -\infty & \text{and } \overline{b_1} = 0 & \text{for } y_i = 1 \\
\underline{b_2} = & 0 & \text{and } \overline{b_2} = \phi_2 & \text{for } y_i = 2 \\
& \vdots & & \\
\underline{b_J} = & \phi_{J-1} & \text{and } \overline{b_J} = +\infty & \text{for } y_i = J
\end{array}
\tag{10.14}
$$

Given these assumptions, this model and associated MCMC procedure represents a relatively simple extension of the spatial probit model. After making a series of m passes through the n-observation vector z, we generate $y^* = \mu + z^{(m)}$. The vector y^* can then be used to produce draws from the conditional distributions for the remaining model parameters β, ρ, ϕ as in the binary spatial probit model.

10.3 Spatial Tobit models

These models deal with situations where a subset of the observations are believed to represent censored values, which result in a truncated distribution for the dependent variable observations. We might argue that censoring of some sample observations arose because the utility is negative for an action measured by our dependent variable observations y. For example, if y measures the number of persons in a sample of census tracts who commute to work by walking, there may be a number of zero observations. These could be argued to represent census tracts where the utility associated with walking as a mode of travel to work is negative.

This same line of argument was used to motivate the truncated multivariate normal conditional posterior distributions for the latent unobserved utilities in the case of spatial probit. Indeed the same approach as taken by Albert and Chib (1993) works here, as was pointed out by Chib (1992) for the case of non-spatial Tobit models. In terms of motivation for spatial dependence in censored observations, it seems likely that census tracts located nearby in large cities would have similar numbers of persons who walk to work. It should also be clear that large rural census tracts, or those in outlying suburbs are not likely to see commuters walking to work as the utility is probably negative. Another example used earlier is the case of regional trade flows, where we might expect to see zero flows between regions where the trade costs exceed a threshold level. Since trade costs are thought to be related to distance, zero trade flows between origin and destination regions might imply zero flows for neighbors to the origin to the same destination, and vice versa. Similar situations arise in observed and unobserved home selling prices. There may be a number of homes that do not sell because the utility associated with owning a house in a central city location plagued by crime and negative externalities from neighboring homes that are abandoned or in poor condition might be below a threshold level required to undertake the transaction costs.

The latent regression model motivation for this model when censoring occurs at zero takes the form in (10.15), where y_2 denotes a vector of non-censored observations.

$$
\begin{aligned}
y^* &= S^{-1}X\beta + S^{-1}\varepsilon \qquad &&(10.15)\\
y^* &= y_1^* \quad \text{if} \quad y^* \leq 0\\
y^* &= y_2 \quad \text{otherwise}\\
S &= I_n - \rho W
\end{aligned}
$$

For the case of Tobit, where we have a *block* of n_1 censored observations and another set of n_2 observed values, we need only produce latent y_1^* for the n_1 censored observations. We construct the mean and variance-covariance matrix for the block of n_1 censored observations *conditional on* the n_2 uncensored observations y_2. We assume the locations of all observations are known, so the $n \times n$ weight matrix W can be formed. The conditional posterior distribution for the n_1 censored observations can be expressed as a multivariate truncated normal distribution $y_1^* \sim TMVN(\mu_1^*, \Omega_{1,1}^*)$, where the mean and variance-covariance are set forth in (10.16) and (10.17).

$$\mu_1^* = E(y_1^*|y_2, X, W, \beta, \rho, \sigma_\varepsilon^2)$$
$$= \mu_1 - (\Psi_{1,1})^{-1}\Psi_{1,2}(y_2 - \mu_2) \tag{10.16}$$
$$\Omega_{1,1}^* = \text{var-cov}(y_1^*|y_2, X, W, \beta, \rho, \sigma_\varepsilon^2)$$
$$= \Omega_{1,1} + (\Psi_{1,1})^{-1}\Psi_{1,2}\Omega_{2,1} \tag{10.17}$$
$$\Omega = \sigma_\varepsilon^2[(I_n - \rho W)'(I_n - \rho W)]^{-1}$$
$$\Psi = \Omega^{-1}$$
$$\mu_1 = (I_n - \rho W)_{1,1}^{-1}X_1\beta$$
$$\mu_2 = (I_n - \rho W)_{2,2}^{-1}X_2\beta$$

We use the subscripts $1, 2$ to denote an $n_1 \times n_2$ matrix, and matrices such as $\Omega_{1,1}$ would contain n_1 rows and columns, whereas $\Omega_{2,2}$ would be of dimension $n_2 \times n_2$. The term $(I_n - \rho W)_{1,1}^{-1}$ refers to the $n_1 \times n_1$ block of the matrix inverse $(I_n - \rho W)^{-1}$, and a similar definition applies to the $n_2 \times n_2$ block used in μ_2. Note that we have a scalar noise variance parameter σ_ε^2 in the model, which appears in the expressions for Ω and Ψ.

Fortunately, calculating $\Omega = \Psi^{-1}$ is *not necessary*. This would be required on each pass through the MCMC sampler because the parameters ρ and σ_ε^2 change, making estimation of this model computationally challenging. The Geweke m-step Gibbs sampler used to produce draws from the multivariate truncated normal distribution works with the *precison matrix* $\Psi = \Omega^{-1}$. We can calculate the matrices $W + W'$ and $W'W$ prior to beginning the MCMC sampling loop and produce $\Psi = (1/\sigma_\varepsilon^2)[I_n - \rho(W + W') + \rho^2 W'W]$ using the current values of ρ and σ_ε^2 obtained during sampling. The terms needed to form the n_1-dimensional mean and variance-covariance for the the block of n_1 TMVN censored observations are:

$$\mu_1^* = \mu_1 - (\Psi_{1,1})^{-1}\Psi_{1,2}(y_2 - \mu_2)$$
$$\Psi_{11}^* = (I_{n1} - \rho(W + W')_{1,1} + \rho^2(W'W)_{1,1})/\sigma_\varepsilon^2$$

where $(W + W')_{1,1}$ and $(W'W)_{1,1}$ refer to the block of $n_1 \times n_1$ observations taken from the matrices $(W + W')$ and $(W'W)$.

Given the $n_1 \times 1$ vector μ_1^* of means and associated precision matrix Ψ_{11}^*, we use the Geweke m-step Gibbs sampling procedure to carry out a sequence of m draws for each censored observation $i = 1, \ldots, n_1$ conditional on all other *censored observations* which we label $-i$. Of course, we follow the scheme detailed in our discussion of the spatial probit model, where previously sampled censored observation values are used when sampling the ith observation during each pass through the Gibbs sampler. This can be used to build up the joint conditional multivariate posterior distribution for the $n_1 \times 1$ latent vector y_1^* which is used to replace the censored observations. Specifically, we generate $y_1^* = \mu_1^* + z^{(m)}$, where m denotes the number of passes made and the vector z is for censored observations. These are sampled from the complete sequence of univariate (conditional) distributions, $z_i|z_{-i}$.

This procedure represents a Gibbs sampling scheme that takes into account dependence between observed and unobserved observations when calculating μ_1^* and $\Omega_{1,1}^*$, and then uses the m-step Gibbs sampler to build up the joint distribution for the $n_1 \times 1$ vector z of censored observations. This latter procedure takes into account spatial dependence between the censored observations.

These latent parameters are then used to produce a full-sample of observations $y^* = \left(y_1^{*\prime} \ y_2^\prime \right)^\prime$, some of which are observed values and others represent sampled unobserved latent variables. The full-sample vector y^* is then used when sampling from the conditional posterior distributions for the remaining model parameters $\beta, \rho, \sigma_\varepsilon^2$.

For the standard non-spatial tobit model with censoring at zero and normally distributed disturbances the marginal effects for the censored regression model take the form, Greene (2000).

$$\partial E[y_i|x_r]/\partial x_r = \beta \Phi \left(x_r' \beta / \sigma \right) \tag{10.18}$$

where $\Phi(\cdot)$ represents the normal CDF function and the subscript r references a variable. Intuitively, this expression indicates that the maximum likelihood Tobit coefficient estimates are adjusted versions of least-squares estimates, where the adjustment involves the proportion of the sample that is censored.

We have already explored expressions like $\partial E[y|x_r]/\partial x_r'$ for our SAR model in the context of interpreting the parameter estimates, as well as calculating SAR probit marginal effects. A similar approach can be used here where the missing or censored observations are replaced with the posterior mean of the draws for the latent values y_1^*.

10.3.1 An example of the spatial Tobit model

A data-generated experiment from Koop (2003), which allows us to control the degree of sample censoring was adapted to generate a sample of 1,000 observations. We draw independent observations x_i from a uniform distribution, $U(a, 1)$ and a disturbance term $\varepsilon_i \sim N(0, 0.5)$. These are used to construct:

$$y^* = (I_n - \rho W)^{-1} X\beta + (I_n - \rho W)^{-1} \varepsilon \tag{10.19}$$

A value of $\rho = 0.7$ was used in conjunction with a spatial weight matrix W generated using random locational coordinates and six nearest neighbors. The degree of censoring that occurs can be controlled using different settings for the parameter a. Negative generated values from the vector y^* are set to zero to reflect sample truncation at zero. We report results for an experiment where 51.3 percent of the 1,000 observation sample was censored.

These are shown in Table 10.5 where we also report Bayesian MCMC SAR model estimates based on the true non-censored values. Ideally, we would like to produce estimates close to those based on the uncensored sample containing the true underlying utilities, which are of course unknown in applied settings. The value of $m = 1$ was used in the Gibbs sampler for the censored observations when constructing the parameters $z_i|z_{-i}$, and values of the vector z from previous passes through the MCMC sampler were used. The results are based on 1,000 retained draws from a sample of 1,200.

TABLE 10.5: SAR and SAR Tobit estimates

Variables		SAR model y_1 mean $\hat{\beta}$	std dev.	SAR Tobit model mean $\hat{\beta}$	std dev.
Constant	($\alpha = 0$)	0.0123	0.0222	0.0280	0.0234
Slope	($\beta = 2$)	1.9708	0.0419	1.9521	0.0619
Wy	($\rho = 0.7$)	0.7089	0.0157	0.7030	0.0181
σ_ε^2	($= 0.5$)	0.4972		0.5072	
R^2		0.7334			

From the table, we see estimates close to the true parameter values and those based on the uncensored sample data. Of course, the standard deviations for the Tobit estimates will be larger than those based on the full sample of data. This reflects additional parameter uncertainty arising from the censoring process.

Experiments indicated that the model worked well in situations where censoring involved up to seventy percent of the sample data. Of course, success requires a model that is good enough to produce accurate imputations of the censored observations. The signal-to-noise ratio used in our experiments resulted in R^2 estimates between 0.7 and 0.8, representing a level of fit consistent with applied practice. An important consideration when contemplating use of the spatial Tobit model is whether the underlying sample data can be plausibly assumed to represent a censored (multivariate) normal distribution. In some cases where we observe excessive censoring, this is an indication that a zero-inflated Poisson process may be more consistent with the underlying data generating process (Agarwal, Gelfand and Citron-Pousty, 2002; Rathbun and Fei, 2006).

The SAR Tobit model may be useful in modeling origin-destination (OD) flows. One of the problems encountered with OD flows is that a number of elements in the flow matrix are often zero. As noted in our Chapter 8 discussion of these models, we can view the zero flows as indicative of negative utility associated with say commodity or migration flows between particular origin-destination pairs. Since positive utility is required to produce flows, we have an observed sample truncation situation.

As an example a sample of 60 origin-destination commuting flows from districts in Toulouse, France were used. Of the 3,600 OD flows 15 percent represented zero values. A series of explanatory variables representing destination-specific, origin-specific and intra-regional variables were used to formulate a model (see Section 8.3). These are labeled in Table 10.6 with prefixes D_- , O_- and I_- respectively. Explanatory variables consisted of employment and housing characteristics for each of the 60 districts. Employment characteristics were: the number of workers employed and unemployed, independent (non-salaried) workers, and the number of employers in each district. Housing characteristics consisted of owner-occupied versus tenants in private and public rental units, and holiday housing. Both the dependent variable number of commuting flows as well as the explanatory variables were transformed using logs.

Posterior means and standard deviations based on a SAR model that ignores the 15 percent sample censoring are presented alongside the SAR Tobit estimates. A value of $m = 1$ was used for the Gibbs sampler and a series of 1,200 draws were produced with the first 200 excluded for burn-in. Of course, in applied practice one might rely on a small number of draws such as this during exploratory analysis. However, when reporting final results for publication, a larger sample of MCMC draws as well as a larger number of excluded burn-in draws should be used. In addition, diagnostics for convergence of the MCMC sampler should be examined.

There is a clear pattern of Tobit estimates being larger (in absolute value terms) than the non-Tobit estimates, which suggests systematic downward bias in estimates that ignore sample censoring. Given the double-log transformation, we can compare the relative magnitudes of the coefficients, which suggests the most important influence was distance. For the other explanatory variables, the number of employed workers within a district (I_-employed workers) exhibited the largest coefficient. The number of employed workers at the origin exhibited a negative influence on interregional commuting flows whereas those at the destination had a positive influence. It may seem somewhat surprising that the number of unemployed workers at both the origin and destination exhibit a positive influence on interregional commuting flows. However, the sum of (logged) employed and unemployed workers reflects the labor force at the origin and destination regions, so this may be a size effect. The sum of the two coefficients for both origin and destination employed and unemployed workers is positive, as is the intraregional coefficient on employed workers. Of course, to fully analyze the posterior means we would need to carry out the direct, indirect and total effects scalar summary calculations to determine the magnitude of spatial spillovers. This can be accomplished using the same summary measures discussed for the non-censored SAR model, where we use our matrix expressions to replace the partial derivative $\partial E[y_i | x_r]/\partial x_r$ in (10.18).

TABLE 10.6: OD SAR and OD SAR Tobit estimates

| | SAR model | | SAR Tobit model | |
Variables	$\hat{\beta}$	std dev.	$\hat{\beta}$	std dev.
Constant	−5.2154	0.4044	−6.1607	0.4705
D_employed workers	0.6928	0.0812	0.7635	0.0977
D_unemployed workers	0.9521	0.2206	1.1150	0.2623
D_independent workers	0.0586	0.0516	0.0946	0.0579
D_employers	−0.0675	0.0373	−0.0817	0.0430
D_holiday housing units	−0.0192	0.0172	−0.0275	0.0204
D_owner occupier	0.0121	0.0236	0.0261	0.0288
D_tenant private housing	−0.0484	0.0384	−0.0438	0.0474
D_tenant social housing	−0.0416	0.0160	−0.0311	0.0189
D_area	0.1451	0.0382	0.1368	0.0484
O_employed workers	−0.3210	0.0824	−0.3790	0.0948
O_unemployed workers	1.2306	0.2335	1.5633	0.2556
O_independent workers	−0.2392	0.0528	−0.2162	0.0576
O_employers	−0.0786	0.0362	−0.1053	0.0407
O_holiday housing units	0.1325	0.0182	0.1469	0.0180
O_owner occupier	−0.0772	0.0238	−0.0854	0.0259
O_tenant private housing	0.6049	0.0415	0.6873	0.0423
O_tenant social housing	−0.0665	0.0162	−0.0859	0.0194
O_area	0.8222	0.0418	0.9543	0.0448
L_employed workers	1.4038	0.3215	1.6420	0.3758
L_unemployed workers	0.6099	1.7787	1.0200	2.0277
L_independent workers	−0.0833	0.3789	−0.1261	0.4082
L_employers	0.0250	0.2903	0.0149	0.3313
L_holiday housing units	−0.0470	0.1262	−0.0680	0.1467
L_owner occupier	−0.1996	0.1659	−0.2281	0.1893
L_tenant private housing	−0.1393	0.2839	−0.1584	0.3109
L_tenant social housing	−0.0907	0.1020	−0.1386	0.1121
L_area	0.1625	0.3006	0.1900	0.3308
Distance	−39.5906	6.6526	−47.8385	8.0114
ρ	0.7646	0.0215	0.7328	0.0215
σ_ε^2	0.7007		0.9125	

10.4 The multinomial spatial probit model

This model is a modification of the basic spatial probit model where y_i can take values $\{j = 0, 1, \ldots, J\}$, representing $J + 1$ alternative choices. The same random utility framework can be used, where we consider utility of observation/region i for choice j, (U_{ji}) relative to some other base choice alternative, (U_{0i}). If we use choice alternative 0 as the base, then our latent utility differences take the form:

$$y_{ji}^* = U_{ji} - U_{0i}, \ j = 1, \ldots, J \tag{10.20}$$

We can treat this as a system of J seemingly unrelated (SURE) SAR regression equations (Wang and Kockelman, 2007). This involves stacking J observations for each set of observed choices as shown in (10.21), where X_{ji} represent $1 \times k$ vectors of explanatory variables associated with each choice. For simplicity we assume k is the same for all choices.

$$\tilde{y}_i = \begin{pmatrix} y_{1i} \\ y_{2i} \\ \vdots \\ y_{Ji} \end{pmatrix}, \quad \tilde{y} = \begin{pmatrix} \tilde{y}_1 \\ \tilde{y}_2 \\ \vdots \\ \tilde{y}_n \end{pmatrix} \tag{10.21}$$

$$X_i = \begin{pmatrix} X_{1i} & 0_k & \cdots & 0_k \\ 0_k & X_{2i} & & \vdots \\ \vdots & & \ddots & 0_k \\ 0_k & \cdots & 0_k & X_{Ji} \end{pmatrix}, \quad \tilde{X} = \begin{pmatrix} X_1 \\ X_2 \\ \vdots \\ X_n \end{pmatrix}$$

Values of $y_{j,i}$ are such that: $y_{j,i} = \delta(\max(\tilde{y}_i^*))$ or 0 if $\max(\tilde{y}_i^*) < 0$.[3] Given the arrangement of the dependent variable vector, we must re-arrange the conventional $n \times n$ spatial weight matrix to produce spatial lags of the dependent variable $\tilde{y}$. This can be done by repeating each row from the $n \times n$ conventional weight matrix J times, which can be expressed as $W \otimes I_J$.

As an example, for the case of three choices where $J = 2$ we would have:

$$\tilde{W} = W \otimes I_2 = \begin{pmatrix} \tilde{W}_1 \\ \vdots \\ \tilde{W}_n \end{pmatrix}$$

$$\tilde{W}_i = \begin{pmatrix} W_{i,1} & 0 & W_{i,2} & 0 & \cdots W_{i,n} & 0 \\ 0 & W_{i,1} & 0 & W_{i,2} \cdots & 0 & W_{i,n} \end{pmatrix} \tag{10.22}$$

---

[3] We note that the indicator function $\delta(\max(\tilde{y}_i^*))$ returns $\{0, 1, \ldots, J\}$.

The matrix product $\tilde{W}\tilde{y}$ results in a spatial lag representing an average of neighboring region observations for each choice. For example, the spatial lag for choice 1, observation/region 1 will consist of a weighted average of neighboring regions choice 1, and for choice 2 we also have an average of neighboring regions choice 2. We note the spatial lag formed in this fashion does not allow for choice 1 in region 1 to depend directly on observed choice 2 outcomes from neighboring regions. That is, the model for utility associated with the vector of say choices 2, made in all regions 1 to n are specified to depend only on choices 2 made in neighboring regions.

We accommodate cross-choice covariance in the conventional multinomial probit fashion by allowing for cross-equation/choice covariance in the disturbances of the J-equation system, where we estimate a single $J \times J$ variance-covariance matrix using all observations. This leads to a SURE SAR model shown in (10.23), where we have assumed for simplicity that the same scalar dependence parameter ρ applies to all J choices.

$$\tilde{y} = \rho\tilde{W}\tilde{y} + \tilde{X}\beta + \tilde{\varepsilon} \tag{10.23}$$

$$\text{var-cov}(\tilde{\varepsilon}_i) = \Sigma = \begin{pmatrix} \sigma_{1,1} & \sigma_{1,2} & \cdots & \sigma_{1,J} \\ \sigma_{2,1} & \sigma_{2,2} & & \\ \vdots & & \ddots & \\ \sigma_{J,1} & & & \sigma_{J,J} \end{pmatrix} \tag{10.24}$$

In (10.23), we allow for different $\beta_1, \ldots, \beta_J$ associated with each of the J equations being modeled.

10.4.1   The MCMC sampler for the SAR MNP model

In the univariate SAR probit model, we implemented an MCMC sampling scheme with *data augmentation*, where y^* were treated as parameters to be estimated. The sampler drew sequentially from the conditional posteriors:

$p(\beta|\rho, y^*),$

$p(\rho|\beta, y^*),$

$p(y^*|\beta, \rho, y)$

to produce estimates for inference. For this multinomial setup we require conditional posteriors taking the form:

$p(\beta|\rho, \Sigma, \tilde{y}^*),$

$p(\rho|\beta, \Sigma, \tilde{y}^*),$

$p(\Sigma|\beta, \rho, \tilde{y}^*),$

$$p(\tilde{y}^*|\beta, \rho, \Sigma, y)$$

The nature of these conditional posteriors as well as methods for sampling from each are discussed in the following sections.

10.4.2   Sampling for β and ρ

Samples from the conditional posterior distributions $p(\beta|\rho, \Sigma, \tilde{y}^*, \tilde{X})$ take the same form as would be used for a seemingly unrelated (SURE) SAR model, containing J equations with a common spatial autoregressive structure involving the same parameter ρ, and spatial weight matrix $\tilde{W}$. This can easily be seen by considering $\tilde{y}^*$ as a set of continuous dependent variables, and noting that we are conditioning on these latent values treated as parameters of the model. If an independent Normal-Wishart prior is used for the parameters β and Σ (and a uniform prior for ρ), the conditional posterior for the parameters β take the form of a normal distribution (Koop, 2003).

If we rely on a non-informative prior for the parameters β, the conditional posterior for the parameters $\beta = \left(\beta_1 \ \beta_2, \ldots, \beta_J \right)'$ take the form in (10.25).

$$p(\beta|\rho, \Sigma, \tilde{y}^*) \sim N(c^*, T^*)$$
$$c^* = T^*(\tilde{X}'(I_{nJ} - \rho\tilde{W})\tilde{y}^*)$$
$$T^* = (\tilde{X}'(I_n \otimes \Sigma^{-1})\tilde{X})^{-1} \tag{10.25}$$

For the parameter ρ, we can sample $p(\rho|\beta, y^*)$, using either the M-H or integration and draw by inversion approach set forth in Chapter 5. This requires evaluating the expression in (10.26)

$$p(\rho|\beta, \Sigma, \tilde{y}^*) \propto |I_n - \rho W|^J \ |\Sigma|^{-n/2} \tag{10.26}$$
$$\cdot \exp\left(-\frac{1}{2}[H\tilde{y}^* - \tilde{X}\beta]'H'(I_n \otimes \Sigma^{-1})H[H\tilde{y}^* - \tilde{X}\beta] \right)$$
$$H = I_{nJ} - \rho\tilde{W}$$

10.4.3   Sampling for Σ

When sampling the conditional distribution for the covariance matrix Σ, the conventional MNP model has an additional identification problem. As noted in the case of the univariate probit model, a scale shift will not change the observed choices. Typically, the MNP model is identified by setting the first diagonal element of the covariance matrix (σ_{11}) to unity.

From a strictly Bayesian viewpoint we can work with what has been labeled a non-identified model by McCulloch, Polson and Rossi (2000). This model specifies a prior for the full set of parameters (β, Σ, ρ), computes the full posterior over β, Σ, ρ and reports the marginal posterior distribution of the identified parameters $(\beta/\sqrt{\sigma_{11}}, \Sigma/\sqrt{\sigma_{11}}, \rho/\sqrt{\sigma_{11}})$. A drawback to this approach is that we cannot rely on improper priors for β, Σ, ρ (McCulloch, Polson and Rossi, 2000).

McCulloch, Polson and Rossi (2000) present an alternative approach that produces an identified model. A detailed discussion of this approach is set forth in Koop (2003), which we follow here. A prior is placed on Σ such that the first diagonal element takes a value of unity. To use this approach, Σ must be re-parameterized. This is accomplished by working with the joint distribution of $g_i = \varepsilon_{1i}$ and $G_{-i} = (\varepsilon_{2i}, \ldots, \varepsilon_{Ji})$, where the $(J \times 1)$ vector $\varepsilon_i = \begin{pmatrix} g_i & G_{-i} \end{pmatrix}'$. This leads to a partitioning of Σ as shown in (10.27).

$$\Sigma = \begin{pmatrix} \sigma_{11} & \eta' \\ \eta & \Sigma_G \end{pmatrix} \tag{10.27}$$

They exploit the fact that any joint distribution can be expressed as the product of a marginal and conditional distribution, in this case: $p(g_i, G_{-i}) = p(g_i)p(G_{-i}|g_i)$. This is of course similar to the approach taken by Geweke (1991) for sampling from the multivariate truncated normal distribution by exploiting a sequence of univariate distributions based on conditionals. The (multivariate) normal distribution for the vector ε_i, leads to a univariate normal distribution for $g_i \sim N(0, \sigma_{11})$, and a $(J-1)$-variate normal distribution for $G_{-i}|g_i \sim N(\eta g_i/\sigma_{11}, \Phi)$. The $(J-1 \times J-1)$ matrix $\Phi = \Sigma_G - \eta\eta'/\sigma_{11}$.

This allows imposing the restriction $\sigma_{11} = 1$, while assigning priors for the remaining unrestricted parameters of the covariance matrix, η and Σ_G. A combination of a normal prior for the $(1 \times J-1)$ vector η and Wishart prior for the $(J-1 \times J-1)$ covariance matrix Φ^{-1} is convenient, since these priors lead to conditional posterior distributions taking known forms that are amenable to Gibbs sampling. These conditional distributions take the form of a normal and Wishart distribution (see Koop (2003) for the detailed expressions).

Nobile (2000) discusses how to generate directly from Wishart and inverted Wishart random matrices conditional on one of the diagonal elements. This provides an alternative way of imposing the normalization constraint in a Bayesian MNP model, where we simply need to assign a Wishart prior to the $(J \times J)$ covariance matrix Σ. Use of the normal priors for the parameters β and a Wishart prior for the covariance matrix Σ then leads to a (multivariate) normal distribution for β conditional on the other model parameters, and a Wishart prior for Σ conditional on the other parameters. Draws from the Wishart prior subject to the restriction $\sigma_{11} = 1$ can be obtained directly, allowing us to impose identification.

The conventional algorithm based on the Bartlett decomposition for producing draws for $E(\Sigma) = \nu V^{-1}$ from the Wishart dimension J distribution $W_J(\nu, V)$ takes the form:

1. Let $V^{-1} = LL'$.

2. where $L = \ell_{ij}, i > j$, a lower triangular matrix.

3. Construct a lower triangular matrix A with a_{ii} equal to the square root of $\chi^2(\nu + 1 - i)$ deviates, $i = 1, \ldots, J$.

4. Set a_{ij} equal to $N(0, 1)$ deviates, for $i > j$.

5. Return $\Sigma = LAA'L'$.

Nobile (2000) proposes modifying this algorithm to allow setting $\sigma_{11} = 1$.

1. Construct a lower triangular matrix A with $a_{11} = 1/\ell_{11}$.

2. Set a_{ii} equal to the square root of $\chi^2(\nu + 1 - i)$ deviates, $i = 2, \ldots, J$.

3. Set a_{ij} equal to $N(0, 1)$ deviates, for $i > j$.

4. Return $\Sigma = LAA'L'$.

Either approach for sampling Σ should work, but we found the approach of Nobile (2000) to have better MCMC sampling properties.

10.4.4   Sampling for $\tilde{y}^*$

The final conditional distribution from which we need to sample is that for $\tilde{y}^*$. We accomplish this using a modification of the Geweke (1991) m-step Gibbs sampling scheme that was applied to the univariate SAR probit model. For the MNP SAR model we have a mean vector and variance-covariance matrix shown in (10.28).

$$
\begin{aligned}
\tilde{y}^* &\sim TMVN\{H^{-1}\tilde{X}\beta, [H'(I_n \otimes \Sigma^{-1})H]^{-1}\} \\
\tilde{y}^* &\sim TMVN(\mu, \Omega) \\
\mu &= H^{-1}\tilde{X}\beta \\
\Omega &= [H'(I_n \otimes \Sigma^{-1})H]^{-1} \\
H &= I_{nJ} - \rho\tilde{W}
\end{aligned}
\tag{10.28}
$$

We can use the method of Geweke (1991) to produce an m-step Gibbs sampler to produce draws from this nJ-variate truncated normal distribution. As before, the method of Geweke (1991) works with a *precision* matrix, which in this case takes the form of an $nJ \times nJ$ matrix: $\Psi = D^{-1'}H'(I_n \otimes \Sigma^{-1})HD^{-1}$, with details regarding the $nJ \times nJ$ block diagonal matrix D provided shortly.

Samples for $v \sim N(0, D^{-1'}H'(I_n \otimes \Sigma^{-1})HD^{-1})$ subject to linear restrictions: $a < D\tilde{y}^* < b$, where D is an $nJ \times nJ$ matrix that restricts $y^*_{j,i}$ to be the largest component of $\tilde{y}^*_i$ if $y_{j,i} = j$ or assures that each component of $\tilde{y}^*_i$ is negative if $\max(y_{j,i}) = 0$.

As in the case of the SAR probit model, the samples v are used to produce the series of $z_i | z_{-i}$ needed to build up the joint posterior for z. This is equivalent to constructing samples from the nJ-variate normal distribution $z \sim N(0, D\Omega D')$ subject to the linear restrictions: $\underline{b} \leq z \leq \overline{b}$. Where $\underline{b} = a - D\mu, \overline{b} = b - D\mu$. The sampled z are used to obtain: $y^* = \mu + D^{-1}z$.

The restrictions applied to samples from $v_i \sim N(0,1)$ are shown in (10.29), where $y_{j,i}$ represents the jth element from $\tilde{y}_i$. These are used to produce an nJ-vector of values for all observations and choices, where previously sampled values $z_1, z_2, \ldots, z_{i-1}, z_{i+1}, \ldots, z_n$ are used during sampling of element $z_i, i = 1, \ldots, nJ$. The same definition for Ψ_{-i} as the ith row of Ψ excluding the ith element applies here as in the case of probit, but Ψ represents the larger $(nJ \times nJ)$ precision matrix.

$$(\underline{b_i} - \gamma_{-i}z_{-i})/r_i < v_i < (\overline{b_i} - \gamma_{-i}z_{-i})/r_i \qquad (10.29)$$
$$\gamma_{-i} = -\Psi_{-i}/\Psi_{i,i}$$
$$r_i = (\Psi_{i,i})^{-1/2}$$

$$\underline{b_i} = -\infty \qquad \text{and } \overline{b_i} = -D^{(0)}\mu_i \text{ for } y_{j,i} = 0$$
$$\underline{b_i} = -D^{(0)}\mu_i \text{ and } \overline{b_i} = -D^{(1)}\mu_i \text{ for } y_{j,i} = 1$$
$$\vdots$$
$$\underline{b_i} = -D^{(J)}\mu_i \text{ and } \overline{b_i} = +\infty \qquad \text{for } y_{j,i} = J$$

An important difference between this model and the SAR probit is introduction of the matrices $D^{(j)}, j = 0, \ldots, J$. The $nJ \times nJ$ matrix D contains a series of $J \times J$ matrices on the diagonal and zeros elsewhere. For each observation i we determine an appropriate $J \times J$ matrix D_i that is placed on the ith diagonal of the matrix D. For the case where we have three choices so $J = 2$, the matrices D_i take the form:[4]

$$D_i = D^{(0)} = \begin{pmatrix} +1 & 0 \\ 0 & +1 \end{pmatrix}, \text{ if } y_{j,i} = 0$$
$$= D^{(1)} = \begin{pmatrix} +1 & 0 \\ +1 & -1 \end{pmatrix}, \text{ if } y_{j,i} = 1$$
$$= D^{(2)} = \begin{pmatrix} -1 & +1 \\ 0 & +1 \end{pmatrix}, \text{ if } y_{j,i} = 2$$

These take a similar form for models involving more choices, for example when there are four choices and $J = 3$ we have:

---

[4] Recall that we work with $J + 1$ choices, so $J + 1 = 3$ choices leads to $J = 2$.

$$D_i = D^{(0)} = \begin{pmatrix} +1 & 0 & 0 \\ 0 & +1 & 0 \\ 0 & 0 & +1 \end{pmatrix}, \text{ if } y_{j,i} = 0$$

$$= D^{(1)} = \begin{pmatrix} +1 & 0 & 0 \\ +1 & -1 & 0 \\ +1 & 0 & -1 \end{pmatrix}, \text{ if } y_{j,i} = 1$$

$$= D^{(2)} = \begin{pmatrix} -1 & +1 & 0 \\ 0 & +1 & 0 \\ 0 & +1 & -1 \end{pmatrix}, \text{ if } y_{j,i} = 2$$

$$= D^{(3)} = \begin{pmatrix} -1 & 0 & +1 \\ 0 & -1 & +1 \\ 0 & 0 & +1 \end{pmatrix}, \text{ if } y_{j,i} = 3$$

While this appears to be a relatively straightforward extension from the SAR probit model, in practice this approach can be slow. There is room for a number of computational improvements.

10.5 An applied illustration of spatial MNP

A data generated experiment was conducted using three choices and $J = 2$, with $n = 400$. A set of continuous y-values representing utilities were generated using:

$$y_1^* = (I_n - \rho W)^{-1} X_1 \beta_1 + (I_n - \rho W)^{-1} \varepsilon_1 \qquad (10.30)$$
$$y_2^* = (I_n - \rho W)^{-1} X_2 \beta_2 + (I_n - \rho W)^{-1} \varepsilon_2 \qquad (10.31)$$
$$(\varepsilon_{1i}, \varepsilon_{2i})' \sim N\left[\begin{pmatrix} 0 \\ 0 \end{pmatrix}, \begin{pmatrix} 1.0 & -0.5 \\ -0.5 & 1.0 \end{pmatrix}\right]$$
$$X_{1i} \sim N(0, 2), \quad i = 1, \ldots, n$$
$$X_{2i} \sim N(0, 2), \quad i = 1, \ldots, n$$

A value of $\rho = 0.7$ was used, and a spatial weight matrix constructed using random vectors of locational coordinates, based on six nearest neighbors. The continuous dependent variables y_1^*, y_2^* were converted to values of 0, 1, and 2 based on:

$$y_{j,i} = 0, \qquad \text{if } \max(\tilde{y}_i^*) < 0$$
$$y_{j,i} = \delta[\max(\tilde{y}_i^*)], \text{ if } \max(\tilde{y}_i^*) \geq 0$$

where $\tilde{y}_i^* = \left(y_{1i}^* \ y_{2i}^* \right)'$ is based on the ith observation from equations (10.30) and (10.31) above.

Table 10.7 shows estimation results based on 1,200 draws with 200 omitted for burn-in. A value of $m = 1$ was used for the Gibbs sampler. The results in the table were based on the Nobile (2000) procedure for sampling the covariance matrix Σ. The table shows estimates based on the continuous observations y_1^*, y_2^* used to produce the experimental choices. Spatial MNP estimates close to these would be an indication of success. From the table we see estimates that are within one standard deviation of the true values. As we would expect, having discrete choice values for the dependent variable in the model greatly increases uncertainty associated with the parameter estimates for β, reflected in the standard deviations that are around four times as large as those for the model based on the continuous dependent variables.

TABLE 10.7: SAR and SAR MNP estimates

Variables	SAR model y_1 $\hat{\beta}$	std dev.	SAR model y_2 $\hat{\beta}$	std dev.	SAR MNP model $\hat{\beta}$	std dev.
$X_{11}, (\beta = 1.0)$	0.9733	0.0246			1.0094	0.1554
$X_{12}, (\beta = 0.5)$	0.4893	0.0245			0.5160	0.0844
$X_{21}, (\beta = 0.5)$			0.5042	0.0262	0.5546	0.0806
$X_{22}, (\beta = 1.0)$			0.9315	0.0254	1.1084	0.1384
$W y_1, (\rho_1 = 0.7)$	0.6741	0.0290				
$W y_2, (\rho_2 = 0.7)$			0.7228	0.0201		
$\tilde{W} y, (\rho = 0.7)$					0.7249	0.0269
$\sigma_{11}^2 = 1$	1.0360				1.0000	
$\sigma_{22}^2 = 1$			1.0690		1.1479	0.4004
$\sigma_{12}^2 = -0.5$					-0.5172	0.2747
R^2	0.7975		0.8518			

Summarizing, there is a great deal of work to be done on MNP models involving spatial lags. The approach set forth here represents a rudimentary approach that does not attempt to exploit special structure of the variance-covariance matrix $\Omega = [(H'(I_n \otimes \Sigma^{-1})H]^{-1}$. It may be possible to exploit the Kronecker product nature of $H = I_J \otimes (I_n - \rho W)$ in conjunction with that of $(I_n \otimes \Sigma^{-1})$ to produce a more computationally efficient approach to estimation. Wang and Kockelman (2007) take this approach when implementing a spatiotemporal seemingly unrelated regression model.

The approach set forth here samples each observation (i) conditional *on all others* $(-i)$, which may be unnecessary. Intuitively, choices observed in region i should only depend on those from nearby regions, which we might define as $i \in \aleph$. This raises the prospect of adopting Geweke's procedure to sample the distribution of each region's z_i conditional only on a limited set of neighboring regions $z_i | z_{i \in \aleph}$.

The matrix H could be generalized to allow for different spatial dependence parameters associated with each choice. This variant of the model is implemented in Autant-Bernard, LeSage and Parent (2008), where:

$$H = \begin{pmatrix} I_n - \rho_1 W & & \\ & \ddots & \\ & & I_n - \rho_J W \end{pmatrix} \tag{10.32}$$

Another avenue for exploration is the relative importance of spatial dependence versus dependence across choices. It may be that cross-choice covariance is unimportant relative to spatial dependence, allowing the model to be simplified by restricting $\Sigma = I_J$. The MNP model is often criticized as over-parameterized, so prior information such as this that reduces the number of parameters to be estimated would be helpful.

10.5.1   Effects estimates for the spatial MNP model

Effects estimates for non-spatial MNP models estimated using maximum likelihood methods are typically calculated by post-estimation simulation of the model with all but a single explanatory variable fixed. The impact of changing the single explanatory variable on the choice probabilities is used to assess the marginal impact of changes in the explanatory variables of interest.

Our Bayesian MCMC estimation procedure produces a set of continuous latent dependent variable values that represent a proxy for unobserved utility associated with the J choices. The posterior mean of these latent variable values can of course be used to produce J posterior choice probabilities, with the excluded $J+1$ choice also recovered.

Most of the same insights regarding interpretation of marginal effects for the SAR probit model apply to the SAR MNP model as well. For example, we can extend our example of burglaries on the decision to purchase a security system to include purchasing a dog, or installing security lights. The SAR MNP model implies that a change in burglaries (an explanatory variable in the model) of neighboring homes j would have an *effect* on the probability that homeowner i purchases a security system, a dog, or security lighting. The effect would depend on spatial proximity of homeowner i to j, captured by the spatial weight matrix W as well as the strength of spatial dependence measured by the parameter ρ. A burglary at home j would have both a direct effect on the probability that homeowner j purchases a security system, a dog or security lighting, as well as an indirect or spatial spillover effect on neighbors' choices regarding these three alternatives. The total effect is the sum of these two effects for each choice alternative.

There are also some important qualifications that may apply to interpretation based on the particular model specification. For example, use of the block diagonal dependence structure: $I_{nJ} - \rho \tilde{W}$, or that shown in (10.32), restricts us to a situation where the utility of choice j by an individual located in region i is not directly influenced by the utility of choice $k \neq j$ by an individual located in a neighboring region $h \neq i$. Cross-choice influence works through dependence captured by the covariance structure Σ, rather than through spatial lags that embody cross-choice spatial dependence. This simplifies calculation of the marginal effects estimates, since changes in $x_{kh,r}$ will not impact y_{ji}, when $k \neq j$, and when $h \neq i$.

The model could of course be extended along these lines by directly introducing spatial lags for utility associated with the J choices being modeled in each equation of the model. For example in the case of three choices where $J = 2$, we might use:

$$y_1 = \rho_{11} W y_1 + \rho_{12} W y_2 + X_1 \beta_1 + \varepsilon_1$$
$$y_2 = \rho_{21} W y_1 + \rho_{22} W y_2 + X_2 \beta_2 + \varepsilon_2$$

In this model, y_1 and y_2 represent $n \times 1$ vectors containing indicators for observed choices 1 and 2 across the n regions. The $n \times k$ matrices X_1, X_2 are also arranged according to regions.

This type of model might be appropriate in situations where we are modeling the utility of a local government choosing to impose a payroll income tax (y_1) versus a sales tax (y_2). The utility of imposing a payroll tax might depend directly on whether neighboring governments have chosen to implement a payroll tax ($\rho_{11} W y_1$) or sales tax ($\rho_{12} W y_2$). Similarly, the utility associated with the sales tax might depend on the existence of both payroll taxes in neighboring governments, ($\rho_{21} W y_1$) as well as sales taxes by neighbors ($\rho_{22} W y_2$). Intuitively, this might arise because firms could relocate to avoid the payroll taxes, and local residents could alter shopping behavior to avoid sales taxes. Imposing the restriction that $\Sigma = I_J$ for this model seems reasonable, since cross-choice dependence is modeled in the mean part of the model.

Calculating marginal effects for this type of model would require altering our basic expression used to produce the scalar summary measures. If we define:

$$\tilde{H} = \begin{pmatrix} I_n - \rho_{11} W & -\rho_{12} W \\ -\rho_{21} W & I_n - \rho_{22} W \end{pmatrix}$$
$$\tilde{X} = \begin{pmatrix} X_1 & 0_k \\ 0_k & X_2 \end{pmatrix}$$
$$\tilde{y} = \begin{pmatrix} y_1 \\ y_2 \end{pmatrix}$$

then, assuming $\Sigma = I_J$, we have a variance-covariance matrix: $\Omega = (\tilde{H}'\tilde{H})^{-1}$, with the associated precision matrix that would be subjected to the m-step Gibbs sampling procedure.

For this two-equation example, the SAR MNP effects are such that changes in an explanatory variable associated with say, equation $j = 2$ will impact observations $y_{1i}, i = 1, \ldots, n$ as well as $y_{2i}, i = 1, \ldots, n$. That is, $\partial y_1 / \partial x_{2r} \neq 0$ and $\partial y_2 / \partial x_{1r} \neq 0$. This of course follows from allowing for cross-choice impacts in the mean component of the model, reflected by the spatial lag structure, and the resulting: $\mu = \tilde{H}^{-1} \tilde{X} \beta$.

For the payroll versus sales tax example, we could use the model to analyze the impact of changes in an explanatory variable such as levels of payroll income in each location on the long-run steady state probabilities that local governments would rely on payroll versus sales taxes. The impact of an increase in payroll income of one local government jurisdiction on the probability of the own- and other-local governments adopting a payroll income tax versus sales tax, plus spatial spillover (indirect) effects on the probability that neighboring governments adopt both payroll and sales taxes could be analyzed using the scalar summary measures.

10.6   Spatially structured effects probit models

An alternative to SAR models for situations involving limited dependent variables is a model that introduces a spatially structured random effects vector (Smith and LeSage, 2004). This type of model was already introduced in the context of interregional trade flows in Chapter 8.

This hierarchical Bayesian model is shown in (10.33), where U_{ik}, indexes utility in regions $i = 1, \ldots, m$ for individuals $k = 1, \ldots, n_i$ within each region. There are $N = \sum_{i=1}^{m} n_i$ observations in the model, and we use w_{ij} to denote the i, jth elements of the $m \times m$ spatial weight matrix W.

$$U_{ik} = X_{ik}\beta + \xi_{ik} \qquad (10.33)$$
$$\xi_{ik} = \theta_i + \varepsilon_{ik}$$
$$\theta_i = \rho \sum_{j=1}^{m} w_{ij}\theta_j + u_i$$

This model treats the unobserved component ξ_{ik} as consisting of a region-specific effect θ_i as well as an individual effect ε_{ik}. The regional effect parameter θ_i captures unobserved common features for observations located within region i. The regional effects parameters are modeled using a SAR process: $\theta_i = \rho \sum_{j=1}^{m} w_{ij}\theta_j + u_i$ which imposes a restriction that individuals located within region i are likely to be similar to those from neighboring regions. The individualistic effects parameters are then assumed to be conditionally independent given the regional effects θ_i.

This model is a variation on a fixed effects model. Using matrix notation we can express the model as in (10.34), where the $m \times 1$ vector θ represents the spatially structured effects.

$$y = X\beta + \Delta\theta + \varepsilon \tag{10.34}$$
$$\theta = \rho W\theta + u$$
$$u \sim N(0, \sigma_u^2 I_m)$$
$$\varepsilon|\theta \sim N(0, V)$$

$$V = \begin{pmatrix} v_1 I_{n_1} & & \\ & \ddots & \\ & & v_m I_{n_m} \end{pmatrix} \tag{10.35}$$

$$\Delta = \begin{pmatrix} \mathbf{1}_1 & & \\ & \ddots & \\ & & \mathbf{1}_m \end{pmatrix} \tag{10.36}$$

We accommodate heterogeneity across regions using a set of variance scalars $v_i, i = 1, \ldots, m$, and use $\mathbf{1}_i, i = 1, \ldots, m$ to denote an ($n_i \times 1$) vector of ones. The effects parameters need not be applied to all regions, for example when working with counties we might estimate state-level effects parameters. The $N \times m$ matrix Δ works to assign the same effect parameter to each of the n_i counties in state i. Specifically, Δ contains row-elements $i = 1, \ldots, m$ that equal 1 if region (county) i is located in state m and zero otherwise. This model interprets the parameters in the $m \times 1$ vector θ as latent indicators for unobservable/unmeasured state-level influences. These are restricted to follow a SAR process, so neighboring states will exhibit similar effects levels.

The model also accommodates heterogeneity across the m broader regions (e.g. states) allowing for different variance scalars v_i to be associated with each of these regions. This is accomplished using the independent identically distributed chi-squared prior discussed in Chapter 5. All observations in each state (broader region) are assigned the same variance scalar parameter.

The spatial autoregressive structure placed on the effects parameters reflects an implied prior for the vector θ conditional on ρ, σ_u^2, V shown in (10.37).

$$\pi(\theta|\rho, \sigma_u^2) \propto (\sigma_u^2)^{-m/2}|B|\exp\left(-\frac{1}{2\sigma_u^2}\theta' B' B\theta\right) \tag{10.37}$$
$$B = I_m - \rho W$$

Estimation of the spatially structured effects vector θ requires introduction of two additional parameters (ρ, σ_u^2) to the model. One of these controls the strength of spatial dependence between regions and the other controls the variance/uncertainty of the prior spatial structure. Given these two scalar parameters along with the spatial structure, the m effects parameters are completely determined. One could view the spatial connectivity matrix W as introducing additional exogenous information that augments the sample data information. In contrast, the conventional fixed effects approach introduces m additional parameters to be estimated without augmenting the sample data information.

There is also an implied prior density for ε conditional on θ, V which takes the form:

$$\pi(\varepsilon|V) \propto |V|^{-1/2}\exp\left(-\frac{1}{2}\varepsilon'V^{-1}\varepsilon\right) \tag{10.38}$$

Smith and LeSage (2004) provide details regarding Bayesian MCMC estimation of this hierarchical linear model, and the model is discussed in detail by Rossi, Allenby and McCulloch (2006). The binary dependent variables are treated as choice outcomes that reflect latent underlying utilities following Albert and Chib (1993). The conditional posterior of z_{ik} for individual k in region i takes the form of a normal distribution truncated at zero:

$$p(z_{ik}|z_{-ik}, \rho, \beta, \theta, \sigma_u^2, V, y) \sim \begin{cases} N(x_i'\beta + \theta_i, v_i) \text{ left-truncated,} & \text{if } y_i = 1 \\ N(x_i'\beta + \theta_i, v_i) \text{ right-truncated,} & \text{if } y_i = 0 \end{cases}$$

This model is considerably faster to estimate than the SAR probit model because it relies on a smaller $m \times m$ spatial component. Smith and LeSage (2004) decompose the $m \times m$ multivariate normal distribution for the effects vector θ into a sequence of univariate normal distributions which are sampled to produce the effects parameter estimates. This is of course similar to the approach of Geweke (1991) outlined here, but does not involve sampling from a truncated normal distribution, just a sequence of univariate normals conditional on other elements of θ_{-i}.

Interpreting the parameters β for this model is similar to that from an ordinary probit model, so there are no spatial spillover effects in this model. However, we can use the spatially structured effects estimates for each region to draw inferences regarding spatial variation in the model relationship. The effects parameters θ are centered on zero, so regions with negative and positive and significant effects point to latent factors at work that are not included in the explanatory variables matrix X.

An interesting extension of this model can be found in Wang and Kockelman (2008a,b), who extend this model to allow for a set of $\sum_{i=1}^{M} n_i = N$, individuals located in M regions across time periods $t = 1, \ldots, T$. Their model is dynamic, taking the form:

$$y_{ikt}^* = \lambda y_{ikt-1}^* + X_{ikt}\beta + \theta_{it} + \varepsilon_{ikt} \tag{10.39}$$

where t indexes time periods, k individuals and i regions. This model allows for temporal dependence governed by the parameter λ. Each individual makes a decision that is observed T times, so we have a balanced panel containing NT observations. The parameters θ_{it} and ε_{ikt} are assumed *iid* distributed over time conditional on controlling for the influence of the lagged dependent variable y^*_{ikt-1}. The argument is that after controlling for time dependence in decisions, $\theta_{it} = \theta_i$ and $\varepsilon_{ikt} = \varepsilon_{ik}$, for all $t = 1, \ldots, T$.

The motivating example for this type of model given by Wang and Kockelman (2008b) is an application to land development decisions. They argue that land usage patterns depend strongly on pre-existing as well as existing conditions, *and* owner/developer expectations of future conditions (such as local and regional congestion, population, and school access). Future expectations are approximated using contemporaneous measures of access and land use intensity, but Wang and Kockelman (2008b) argue that spatial correlation in unobserved factors is likely to remain.

They argue that land use conversion decisions can be viewed as an *ordered probit* situation if we consider varying intensity levels of land development. As already noted, ordered probit models describe situations where there are more than two choice outcomes, but the alternatives exhibit a natural or logical ordering. As noted, in the simple cross-sectional case where individual i's choices are independent from those of other individuals in the non-spatial model, the cut-point values ϕ can be determined by examining the maximum (and minimum) values of the latent data y^*_i over all individuals $i = 1, \ldots, n$ who have chosen alternative j.

$$\bar{\phi}_{j-1} = \max\{\max\{y^*_i : y_i = j\}, \phi_{j-1}\}$$
$$\bar{\phi}_{j+1} = \min\{\min\{y^*_i : y_i = j + 1\}, \phi_{j+1}\}$$

Wang and Kockelman (2008a) point out that in a spatial model setting where choices of individuals are not independent, but exhibit both space as well as time dependence, this line of argument no longer holds. To pursue this, we consider the (multivariate) normal prior placed on the cut-point parameters by Wang and Kockelman (2008a).

$$\phi \sim N(g, Q)\delta(\phi_1 < \phi_2 < \ldots, \phi_{J-1}) \tag{10.40}$$

where g is a vector of prior means (with elements g_j) and Q is a (diagonal) matrix containing prior variances, which we label q_j. Recall, $\delta(A)$ is an indicator function for each event A, so $\delta(A) = 1$ for outcomes where A occurs and $\delta(A) = 0$ otherwise. This acts as a constraint to ensure probabilities derived from the thresholds are positive. Of course, in the limit with all elements of $g_j = 0$ and the variances q_j infinite, we have the flat prior used in our discussion of the ordered probit model in Section 10.2. The conditional posterior for these parameters takes the form in (10.41), where we use $\diamond$ to denote conditioning arguments other than ϕ_{-j} consisting of other parameters in the model.

$$p(\phi_j|\phi_{-j},\diamond) \propto \delta(\bar{\phi}_{j-1} < \phi_j < \bar{\phi}_{j+1})\exp\left(-\frac{1}{2q_j}(\phi_j - g_j)^2\right) \qquad (10.41)$$

Of course, in the limiting case of a flat prior where $q_j \to \infty$, this collapses to our Uniform distribution:

$$p(\phi_j|\phi_{-j},\diamond) \propto U(\bar{\phi}_{j-1} < \phi_j < \bar{\phi}_{j+1}), j = 2,\ldots,J-1 \qquad (10.42)$$

The bounding values are determined by examining the maximum (and minimum) values of the latent data y^*_{ikt} over all individuals $k = 1,\ldots,n_i$, and all regions $i = 1,\ldots,M$ who have chosen alternative j at all times $t = 1,\ldots,T$. In this general spatial model, individuals' choices are spatially dependent on those of individuals in nearby regions and past time periods leading to:

$$\bar{\phi}_{j-1} = \max\{\max\{y^*_{ikt} : y_{ikt} = j\},\phi_{j-1}\}$$
$$\bar{\phi}_{j+1} = \min\{\min\{y^*_{ikt} : y_{ikt} = j+1\},\phi_{j+1}\}$$

Because of the dependence of y^*_{ikt} on other time periods and regions, the lower and upper bounds in this model also exhibit dependence. This can lead to a multimodal posterior distribution for these parameters.

10.7   Chapter summary

We have seen that Bayesian treatment of observable binary and polychotomous dependent variables y as indicators of latent underlying utilities y^* can be useful in modeling limited dependent variables that exhibit spatial dependence. Albert and Chib (1993) argue that if the vector of latent utilities y^* were known, $p(\beta,\rho,|y^*) = p(\beta,\rho|y^*,y)$, allowing us to view y^* as an additional set of parameters to be estimated. This leads to a (joint) conditional posterior distribution for our model parameters (conditioning on both y^*,y) that takes the same form as the Bayesian estimation problem from Chapter 5.

For models involving spatial dependence, we need to perform some computational work to sample from the conditional posterior distribution for the parameters y^* that we introduce in the model. The spatial dependence structure leads to a multivariate truncated normal distribution for these parameters, rather than the simple univariate truncated normal distribution that arises in the case of independent sample data. However, we showed that a procedure proposed by Geweke (1991) can be used to successfully sample from this conditional distribution. The procedure samples from this multivariate truncated normal distribution by breaking the task into an m-step

Gibbs sampler that carries out m-draws from a series of n univariate conditional distributions. These m-draws provide an asymptotically consistent estimate for the parameters y^*. These are then used when sampling from the conditional distributions of the remaining model parameters, β, ρ.

Despite the drawback arising from the computational intensity of this approach, there are a number of desirable aspects as well. One point is that implementation of the method is quite simple from a coding standpoint. We simply need to add code to our existing routine for MCMC estimation of the Bayesian SAR model to implement the m-step Gibbs sampler, with the remaining code unchanged. This amounts to a few lines of code that calls a specialized function to carry out the truncated multivariate normal sampling task.

References

Abreu, M., H.L.F. de Groot and R.J.G.M. Florax (2004). "Space and Growth: A Survey of Empirical Evidence and Methods," Tinbergen Institute Working Paper No. TI 04-129/3. Available at SSRN: http://ssrn.com/abstract=631007

Agarwal, D.K., A.E. Gelfand, and S. Citron-Pousty (2002). "Zero-inflated models with application to spatial count data," *Environmental and Ecological Statistics* 9: 341-355.

Albert, J. H. and S. Chib (1993). "Bayesian Analysis of Binary and Polychotomous Response Data," *Journal of the American Statistical Association*, 88:422, 669-679.

Allenby, G., G. Fennell, A. Bemmaor, V. Bhargava, F. Christen, J. Dawley, P. Dickson, Y. Edwards, M. Garratt, J. Ginter, A. Sawyer, R. Staelin, and S. Yang (2002). "Market Segmentation Research: Beyond Within and Across Group Differences," *Marketing Letters*, 13:3, 231-241.

Anderson, J.E. (1979). "A Theoretical Foundation for the Gravity Model," *American Economic Review*, 69:1, 106-116.

Anderson, J.E. and E. van Wincoop (2004). "Trade Costs," *Journal of Economic Literature*, 2004, 42:3, 691-751.

Anderson, T.W. and H. Rubin (1949). "Estimation of the Parameters of a Single Equation in a Complete System of Stochastic Equations," *Annals of Mathematical Statistics*, 20, 46-63.

Anselin, L. (1988). *Spatial Econometrics: Methods and Models*, Dordrecht: Kluwer Academic Publishers.

Anselin, L. (1988b). "Lagrange multiplier test diagnostics for spatial dependence and spatial heterogeneity," *Geographical Analysis*, 20, 1-17.

Anselin, L. (2003). "Spatial Externalities, Spatial Multipliers and Spatial Econometrics," *International Regional Science Review*, 26, 153-166.

Anselin, L. and A. Bera (1998). "Spatial dependence in linear regression models with an introduction to spatial econometrics," in *Handbook of Applied Economic Statistics*, A. Ullah and D.E. Giles (eds). New York: Marcel Dekker, 237-289.

Anselin, L., A. Bera, R. Florax, and M. Yoon (1996). "Simple diagnostic tests for spatial dependence," *Regional Science and Urban Economics*, 26, 77-104.

Anselin, L. and J. LeGallo (2006). "Interpolation of Air Quality Measures in Hedonic House Price Models: Spatial Aspects," *Spatial Economic Analysis*, 1:1, 31-52.

Anselin L., A. Varga and Z. Acs (1997). "Local geographic spillovers between university research and high technology innovations," *Journal of Urban Economics*, 42: 422-448.

Autant-Bernard, C. (2001). "The geography of knowledge spillovers and technological proximity," *Economics of Innovation and New Technology*, 10, 237-254.

Autant-Bernard, C., J.P. LeSage and O. Parent (2008). "Firm innovation strategies: a spatial cohort multinomial probit approach," forthcoming in *Annales d'Economie et de Statistique*.

Autant-Bernard, C., J. Mairesse and N. Massard (2007). "Spatial knowledge diffusion through collaborative networks," *Papers in Regional Science* 86:3, 341-350.

Banerjee, S., B.P. Carlin and A.E. Gelfand (2004). *Hierarchical Modeling and Analysis for Spatial Data*, Boca Raton: Chapman & Hall/CRC.

Barry, R.P. and R.K. Pace (1997). "Kriging with Large Data Sets Using Sparse Matrix Techniques," *Communications in Statistics: Computation and Simulation*, 26, 619-629.

Barry, R.P. and R.K. Pace (1999). "A Monte Carlo Estimator of the Log Determinant of Large Sparse Matrices," *Linear Algebra and its Applications*, 289, 41-54.

Bavaud, F. (1998). "Models for Spatial Weights: A Systematic Look," *Geographical Analysis*, 30, 153-171.

Behrens, K., C. Ertur and W. Koch (2007). "Dual Gravity: Using Spatial Econometrics to Control for Multilateral Resistance," *CORE Discussion Paper* No. 2007/59.

Behrens, K. and J.-F. Thisse (2007). "Regional economics: A new economic geography perspective," *Regional Science and Urban Economics*, 37:4, 457-465.

Belsley, D.A., E. Kuh, and R.E. Welch (1980). *Regression Diagnostics: Identifying Influential Data and Source of Collinearity*, New York, John Wiley.

Bernstein, D.S. (2005). *Matrix Mathematics*, Princeton, New Jersey: Princeton University Press.

Beron, K.J. and W.P.M. Vijverberg (2000). "Probit in a Spatial Context: A Monte Carlo Approach," in *Advances in Spatial Econometrics*, L. Anselin and R. Florax (eds.) Heidelberg: Springer-Verlag.

Besag, J., J.C. York and A. Mollie (1991). "Bayesian Image Restoration, with Two Principle Applications in Spatial Statistics," *Annals of the Institute of Statistical Mathematics*, 43, 1-59.

Bolduc, D., B. Fortin and S. Gordon (1997). "Multinomial Probit Estimation of Spatially Interdependent Choices: An Empirical Comparison of Two New Techniques," *International Regional Science Review*, 20, 77-101.

Bonacich, P.B. (1987). "Power and centrality: a family of measures," *American Journal of Sociology*, 92, 1170-1182.

Brasington, D.M. and D. Hite (2005). "Demand for Environmental Quality: A Spatial Hedonic Analysis," *Regional Science and Urban Economics*, 35, 57-82.

Burridge, P. (1980). "On the Cliff-Ord test for spatial autocorrelation," *Journal of the Royal Statistical Society B*, 42, 107-108.

Byron, R.P. (1992). "Polynomial Approximations in Cross-Sectional Models," *Journal of Applied Econometrics*, 7, 309-322.

Chen, J. and R. Jennrich (1996). "The Signed Root Deviance Profile and Confidence Intervals in Maximum Likelihood Analysis," *Journal of the American Statistical Association*, 91, 993-998.

Chib, S. (1992). "Bayes inference in the Tobit censored regression model," *Journal of Econometrics*, 51, 79-99.

Chib, S. (1995). "Marginal likelihoods from the Gibbs Sampler," *Journal of the American Statistical Association*, 90, 1313-1321.

Chib, S. and I. Jeliazkov (2001). "Marginal Likelihood from the Metropolis-Hastings Output," *Journal of the American Statistical Association*, 96, 270-281.

Chiu, Y.M., T. Leonard and K. Tsui (1996). "The Matrix-Logarithmic Covariance Model," *Journal of the American Statistical Association*, 91, 198-210.

Christensen, O., G. Roberts and M. Sköld (2006). "Robust Markov chain Monte Carlo methods for spatial generalised linear mixed models," *Journal of Computational and Graphical Statistics* 15, 1-17.

Cordy, C.B. and D.A. Griffith (1993). "Efficiency of Least Squares Estimators in the Presence of Spatial Autocorrelation," *Communications in Statistics – Simulation and Computation*, 22, 1161-1179.

Cressie, N. (1993). *Statistics for Spatial Data*, Revised edition, New York: John Wiley.

Cressie, N. (1995). "Bayesian smoothing of rates in small geographic areas," *Journal of Regional Science*, 35, 659-673.

Curry, L. (1972). "A spatial analysis of gravity flows," *Regional Studies*, 6, 137-147.

Dall'erba, S. and J. LeGallo (2007). "Regional Convergence and the Impact of European Structural Funds over 1989-1999: A Spatial Econometric Analysis," forthcoming in *Papers in Regional Science*.

Davidson, R. and J. MacKinnon (2004). *Econometric Theory and Methods*. New York: Oxford University Press.

Davidson, R. and J. MacKinnon (1993). *Estimation and Inference in Econometrics*. New York: Oxford University Press.

Debreu, G. and I. N. Herstein (1953). "Nonnegative Square Matrices," *Econometrica*, 21, 597-607.

Dennison, D.G.T., C.C. Holmes, B.K. Mallick and A.F.M. Smith (2002). *Bayesian Methods for Nonlinear Classification and Regression*. West Sussex, England: John Wiley & Sons.

Dong, P. (2008). "Generating and updating multiplicatively weighted Voronoi diagrams for point, line and polygon features in GIS," *Computers & Geosciences*, 34:4, 411-421.

Dubin, R. (1988). "Estimation of Regression Coefficients in the Presence of Spatially Autocorrelated Error Terms," *Review of Economics and Statistics*, 70, 466-474.

Duranton, G. and D. Puga (2001). "Nursery cities: Urban diversity, Process Innovation, and the life cycle of product," *American Economic Review*, 91:5, 1454-1477.

Elhorst, J.P. (2001). "Dynamic Models in Space and Time," *Geographical Analysis*, 33, 119-140.

Eppstein, D., M.S. Paterson and F.F. Yao (1997). "On Nearest-Neighbor Graphs," *Discrete and Computational Geometry*, 17, 263-282.

Ertur, C. and W. Koch (2007). "Convergence, human capital and international spillovers," *Journal of Applied Econometrics*, 22:6, 1033-1062.

Ertur, C., J. LeGallo and J.P. LeSage (2007). "Local versus Global Convergence in Europe: A Bayesian Spatial Econometric Approach," *Review of Regional Studies*, 37:1, 82-108.

Feenstra, R.C. (2002). "Border Effects and the Gravity Model: Consistent

Methods for Estimation," *Scottish Journal of Political Economy*, 49:5, 491-506.

Fernández, C., E. Ley and M.F.J. Steel (2001). "Benchmark priors for Bayesian model averaging," *Journal of Econometrics*, 100, 381-427.

Fingleton, B. (2001). "Theoretical economic geography and spatial econometrics: dynamic perspectives," *Journal of Economic Geography*, 1, 201-225.

Fischer, M.M. and J.P. LeSage (2009). "Spatial econometric modeling of origin-destination flows," in *Handbook of Applied Spatial Analysis: Software Tools, Methods and Applications*, M. M. Fischer and A. Getis (eds.), Berlin: Springer-Verlag.

Fischer, M.M. and C. Stirbock (2006). "Pan-European regional income growth and club-convergence," *The Annals of Regional Science*, 40:4, 693-721.

Fischer, M.M., T. Scherngell and E. Jansenberger (2006). "The geography of knowledge spillovers between high-technology firms in Europe evidence from a spatial interaction modelling perspective," *Geographical Analysis*, 38:3, 288-309.

Fischer, M.M., T. Scherngell and M. Reismann (2008). "Knowledge spillovers and total factor productivity. Evidence using a spatial panel data model," forthcoming in *Geographical Analysis*.

Flemming, M.M. (2004). "Techniques for estimating spatially dependent discrete choice models," in *Advances in spatial econometrics: Methodology, tools and applications*, L. Anselin, R.J.G.M. Florax and S.J. Rey (eds.), New York: Springer-Verlag, 145-168.

Florax, R.J.G.M., H. Folmer and S.J. Rey (2003). "Specification searches in spatial econometrics: the relevance of Hendry's methodology," *Regional Science and Urban Economics*, 33, 557-579.

Fotheringham, A.S. and C. Andrew (1999). "Regularities in Spatial Information Processing: Implications for Modeling Destination Choice," *The Professional Geographer*, 51:2, 227-239.

Frühwirth-Schnatter, S., and H. Wagner (2006). "Auxiliary mixture sampling for parameter-driven models of time series of counts," *Biometrika*, 93: 827-841.

Garofalo, G. and S. Yamarik (2002). "Regional Convergence: Evidence from a New State-By-State Capital Stock Series", *Review of Economics and Statistics*, 84:2, 316-323.

Gelfand, A.E., S.E. Hills, A. Racine-Poon and A.F.M. Smith. (1990). "Illustration of Bayesian Inference in Normal Data Models Using Gibbs Sampling," *Journal of the American Statistical Association*, 85, 972-985.

Gelfand, A.E., S. Sahu and B.P. Carlin (1995). "Efficient parameterizations for normal linear mixed models," *Biometrika*, 82:3, 479-488.

Gelfand, A.E. and A.F.M Smith (1990). "Sampling-Based Approaches to Calculating Marginal Densities," *Journal of the American Statistical Association*, 85, 398-409.

Gelman, A., J.B. Carlin, H.S. Stern and D.B. Rubin (1995). *Bayesian Data Analysis*. New York: Chapman & Hall.

Geman, S. and D. Geman (1984). "Stochastic relaxation, Gibbs distributions, and the Bayesian restoration of images," *IEEE Transactions on Pattern Analysis and Machine Intelligence,* 6, 721-741.

Geweke, J. (1991). "Efficient Simulation from the Multivariate Normal and Student-t Distributions Subject to Linear Constraints and the Evaluation of Constraint Probabilities," in *Proceedings of 23rd Symposium on the Interface Between Computing Science and Statistics*, E. Kermanidas, (ed.), Interface Foundation Of North America Inc., Fairfax, Va., 571-578.

Geweke, J. (1993). "Bayesian Treatment of the Independent Student *t* Linear Model," *Journal of Applied Econometrics*, 8, 19-40.

Gilley, O.W. and R.K. Pace (1996). "The Harrison and Rubinfeld Data Revisited," *Journal of Environmental and Economic Management*, 31, 403-405.

Girard, D.A. (1989). "A Fast 'Monte Carlo Cross-Validation' Procedure for Large Least Squares Problems with Noisy Data," *Numerische Mathematik*, 56, 1-23.

Golub, G. H. and C.F. Van Loan (1996). *Matrix Computations*, Baltimore: John Hopkins University Press.

Goodman, J.E. and J. O'Rourke (1997). *Handbook of Discrete and Computational Geometry*, Boca Raton: CRC Press.

Gradshteyn, I.S. and I.M. Ryzhik (1980). *Table of Integrals, Series, and Products*, Corrected and Enlarged Edition, Orlando, Florida: Academic Press.

Granger, C.W.J. (1980). "Long-memory Relationships and the Aggregation of Dynamic Models," *Journal of Econometrics*, 14, 227-238.

Greene, W. (1997). *Econometric Analysis*, third edition, Upper Saddle River, New Jersey: Prentice-Hall.

Greene, W. (2000). *Econometric Analysis*, fourth edition, Upper Saddle River: Prentice-Hall.

Griffith, D.A. (2000). "Eigenfunction Properties and Approximations of Selected Incidence Matrices Employed in Spatial Analysis," *Linear Algebra and its Applications*, 321, 95-112.

Griffith, D.A. (2007). "Spatial Structure and Spatial Interaction: 25 Years Later," *The Review of Regional Studies*, 37:1, 28-38.

Griffith, D.A. and K.G. Jones (1980). "Explorations into the relationship between spatial structure and spatial interaction," *Environment and Planning A*, 12, 187-201.

Haining, R. (1979). "Statistical Tests and Process Generators for Random Field Models," *Geographical Analysis*, 11, 45-64.

Haining, R. (1990). *Spatial Data Analysis in the Social and Environmental Sciences.* Cambridge: Cambridge University Press.

Haining, R. (1994). "Diagnostics for Regression Modeling in Spatial Econometrics," *Journal of Regional Science*, 34, 325-341.

Harrison, D. and D.L. Rubinfeld (1978). "Hedonic Housing Prices and the Demand for Clean Air," *Journal of Environmental Economics and Management*, 5, 81-102.

Hastings, W.K. (1970). "Monte Carlo sampling methods using Markov chains and their applications," *Biometrika*, 57, 97-109.

Haubrich, J.G. (1993). "Consumption and Fractional Differencing: Old and New Anomalies," *The Review of Economics and Statistics*, 75:4, 767-772.

Hausman, J.A. (1978). "Specification Tests in Econometrics," *Econometrica*, 46, 1251-1272.

Hendry, D., A. Pagan, and D. Sargan (1984). "Dynamic Specification," in: Z. Griliches and M. Intrilligator (eds.), *Handbook of Econometrics, Volume 2*, Amsterdam: North-Holland, 1023-1100.

Hepple, L.W. (1995a). "Bayesian techniques in spatial and network econometrics: 1. Model Comparison and posterior odds," *Environment and Planning A*, 27, 447-469.

Hepple, L.W. (1995b). "Bayesian techniques in spatial and network econometrics: 2. Computational methods and algorithms," *Environment and Planning A*, 27, 615-644.

Hepple, L.W. (2004). "Bayesian model choice in spatial econometrics," in *Advances in Econometrics: Volume 18: Spatial and Spatiotemporal Econometrics*, J.P. LeSage and R.K. Pace (eds.), Oxford: Elsevier Ltd., 101-126.

Holloway, G., B. Shankara and S. Rahman (2002). "Bayesian spatial probit estimation: a primer and an application to HYV rice adoption," *Agricultural Economics*, 27:3, 383-402.

Holmes, C.C., D.G.T. Denison and B.K. Mallick (1999). "Accounting for model uncertainty in seemingly unrelated regressions," Technical Report,

Imperial College, London.

Horn, R.A. and C.R. Johnson (1993). *Matrix Analysis.* New York: Cambridge University Press.

Horn, R.A. and C.R. Johnson (1994). *Topics in Matrix Analysis,* Cambridge: Cambridge University Press.

Hosking, J.R.M. (1981). "Fractional Differencing," *Biometrika,* 68, 165-176.

Huang, D. and V.V. Anh (1992). "Estimation of Spatial ARMA Models," *Australian & New Zealand Journal of Statistics,* 34:3, 513-530.

Hutchinson, M.F. (1990). "A Stochastic Estimator of the Trace of the Influence Matrix for Laplacian Smoothing Splines," *Communications in Statistics: Simulation and Computation,* 19, 433-450.

Irwin, E.G. and N.E. Bockstael (2004). "Analysis of Urban Land Markets and the Impact of Land Market Regulation," *Regional Science and Urban Economics,* 34:6, 705-725.

Judge, G., R.C. Hill, and W.E. Griffiths (1982). *The Theory and Practice of Econometrics,* New York: Wiley.

Katz, L. (1953). "A new status index derived from sociometric analysis," *Psychometrika,* 18, 39-43.

Kelejian, H.H. and D.P. Robinson (1995). "Spatial Correlation: A Suggested Alternative to the Autoregressive Model," in *New directions in spatial econometrics,* L. Anselin and R.J.G.M Florax (eds.), Berlin: Springer-Verlag, 75-95.

Kelejian, H.H. and I.R. Prucha (1998). "A Generalized Spatial Two-Stage Least Squares Procedure for Estimating a Spatial Autoregressive Model with Autoregressive Disturbances," *Journal of Real Estate and Finance Economics,* 17:1, 99-121.

Kelejian, H.H. and I.R. Prucha (1999). "A generalized moments estimator for the autoregressive parameter in a spatial model," *International Economic Review,* 40, 509-33.

Kelejian, H.H. and I.R. Prucha (2007). "The relative efficiencies of various predictors in spatial econometric models containing spatial lags," *Regional Science and Urban Economics,* 37:3, 363-374.

Kelejian, H.H. and I.R. Prucha (2007). "HAC estimation in a spatial framework," *Journal of Econometrics,* 140:1, 131-154.

Kelejian, H.H., G.S. Tavlas and G. Hondronyiannis (2006). "A Spatial Modeling Approach to Contagion Among Emerging Economies," *Open Economies Review,* 17:4/5, 423-442.

Kim, C.W., T.T. Phipps and L. Anselin (2003). "Measuring the Benefits of Air Quality Improvement: A Spatial Hedonic Approach," *Journal of Environmental Economics and Management*, 45, 24-39.

Koop, G. (2003). *Bayesian Econometrics*, West Sussex, England: John Wiley & Sons.

Lacombe, D. (2004). "Does Econometric Methodology Matter? An Analysis of Public Policy Using Spatial Econometric Techniques," *Geographical Analysis*, 36, 87-89.

Lange, K.L., R.J.A. Little and J.M.G. Taylor (1989). "Robust Statistical Modeling Using the *t* Distribution," *Journal of the American Statistical Association*, 84, 881-896.

Lee, L.-F. (2004). "Asymptotic Distributions of Quasi-Maximum Likelihood Estimators for Spatial Econometric Models," *Econometrica*, 72, 1899-1926.

Lee, M. and R.K. Pace (2005). "Spatial Distribution of Retail Sales," *Journal of Real Estate Finance and Economics*, 31:1, 53-69.

Le Gallo J., C. Ertur and C. Baumont (2003). "A spatial econometric analysis of convergence across European regions, 1980-1995," in B. Fingleton (ed.) *European Regional Growth*, Berlin: Springer-Verlag.

LeSage, J.P. (1997). "Bayesian Estimation of Spatial Autoregressive Models," *International Regional Science Review*, 20:1&2, 113-129.

LeSage, J.P. (1999). *The Theory and Practice of Spatial Econometrics*, a manual to accompany the Spatial Econometrics Toolbox, available at: www.spatial-econometrics.com

LeSage, J.P. (2000). "Bayesian Estimation of Limited Dependent Variable Spatial Autoregressive Models," *Geographical Analysis*, 32:1, 19-35.

LeSage, J.P. (2007). *Spatial Econometrics Toolbox*, available at: www.spatial-econometrics.com.

LeSage, J.P. and M.M. Fischer (2008). "Spatial Growth Regressions: Model Specification, Estimation and Interpretation," forthcoming in *Spatial Economic Analysis*.

LeSage, J. P., M.M. Fischer and T. Scherngell (2007). "Knowledge Spillovers across Europe, Evidence from a Poisson Spatial Interaction Model with Spatial Effects," *Papers in Regional Science*, 86:3, 393-421.

LeSage, J.P. and C. Llano (2007). "A Spatial Interaction Model With Spatially Structured Origin and Destination Effects," SSRN working paper, available at SSRN: http://ssrn.com/abstract=924603.

LeSage, J.P. and R.K. Pace (2004). "Models for Spatially Dependent Missing

Data," *Journal of Real Estate Finance and Economics*, 29:2, 233-254.

LeSage, J.P. and R.K. Pace (2004a). "Introduction," *Advances in Econometrics: Volume 18: Spatial and Spatiotemporal Econometrics*, J.P. LeSage and R.K. Pace (eds.), Oxford: Elsevier Ltd., 1-32.

LeSage, J.P. and R.K. Pace (2004b). "Using Matrix Exponentials to Estimate Spatial Probit/Tobit Models," in *Recent Advances in Spatial Econometrics*, J. Mur, H. Zoller, and A. Getis (eds.), Palgrave Publishers, 105-131.

LeSage, J.P. and R.K. Pace (2007). "A Matrix Exponential Spatial Specification," *Journal of Econometrics*, 140:1, 190-214.

LeSage, J.P. and R.K. Pace (2008). "Spatial econometric modeling of origin-destination flows," *Journal of Regional Science*, 48:5, 941-967.

LeSage, J.P. and O. Parent (2007). "Bayesian Model Averaging for Spatial Econometric Models," *Geographical Analysis*, 39:3, 241-267.

LeSage, J.P. and W. Polasek (2008). "Incorporating transportation network structure in spatial econometric models of commodity flows," *Spatial Economic Analysis*, 3:2, 225-245.

Lindley, D.V. (1957). "A Statistical Paradox," *Biometrika*, 44, 187-192.

López-Bazo, E., E.Vayá and M. Artís (2004). "Regional externalities and growth: evidence from European regions," *Journal of Regional Science*, 44:1, 43-73.

Madigan, D. and J. York (1995). "Bayesian graphical models for discrete data," *International Statistical Review*, 63, 215-232.

Marcus, M. and H. Minc (1992). *A Survey of Matrix Theory and Matrix Inequalities*. New York: Dover.

Mardia, K.V. and R.J. Marshall (1984). "Maximum likelihood estimation of models for residual covariance in spatial regression," *Biometrika*, 71, 135-146.

Marsh, T.L. and R.C. Mittelhammer (2004). "Generalized maximum entropy estimation of a first order spatial autoregressive model," in *Advances in Econometrics, Volume 18, Spatial and Spatiotemporal Econometrics*, J.P. LeSage and R.K. Pace (eds.), Oxford: Elsevier Ltd., 199-234.

Martellosio, F. (2006). "Some Correlation Properties of Spatial Autoregressions on Irregular Lattices," manuscript, University of Reading.

Martin, R.J. (1993). "Approximations to the Determinant Term in Gaussian Maximum Likelihood Estimation of Some Spatial Models," *Communications in Statistics – Theory and Methods*, 22, 189-205.

McCulloch, R.E., N.G. Polson and P.E. Rossi (2000). "A Bayesian analysis of the multinomial probit model with fully identified parameters," *Journal of Econometrics*, 99, 173-193.

McMillen, D.P. (1992). "Probit with spatial autocorrelation," *Journal of Regional Science*, 32:3, 335-348.

Metropolis, N., A.W. Rosenbluth, M.N. Rosenbluth, A.H. Teller and E. Teller. (1953). "Equation of state calculations by fast computing machines," *Journal of Chemical Physics*, 21, 1087-1092.

Moran, P.A.P. (1948). "The interpretation of statistical maps," *Biometrika*, 35, 255-260.

Newton, M.A. and A.E. Raftery (1994). "Approximate Bayesian inference with the weighted likelihood bootstrap," *Journal of the Royal Statistical Society B*, 56, 3-48.

Nobile, A. (2000). "Comment: Bayesian multinomial probit models with a normalization constraint," *Journal of Econometrics*, 99, 335-345.

Ord, J.K. (1975). "Estimation Methods for Models of Spatial Interaction," *Journal of the American Statistical Association*, 70, 120-126.

Pace, R.K. (2007). *Spatial Statistics Toolbox* available at www.spatial-statistics.com.

Pace, R.K. and R.P. Barry (1997). "Quick computation of spatial autoregressive estimators," *Geographical Analysis*, 29, 232-246.

Pace, R.K. and R.P. Barry (1998). "Simulating Mixed Regressive Spatially Autoregressive Estimators," *Computational Statistics*, 13, 397-418.

Pace, R.K., R.P. Barry, O.W. Gilley and C.F. Sirmans (2000). "A Method for Spatial-temporal Forecasting with an Application to Real Estate Prices," *International Journal of Forecasting*, 16, 229-246.

Pace, R.K. and J.P. LeSage (2002). "Semiparametric Maximum Likelihood Estimates of Spatial Dependence," *Geographical Analysis*, 34, 76-90.

Pace, R.K. and J.P. LeSage (2003a). "Likelihood Dominance Spatial Inference," *Geographical Analysis*, 35:2, 133-147.

Pace, R.K. and J.P. LeSage (2003b). "Spatial Autoregressive Local Estimation," in *Recent Advances in Spatial Econometrics*, J. Mur, H. Zoller and A. Getis (eds.) Palgrave Publishers, 105-131.

Pace, R.K. and J.P. LeSage (2004). "Chebyshev Approximation of Log-determinants of Spatial Weight Matrices," *Computational Statistics and Data Analysis*, 45, 179-196.

Pace, R.K. and J.P. LeSage (2006). "Interpreting Spatial Econometric Models," paper presented at the Regional Science Association International

North American meetings, Toronto, Ontario, Canada.

Pace, R.K. and J.P. LeSage (2008). "A Spatial Hausman Test," *Economics Letters*, 101, 282-284.

Pace, R.K. and J.P. LeSage (2009a). "A Sampling Approach to Estimating the Log Determinant Used in Spatial Likelihood Problems," forthcoming in *Computational Statistics and Data Analysis*.

Pace, R.K. and J.P. LeSage (2009b), "Omitted variables biases of OLS and spatial lag models," in *Progress in Spatial Analysis: Theory and Computation, and Thematic Applications*, A. Pez, J. Le Gallo, R. Buliung and S. DallErba (eds.), Berlin: Springer.

Pace, R.K. and J.P. LeSage (2009c). "Simple Bounds for Difficult Spatial Likelihood Problems," unpublished manuscript.

Pace, R.K. and J.P. LeSage (2009d). "Spatial Econometrics," forthcoming in *Handbook of Spatial Statistics*, A.E. Gelfand, P. Diggle, M. Fuentes and P. Guttorp (eds.), Boca Raton, Florida: Chapman & Hall.

Pace, R.K. and D. Zou (2000). "Closed-Form Maximum Likelihood Estimates of Nearest Neighbor Spatial Dependence," *Geographical Analysis*, 32, 154-172.

Parent, O. and J.P. LeSage (2008). "Using the variance structure of the conditional autoregressive specification to model knowledge spillovers," *Journal of Applied Econometrics*, 23:2, 235-256.

Pfeifer, P.E. and S.J Deutsch (1980). "A three-stage iterative procedure for space-time modeling," *Technometrics* 22, 35-47.

Pollack, R.A. and T.J. Wales (1991). "The Likelihood Dominance Criterion: A New Approach to Model Selection," *Journal of Econometrics*, 47, 227-242.

Press, W., S. Teukolsky, W. Vetterling and B. Flannery (1996). *Numerical Recipes in Fortran 77*, second edition, New York: Cambridge University Press.

Ranjan, R. and J.L. Tobias (2007). "Bayesian Inference For The Gravity Model," *Journal of Applied Econometrics*, 22: 817-838.

Rathbun, S.L. and S. Fei (2006). "A spatial zero-inflated poisson regression model for oak regeneration," *Environmental and Ecological Statistics*, 13:4, 406-426.

Richardson, S. and P.J. Green (1997). "On Bayesian analysis of mixtures with an unknown number of components," *Journal of the Royal Statistical Society B*, 59, 731-792.

Ripley, B. (1981). *Spatial Statistics*, New York: Wiley.

Ripley, B. (1988). *Statistical Inference for Spatial Processes*, Cambridge: Cambridge University Press.

Rossi, P.E., G.M. Allenby and R. McCulloch (2006). *Bayesian Statistics and Marketing* (Wiley Series in Probability and Statistics).

Sen, A. and T.E. Smith (1995). *Gravity Models of Spatial Interaction Behavior*, Heidelberg: Springer-Verlag.

Smirnov O.A. (2005). "Computation of the Information Matrix for Models With Spatial Interaction on a Lattice," *Journal Of Computational and Graphical Statistics*, 14:4, 910-927.

Smirnov, O. and L. Anselin (2001). "Fast Maximum Likelihood Estimation of Very Large Spatial Autoregressive Models: A Characteristic Polynomial Approach," *Computational Statistics and Data Analysis* 35, 301-319.

Smith, T.E. and J.P. LeSage (2004). "A Bayesian Probit Model with Spatial Dependencies," in *Advances in Econometrics: Volume 18: Spatial and Spatiotemporal Econometrics*, J.P. LeSage and R.K. Pace (eds.), Oxford: Elsevier Ltd., 127-160.

Snyder, J.P. and P.M. Voxland (1989). *An Album of Map Projections*, Washington: US Government Printing Office.

Strang, G. (1976). *Linear Algebra and its Applications*, New York: Academic Press.

Ter Hofstede, F., M. Wedel and J.E.M. Steenkamp (2002). "Identifying Spatial Segments in International Markets," *Marketing Science*, 21, 160-177.

Tiebout, C.M. (1956). "A Pure Theory of Local Expenditures," *The Journal of Political Economy*, 64:5, 416-424.

Tiefelsdorf, M. (2003). "Misspecifications in interaction model distance decay relations: A spatial structure effect," *Journal of Geographical Systems*, 5, 25-50.

Turnbull, G.K. and G. Geon (2006). "Local government internal structure, external constraints and the median voter," *Public Choice*, 129, 487-506.

Wang, X. and K.M. Kockelman (2007). "Specification and estimation of a spatially and temporally autocorrelated seemingly unrelated regression model: application to crash rates in China," *Transportation*, 34:3, 281-300.

Wang, X. and K.M. Kockelman (2008a). "The Dynamic Spatial Ordered Probit Model: Methods For Capturing Patterns Of Spatial And Temporal Autocorrelation In Ordered Response Data, Using Bayesian Estimation," paper presented at the 54th North American Regional Science Association

International Conference, Savannah, Georgia (2007).

Wang, X. and K.M. Kockelman (2008b). "Application of the Dynamic Spatial Ordered Probit Model: Patterns of Land Development Change in Austin, Texas," paper presented at the 54th North American Regional Science Association International Conference, Savannah, Georgia (2007).

Whittle, P. (1954). "On Stationary Processes in the Plane," *Biometrika*, 41, 434-49.

Wilson, J.D. (1986). "A theory of interregional tax competition," *Journal of Urban Economics*, 19:3, 296-315.

Yang, S. and G.M. Allenby (2003). "Modeling interdependent consumer preferences," *Journal of Marketing Research*, 40, 282-294.

Zellner, A. (1971). *An Introduction to Bayesian Inference in Econometrics*. New York: John Wiley & Sons.

Zellner, A. (1986). "On assessing prior distributions and Bayesian regression analysis with g-prior distributions," In *Bayesian Inference and Decision Techniques: Essays in Honor of Bruno de Finetti*, P. Goel and A. Zellner, (eds.), Amsterdam: North-Holland/Elsevier, 233-243.

Zhang, Y., W.E. Whitehead, D.J. Leith and L. Walshe (2008). "Log-det Approximation based on Uniformly Distributed Seeds and its Application to Gaussian Process Regression," forthcoming in *Journal of Computational and Applied Mathematics*.

Zhou, B. and K.M. Kockelman (2008). "Neighborhood impacts on land use change: a multinomial logit model of spatial relationships," *The Annals of Regional Science*, 42:2, 321-340.

Index

A

Additive error specification, 280
Address matching, 1
Almon distributed lags, 96
Alternative linear regression models, 173
Alternative spatial weight structures, 168
Analytical solutions, to Bayesian spatial model, 127, 130–133
Anselin, Luc, xi
Asymmetric loss function, 157
Asymptotic omitted variable bias, 67
Austrian road/rail network, 230
    interregional flows, 229
    regions in, 231
Autoregressive case, 204–205
Average direct impact, 36
Average total impact, 37
    to and from observations, 36
Averaged impact estimates, 183

B

Band matrices, 86
Bandwidth, reducing in matrices, 86
Bartlett decomposition, 309
Basic determinant computation, 81–84
Bayes' Theorem, 129, 130, 171
Bayesian estimation, ix, 31, 56
    computational competitiveness
        with maximum likelihood
        techniques, 145
    convergence of samplers, 141
    integration problems, 126
    of latent utility, 281
    MCMC draws from, 166
    for multi-model and MESS
        comparisons, 254
    for spatial econometric interaction
        models, 218–222

using MESS model, 250–255
Bayesian hierarchical spatial models, 234
Bayesian inference/learning, 125
Bayesian latent variable treatment, 281–283
    applied illustrations of spatial probit
        model, 289–293
    Gibbs sampling conditional
        distribution, 285–287
    implementation issues, 287–289
    MCMC sampler for SAR probit
        model, 284–285
    SAR probit model and, 283
Bayesian Markov Chain Monte
        Carlo estimation, for spatial
        econometric interaction
        models, 220
Bayesian methodology, 123, 124–127
    analytical approaches to, 127–130
    focus on data and parameter
        distributions, 124
Bayesian model comparison, 126, 155,
        168–169
    applied illustration, 175–178
    illustration of MC^3 and model
        averaging, 178–184
    models based on different variables,
        173–175
    models based on different weights,
        169–173
    posterior probabilities for models
        with different parameter and
        weight matrices, 176
Bayesian model probabilities, 162
Bayesian spatial econometric models,
        123–124
    analytical solution, 130–133
    applied illustration, 142–145
    Bayesian methodology, 124–127
    Beta prior distributions, 143

337

Normal-inverse gamma prior (NIG)
 distributions, 128, 129, 130, 170
Normal-Wishart prior, 308
Numerical integration, 127, 132
 for anti-log of log-marginal posterior
  density, 187
 MCMC draw by inversion using, 140
 of posterior distribution, 123
 univariate, 138, 172

O

OD SAR estimates, 305
OD SAR Tobit estimates, 305
Omitted variables, 45, 68
 bias of least-squares, 63–66
 correlation with explanatory
  variables, 193
 correlation with included variables,
  280
 correlation with technical knowledge
  stocks, 160
 Hausman test for OLS and SEM
  estimates, 61–63
 limited effect on spatial regression
  methods, 67
 with spatial character, 202
 with spatial dependence, 60–61
 spatial dependence in regressor
  with, 65
 in spatiotemporal and spatial
  models, 193
Omitted variables bias, 61
 as function of spatial dependence, 66
 least-squares, 63–66
 SDM model protection against, 157
 for spatial regressions, 67–68
 trade-off with inclusion of redundant
  variables, 174
Omitted variables motivation, 25, 27–28,
  28
Opposite signs, multi-model
  comparisons of significant
  coefficients with, 249
Order of neighbors, partitioning
  impacts by, 40–41
Ordered spatial probit model, 280,
  297–299

for land use conversion decisions,
  319
Ordering algorithms, 86–87
Orderings, and decomposition times for
  sparse matrices, 85–86
Ordinary least squares (OLS), 61
Origin- and destination-centric OD flow
  arrangements, 213
Origin-based spatial dependence, 216,
  219, 225, 229
Origin-centric ordering, 213, 214, 233
Origin-centric schemes, for OD flows,
  213
Origin-destination commuting flows,
  303–304
Origin-destination flow data, 94
Origin-destination flow matrix, 212
Origin-destination (OD) flows, ix, 211
 Hurricane Katrina examples, 279
 SAR and SAR Tobit estimates, 305
 spatial econometric modeling of, 217
 spatial nature of, 212
 zero flow problems in, 223, 303
Origin-destination region contiguity
  relationships, 230
Origin-to-destination dependence, 216,
  225
Out-of-sample forecasting, 5, 124
 and Bayesian estimation, 127
Over-parameterization problem,
  solution to, 8

P

Panel data sets, 29
Parameter dispersion
 estimates of, 54–60
 posterior distribution for, 124
 and sample sizes, 137
Parameter estimation, 124, 125
 calculating summary measures of
  impacts, 39
 direct and indirect impacts in theory,
  34–39
 interpreting, 33–34
 measures of dispersion for impact
  estimates, 39–40